Quinoa

Improvement and Sustainable Production

Quinoa
Improvement and Sustainable Production

Edited by

Kevin Murphy and Janet Matanguihan

WILEY Blackwell

Published by John Wiley & Sons, Inc., Hoboken, New Jersey
Published simultaneously in Canada

For general information on our other products and services or for technical support, please contact our Customer Care Department within the United States at (800) 762-2974, outside the United States at (317) 572-3993 or fax (317) 572-4002.

Wiley also publishes its books in a variety of electronic formats. Some content that appears in print may not be available in electronic formats. For more information about Wiley products, visit our web site at www.wiley.com.

Library of Congress Cataloging-in-Publication Data:

Quinoa : improvement and sustainable production / edited by Kevin Murphy and Glafera Janet Matanguihan.
pages cm
Includes bibliographical references and index.
ISBN 978-1-118-62805-8 (cloth)
1. Quinoa. 2. Crop improvement. 3. Sustainable agriculture. I. Murphy, Kevin (Kevin Matthew), 1972- editor. II. Matanguihan, Glafera Janet, editor.
SB177.Q55Q56 2015
664′.7–dc23

2015006917

Printed in Singapore by C.O.S. Printers Pte Ltd

10 9 8 7 6 5 4 3 2 1

Contents

List of Contributors

Sergio Núñez de Arco
Andean Naturals, Inc.,
Foster City, CA, USA

Didier Bazile
UPR47, GREEN, Centre de Coopération Internationale en Recherche Agronomique pour le Développement
Campus International de Baillarguet
Montpellier France

Carmen Del Castillo
Faculty of Agronomy
Universidad Mayor de San Andres
La Paz, Bolivia

Bruno Condori
Consultative Group on International Agricultural Research – International Potato Center, La Paz, Bolivia

Sayed S.S. Eisa
Agricultural Botany Department, Faculty of Agriculture, Ain Shams University, Cairo, Egypt

Francisco F. Fuentes
Facultad de Agronomía e Ingeniería Forestal, Pontificia Universidad Católica de Chile, Casilla 306–22, Santiago, Chile

Magali Garcia
Faculty of Agronomy,
Universidad Mayor de San Andres
La Paz, Bolivia

Juan Antonio González
Instituto de Ecologia – Area de Botánica
Fundación Miguel Lillo Tucumán
Tucumán, Argentina

Veronica Guwela
International Crops Research Institute for the Semi-Arid Tropics, Lilongwe, Malawi

Sayed Abd Elmonim Sayed Hussin
Agricultural Botany Department, Faculty of Agriculture, Ain Shams University
Cairo, Egypt

Eric N. Jellen
Plant and Wildlife Sciences
Brigham Young University
Provo, UT, USA

Bozena Kolano
Department of Plant Anatomy and Cytology
University of Silesia, Poland

Moses F.A. Maliro
Department of Crop and Soil Sciences, Bunda College Campus, Lilongwe University of Agriculture and Natural Resources
Lilongwe, Malawi

Enrique A. Martínez
Centro de Estudios Avanzados en Zonas Áridas
La Serena and Facultad de Ciencias del Mar
Universidad Católica del Norte
Coquimbo, Chile

Janet B. Matanguihan
Department of Crop and Soil Sciences
Washington State University
Pullman, WA, USA

Peter J. Maughan
Plant and Wildlife Sciences
Brigham Young University
Provo, UT, USA

Florent Mouillot
IRD, UMR 5175 CEFE
Montpellier, France

Kevin M. Murphy
Department of Crop and Soil Sciences
Washington State University
Pullman, WA, USA

Luz Gomez-Pando
Universidad Nacional Agraria La Molina
Agronomy Faculty
Lima, Peru

Adam J. Peterson
Department of Crop and Soil Sciences
Washington State University
Pullman, WA, USA

Milton Pinto
PROINPA Foundation
538 Americo Vespucio St., P.O. Box 1078,
La Paz, Bolivia

Griselda Podazza
Instituto de Ecología, Fundación Miguel Lillo
Tucumán, Argentina

Fernando Eduardo Prado
Facultad de Ciencias Naturales e IML
Fisiología Vegetal
Tucumán, Argentina

Serge Rambal
CNRS, UMR 5175 CEFE
Montpellier, France
Departamento de Biologia
Universidade Federal de Lavras
Lavras, MG, Brazil

Jean-Pierre Ratte
CNRS, UMR 5175 CEFE
Montpellier, France

Carmen Reguilón
Instituto de Entomología, Fundación Miguel
Lillo Tucumán, Argentina

Wilfredo Rojas
PROINPA Foundation
Av. Elias Meneces km 4
El Paso, Cochabamba, Bolivia

Mariana Valoy
Instituto de Ecología, Fundación Miguel Lillo
Tucumán, Argentina

Thierry Winkel
IRD, UMR 5175 CEFE
Montpellier, France

Geyang Wu
School of Food Science
Washington State University
Pullman, WA, USA

Preface

The seeds of this book took root in the summer of 2010, during the first year of our multilocation quinoa trials across three major climatic regions of Washington State. We began growing and evaluating quinoa thanks to generous funding from the Organic Farming Research Foundation, and growers around the state looked on with keen interest. In that first year we tested 44 varieties of quinoa sourced from almost as many diverse geographical locations and we were mildly surprised when only 12 of these actually produced seed in our northern latitude. That first year we were introduced to many of the ongoing challenges we continue to face 5 years later, including susceptibility to preharvest sprouting and downy mildew, photoperiod insensitivity, pollen sterilization resulting from high summer temperatures with little to no rainfall or supplemental irrigation, and the negative effects of aphid and lygus predation. We quickly realized that if quinoa were to become a successfully grown crop in the Pacific Northwest region of the United States, it would require a concerted effort of a transdisciplinary cadre of scientists with a range of expertise, a forward-thinking and risk-taking group of innovative farmers, and a strong supporting cast of distributors, processors, and consumers. From that first year, with only one junior faculty and one undergraduate research intern collaborating with three farmers, the quinoa group at Washington State University has grown into diverse team of over 10 faculty and 10 graduate students, each addressing a key component of quinoa breeding, agronomy, sociology, entomology, or food science. This book is intended to lay the groundwork for the latest quinoa research worldwide and to assist faculty and students new to the crop to gain a foothold of understanding into quinoa genomics and breeding, global agronomy and production, and marketing.

In August 2013, Washington State University hosted the International Quinoa Research Symposium (IQRS). One hundred and sixty enthusiastic participants from 24 countries descended on Pullman, Washington and shared knowledge, questions, obstacles, observations, and ideas on the path forward during an intense, vibrant and thought-provoking 3 days of talks, field visits, poster sessions, and quinoa vodka infused social exploration. Many of the co-authors of the various chapters in this book were attendees and/or presenters at the IQRS, and the symposium provided a safe forum for the open discussion of ideas that have found their way into the chapters of this book. Symposium attendees who have contributed to this book include Didier Bazille, Juan Antonio Gonzalez, Luz Gomez Pando, Rick Jellen, Moses Maliro, Enrique Martinez (in absentia), Jeff Maughan, Sergio Núñez de Arco, Adam Peterson, Wilfredo Rojas, Geyang Wu, and co-editors Janet Matanguihan and Kevin Murphy.

Keynote speakers at the IQRS included Sven-Erik Jacobsen, renowned quinoa researcher from University of Copenhagen, Tania Santivanez from the United Nations Food and Agriculture Organization, and John McCamant, a long-time quinoa farmer and researcher from White Mountain Farms in Colorado, USA. Other esteemed presenters not mentioned included Daniel Bertero from the University of Buenos Aires, Argentina, Morgan Gardner of Washington State University, Frank Morton of Wild Garden Seeds in Oregon, and Hassan Munir of the

University of Agriculture Faisalabad, Pakistan, as well as numerous poster presentations. Finally, the highlight of the symposium for many attendees was the eloquent thoughts delivered by a group of five Bolivian farmers, who traveled to the United States for the first time to join in the international discussion on the many social and political aspects of quinoa cultivation.

This book reflects the many presentations and discussions that took place at the IQRS, and is intended to provide the reader with a comprehensive base knowledge of the current body of knowledge of the ever-expanding, global scientific research of quinoa. In Chapter 1, Gonzalez et al. provide a solid overview of quinoa as an Incan crop, primarily in Peru and Bolivia, now facing a diversity of global challenges. Chapter 2 follows up on this introduction by discussing the origin, domestication, diversification, and cultivation of quinoa from a Chilean perspective.

Chapter 3 by Garcia et al. encapsulates many of the wide-ranging agronomic and agroecological cultural practices of quinoa throughout the major growing regions of South America as a whole. This broad chapter provides a botanical and taxonomical description of quinoa, ecology and phytogeography of quinoa, and many tangible production practices across a wide range of climates, soils, and growing conditions that can be emulated in nontraditional growing regions around the world. Rambal et al. follow this with a description of the historical trends in quinoa yield in the southern Bolivian altiplano, including important lessons from climate and land-use projections in Chapter 4. Valoy et al. then discuss in Chapter 5 the potential of using natural enemies and chemical compounds in quinoa for biological control of pests. This chapter follows up on the agroecological themes discussed in Chapter 3, and compiles and elucidates a vast array of knowledge gained through previous research in this realm of quinoa science, and provides the thoughtful reader many potential ideas for new research in this direction.

In Chapter 6, Peruvian plant breeder Gomez-Pando describes the historical and modern context of quinoa breeding in the Andean regions. Beginning with the effect of farmer selection on seed color, dormancy, seed size and seed coat thickness, salt and drought tolerance, and adaptation to multiple and countless microclimates, Gomez-Pando then moves on to highlight the rise of modern quinoa breeding in the 1960s, the collection of quinoa genetic resources and *in situ* conservation, and the goals and methodology employed by current quinoa breeders.

Matanguihan et al. follow this with an in-depth discussion on the cytogenetics, genomic structure, and diversity of quinoa in Chapter 7. Information on close genetic relatives of *Chenopodium quinoa* are discussed, along with DNA-based molecular genetic tools and linkage maps which can facilitate and accelerate the transfer of exotic genes into *C. quinoa*. Also included in Chapter 7 is a review of phenotypic and genetic diversity studies which show that the genetic variability of quinoa has a spatial structure and distribution. The congruence between genetic differentiation and ecogeography suggests that quinoa all over the southern Andes may be undergoing similar processes of genetic differentiation. Not surprisingly, human activities, specifically seed exchange routes, have significantly affected the genetic structure of quinoa.

In Chapter 8, Rojas and Pinto discuss the *ex-situ* conservation of quinoa genetic resources from a Bolivian perspective. According to Rojas and Pinto, the Bolivian quinoa germplasm collection has the greatest diversity in the world, and this diversity represents the cultural importance of quinoa in Bolivian customs, indigenous consumption, and production. Chapter 8 also provides insight into the center of origin and diversity of quinoa, the geographical distribution of quinoa, and steps needed for the *ex situ* management and conservation of quinoa.

Chapters 9 and 10 discuss quinoa cultivation n two continents, Africa and North America, that are considered nontraditional quinoa production environments. In Chapter 9, Maliro and Guwela describe the necessity of stabilizing food security and alleviating malnutrition in Africa, and the potential for quinoa as a novel crop to make a positive contribution to these efforts. The goals of quinoa breeding in Africa and information from recent quinoa trials in Malawi and Kenya are discussed in an effort to address the challenges and considerations for future quinoa research in Africa. Key among these considerations is the acceptability of quinoa into African diets. In Chapter 10, Peterson and Murphy discuss quinoa introduction to the United States as a crop approximately 30 years ago, and the key breeding, research, and production events in the time period after its introduction. Recent research at Washington State University is highlighted in this chapter.

In Chapter 11, Wu describes the nutritional properties of quinoa that have played an important role in bringing the crop to worldwide attention. Finally, in a refreshing departure from the scientific writing in the previous chapters, Nuñez de Arco provides an insider's view into the marketing of quinoa in Chapter 12. Of particular interest are the personal descriptions and snapshots of the lives of smallholder farmers, of which an estimated 35,000 produce quinoa in Bolivia, who discuss their philosophy of marketing quinoa under the current fluctuations in the supply and demand of this increasingly popular crop.

This book is a reflection of the increasing importance of quinoa in the global market. The roster of contributors – from South America, Europe, Africa and North America – also reflects the expansion of quinoa from its origins to new production areas in the world. It was a pleasure to work with colleagues from countries who have grown quinoa for centuries, and with colleagues from countries which are growing quinoa for the first time. We are indebted to these authors for their willingness to share their expertise and for their cooperation in the process of shaping this book. It is our hope that this book will contribute to quinoa knowledge to benefit growers, students, researchers, and professionals from universities and institutes involved in the improvement of quinoa and its sustainable production.

Kevin M. Murphy
Janet B. Matanguihan

Chapter 1
Quinoa: An Incan Crop to Face Global Changes in Agriculture

Juan Antonio González[1], Sayed S. S. Eisa[2], Sayed A. E. S. Hussin[2], and Fernando Eduardo Prado[3]

[1] *Instituto de Ecologia – Area de Botánica, Fundación Miguel Lillo, Tucumán, Argentina*
[2] *Agricultural Botany Department, Faculty of Agriculture, Ain Shams University (ASU), Cairo, Egypt*
[3] *Facultad de Ciencias Naturales e IML, Fisiología Vegetal, Miguel Lillo 205, 4000 Tucumán Argentina*

INTRODUCTION

Environmental changes have always occurred in the past but in the last decades these have escalated to critical levels, presenting environmental risk to people, especially in terms of food supply, as it affects crop yield, production, and quality. Rapid population growth leads to increase in demand for land and thus to accelerated degradation and destruction of the environment (Alexandratos 2005; IPCC 2007). Probably the most important change driven by human activity is the increasing accumulation of greenhouse gases such as carbon dioxide (CO_2), among others (Wallington et al. 2004; Montzka et al. 2011). Greenhouse gases can absorb and emit infrared radiation, and thus a global earth warming occurs, otherwise known as the greenhouse effect. Many scientists agree that even a small increase in the global temperature would lead to significant climate and weather changes, affecting cloud cover, precipitation, wind patterns, the frequency and severity of storms, and the duration of seasons (Solomon et al. 2009). This scenario will lead to scarce natural resources and the reduction of food production.

The net consequences of global warming on crop physiology and yield are not yet fully understood, but there are some evidences indicating that decrease in yield may be the main response (Parry et al. 2005). Another deleterious effect of global warming is the increase in diseases, especially those caused by fungi and bacteria, as a consequence of higher humidity (Chakraborty et al. 2000; Hunter 2001). As most crops worldwide are well adapted to previous weather conditions, many of these crops will become less productive and may even disappear in a future of increasing climate change. It is therefore necessary to explore plant species as alternative crops or develop new crops to grow under these changing weather patterns. In this sense, it is very important to take into account plant species that grow in different altitudinal levels or those that have thrived in mountain regions for millennia. Mountain plants, especially those adapted and cultivated in different altitudinal levels, may be very important because of the genetic richness that enabled those adaptations.

Quinoa (*Chenopodium quinoa* Willd.), a native grain to the Andean highlands in South America, could be an excellent alternative crop in many regions of the world. Quinoa has been grown

Quinoa: Improvement and Sustainable Production, First Edition. Edited by Kevin Murphy and Janet Matanguihan.

in the Andes about 5,000–7,000 years ago and has been cultivated in different ecological zones from sea level in the northwest region of Chile to altitudes over 4,000 m above sea level (masl) in the Bolivian Altiplano (Fuentes et al. 2009). Owing to this plasticity, quinoa has been introduced to higher latitudes as a new or alternative crop, with reports indicating an acceptable adaptation of this species in the United States, Canada, and Europe (Johnson and Ward 1993; Jacobsen 1997) and recently in Morocco (Jellen et al. 2005), India (Bhargava et al. 2006, 2007), and Italy (Pulvento et al. 2010).

A BRIEF HISTORY OF QUINOA CULTIVATION

Archeological studies provide evidence on the consumption of quinoa as human food thousands of years before the first Spanish conquerors arrived in America. Uhle (1919), taking into account evidences from Ayacucho (Perú), said that quinoa domestication began almost 5,000 years BC. According to Nuñez (1974), quinoa was utilized in the north region of Chile at least 3,000 years BC. Many chronicles and archeological studies provide evidence that quinoa was used by indigenous people for centuries in Colombia, Ecuador, Perú, Bolivia, Chile, and the Argentinean northwest. During pre Columbian times, quinoa seed served as a staple food in the Incan diet, leading the Incas to call it the "mother grain" and considered it as a gift of the sun god, "Inti." It is believed that the Incas considered quinoa to be a sacred plant. Religious festivals including an offering of quinoa in a fountain of gold to the Inti god were held. The Inca Emperor used a special gold tool to make the first furrow of each year's quinoa planting. In Cuzco, ancient Incas worshipped entombed quinoa seeds as the progenitors of the city. The first Spanish conqueror who mentioned quinoa was Pedro de Valdivia. In 1551, he wrote to Carlos I, the Spanish Emperor, about the presence of some crops in the neighboring area of Concepción, Chile and specifically mentioned "*… maize, potatoes and quinuas …*" (Tapia 2009). On the other hand, in the *Comentarios Reales de los Incas*, a book written by Inca Garcilaso de la Vega and published in 1609 in Lisbon, Portugal, Garcilaso mentioned "*quinoa*" as one of the first crops in the Inca Empire (de la Vega 1966). Garcilaso mentioned that there was an intent to export quinoa to Spain but the seeds were nonviable. Other authors had also mentioned the existence of quinoa in Pasto and Quito, Ecuador (Cieza de León 1560), in Collaguas, Bolivia (Ulloa Mogollón 1586), Chiloé island in Chile (Cortés Hogea 1558), and in the Argentinean Northwest and Cordoba province, Argentina (de Sotelo 1583). During the Spanish conquest of South America in the sixteenth century, quinoa was scorned as a "food for Indians" and the conquerors destroyed fields of quinoa, actively suppressing its "non-Christian" production and consumption. The Incan peoples under the yoke of Spanish oppression were forbidden to grow it on pain of death and were forced to grow corn instead. According to Tapia (2009), after the Spanish conquest, the quinoa crop was preserved by Andean peoples in "aynokas" (communal lands) for centuries. This cropping practice also allowed the conservation of quinoa germplasm *in situ* (Tapia 2009). Today, quinoa is cultivated in more than 50 countries beyond the Andes. As a result, the cloud of ambiguity that has enveloped this crop for more than four centuries is beginning to disappear (National Research Council 1989).

NUTRITIONAL VALUE OF QUINOA SEED

There is extensive literature on the chemical composition of quinoa seed (González et al. 1989; Ando et al. 2002; Repo-Carrasco et al. 2003; Abugoch 2009), which cover all nutritional aspects such as chemical characterization of proteins (Brinegar and Goundan 1993; Hevia et al. 2001), fatty acid composition of the seed oil (Wood et al. 1993; Ando et al. 2002), mineral content (Koziol 1992; Konishi et al. 2004; Prado et al. 2010), and nutritional value (Prakash et al. 1993; Ranhotra et al. 1993; Ruales and Nair 1992).

The lipid content of quinoa seed is higher than that in common cereals (Repo-Carrasco-Valencia

2011) and is mainly located in the embryo. The oil of quinoa seed is rich in polyunsaturated fatty acids (linoleic and linolenic) and in oleic acid. Its level of unsaturated fatty acids in relation to human nutrition is better than those in other cereals (Alvarez-Jubete et al. 2009). According to the Food and Agricultural Organization (FAO) recommendations on fats and fatty acids in human nutrition (FAO/WHO 2010), infant food should contain 3–4.5% energy in the form of linoleic acid (LA) and 0.4–0.6% in the form of linolenic acid (ALA), which corresponds to LA/ALA ratio (*n*-6/*n*-3 ratio) between 5 (minimum) and 11.2 (maximum). The LA/ALA ratio of quinoa oil is 6.2 (Alvarez-Jubete et al. 2009) and thus falls within the FAO/WHO (2010) recommended values. Furthermore, a diet with a high *n*-6/*n*-3 ratio promotes the pathogenesis of many degenerative diseases such as cardiovascular disease, cancer, osteoporosis, as well as inflammatory and autoimmune diseases (Simopoulos 2001). The main carbohydrate in quinoa seed is the starch where soluble sugars, that is, sucrose, glucose, and fructose are present at low levels (González et al. 1989). Quinoa starch is located mainly in the perisperm and it occurs both as small individual granules and larger compound granules composed of hundreds of individual granules (Prado et al. 1996). The individual granules are polygonal with a diameter of 1.0–2.5 μm and the compound granules are oval, with a diameter of 6.4–32 μm (Atwell et al. 1983). Quinoa starch is rich in amylopectin and gelatinizes at relatively low temperatures (57–71°C). Moreover, it has excellent freeze-thaw stability attributed to its rich amylopectin content (Ahamed et al. 1996). In comparison with common cereals, quinoa is an excellent source of γ-tocopherol (vitamin E), containing about 5 mg/100 g DM (Ruales and Nair 1993). The content of γ-tocopherol is of particular biological relevance because of its potential anticarcinogenic and anti-inflammatory activities (Jiang et al. 2001). Quinoa also contains significant amounts of riboflavin, thiamine, and, especially, vitamin C that is uncommon in cereals (Koziol 1992; Ruales and Nair 1993; Repo-Carrasco et al. 2003). Recently, it has been demonstrated that quinoa seed also contains high levels of folate (Schoenlechner et al. 2010). The folate content found in quinoa is 132.7 mg/100 g DM, about 10-fold higher than that in wheat seed. Quinoa bran contains a higher amount of folate than flour fraction (Repo-Carrasco-Valencia 2011). Furthermore, quinoa seed does not contain allergenic compounds such as gluten or prolamine or enzyme (protease and amylase) inhibitors present in most common cereals (Zuidmeer et al. 2008) or trypsin and chymotrypsin inhibitors present in soybean seeds (Galvez Ranilla et al. 2009).

Despite its healthy nutritional composition, several cultivars of quinoa contain bitter saponins, glycosylated secondary metabolites in the seed coat that act as antinutrients and deterrents of seed predators such as birds and insects (Solíz-Guerrero et al. 2002). Saponins are concentrated in external layers of the seed (Prado et al. 1996) and include a complex mixture of triterpene glycosides that are derivatives of oleanolic acid, hederagenin, phytolaccagenic acid, serjanic acid, and 3β,23,30-trihydroxy olean-12-en-28-oic acid, which bear hydroxyl and carboxylate groups at C-3 and C-28, respectively (Kuljanabhagavad et al. 2008). Presently, at least 16 different saponins have been detected in quinoa seeds (Woldemichael and Wink 2001). Saponins are reported to be toxic for cold-blooded animals and have been used as fish poison by South American inhabitants (Zhu et al. 2002). They have some adverse physiological effects, as they are membranolytic against cells of the small intestine and possess hemolytic activity (Woldemichael and Wink 2001). Moreover, saponins form complexes with iron and may reduce its absorption.

Although saponins have negative effects, they also have positive effects such as reducing serum cholesterol levels, possessing anti-inflammatory, antitumor, and antioxidant activities, and enhancing drug absorption through the mucosal membrane. Saponins also exhibit insecticidal, antibiotic, antiviral, and fungicidal properties (Kuljanabhagavad and Wink 2009). Furthermore, saponins act as immunological and absorption adjuvant to enhance antigen-specific antibody and mucosal response (Estrada et al. 1998).

Saponin content varies among genotypes, ranging between 0.2 and 0.4 g/kg DM (sweet genotypes) and 4.7 and 11.3 g/kg DM (bitter genotypes). Therefore, selection of sweet genotypes with very low saponin content in the seeds is one of the main breeding goals in quinoa. However, selection for sweet genotypes is retarded by cross-pollination (Mastebroek et al. 2000). The tissue containing saponins is of maternal origin, and the saponin content of the seed reflects the genotype of the plant from which the grain is harvested (Ward 2001). According to Gandarillas (1979), the saponin content trait is controlled by two alleles at a single locus, with the bitter allele (high saponin) dominant to the sweet allele (low saponin). More recently, researchers have observed that saponin content in quinoa seed is a continuously distributed variable and is therefore more likely to be polygenically controlled and quantitatively inherited (Galwey et al. 1990; Jacobsen et al. 1996).

Quinoa seeds must be freed of seed coat saponins before consumption. Saponins can be easily eliminated by water washing or abrasive dehulling. There was no difference in the removal of saponins observed between the two methods (Ridout et al. 1991), although the latter method has the advantage of not generating wastewater. However, some nutrients can be lost when the abrasive dehulling method is used (Repo-Carrasco-Valencia 2011).

Among the nutritional attributes of quinoa seed, prominent is its high-quality protein that is gluten-free and has an exceptional amino acid balance. The presence of essential amino acids such as methionine, threonine, lysine, and tryptophan are very important because they are limiting amino acids in most cereal grains (Gorinstein et al. 2002). The high level of tryptophan found in the seed of the Bolivian cultivar "Sajama" is noteworthy (Comai et al. 2007). Protein quality is determined by its biological value (BV), which is an indicator of protein intake by relating nitrogen uptake to nitrogen excretion. The highest values of BV correspond to whole egg (93.7%) and cow milk (84.5%) (Friedman 1996). The protein of quinoa seed has a BV of 83%, which is higher than that of fish (76%), beef (74.3%), soybean (72.8%), wheat (64%), rice (64%), and corn (60%) protein (Abugoch 2009).

According to the FAO/WHO nutritional requirements for 10- to 12-year-old children, quinoa protein possesses adequate levels of phenylalanine, tyrosine, histidine, isoleucine, threonine, and valine (FAO/WHO 1990). Consequently, there is no need to combine quinoa seed with other protein sources to supply human requirements for essential amino acids. This nutritional aspect of quinoa is very significant as it can provide a new protein source for a good diet. Quinoa may also be an important alternative crop for mountainous regions of the world, where many people live. In these regions, there are severe constraints in obtaining good quality food and quinoa will be able to supply the nutrient requirements that other crops cannot, especially for children.

The nutritional composition of quinoa seed is determined by both the genotype and the environment. The metabolism of nitrogen-containing compounds, that is, proteins and amino acids, may be strongly affected by environmental conditions (Triboi et al. 2003). In a recent ecophysiological study carried out on 10 quinoa cultivars from the Bolivian highland region (Patacamaya site, 3,600 masl) and northwest Argentinean lowland region (Encalilla site, 2,000 masl), González et al. (2011) demonstrated that in six cultivars (Amilda, Kancolla, Chucapaka, Ratuqui, Robura, and Sayaña) the protein content showed an increment in the lowland growing site when compared with seeds from the highland site. In contrast, four cultivars (CICA, Kamiri, Sajama, and Samaranti) showed a decreased content (Table 1.1). Similarly, it has also been demonstrated that both the content and the composition of quinoa saponins are affected by environmental conditions. Both drought and salinity decreased the content and profile of saponins of quinoa cultivars (Solíz-Guerrero et al. 2002; Dini et al. 2005; Gómez-Caravaca et al. 2012). In effect, many metabolic and physiological aspects of crops are affected by agroecological conditions (Triboi et al. 2003). Soil type and climatic conditions also play a crucial role in the success of crops. These are important results and should be taken into account when choosing a commercial cultivar.

Table 1.1 Protein content (g/100 g DW) of quinoa seeds cultivated in two agroecological sites (Patacamaya, 3,600 masl and Encalilla, 2,000 masl).

Cultivar	Patacamaya (g/100 g DW)	Encalilla (g/100 g DW)	Difference (%)
Amilda	11.41	12.5	8.7
Kancolla	14.44	15.17	4.8
Chucapaka	11.67	14.34	18.6
CICA	15.46	13.46	−14.9
Kamiri	13.98	13.12	−6.6
Ratuqui	10.38	15.53	33.2
Robura	9.62	10.43	7.8
Sajama	12	9.15	−31.1
Samaranti	12.26	9.34	−31.3
Sayaña	11.36	13.85	18.0

Quinoa may be considered as a potential alternative crop in many regions of the world due to the nutritional quality of its seed and its good potential for adaptation (González et al. 1989, 2012; Dini et al. 2005; Comai et al. 2007; Thanapornpoonpong et al. 2008). Probably all these aspects were taken into account by the FAO when it included quinoa in the list of most promising crops for world food security and human nutrition in the twenty-first century (FAO 2006). The National Aeronautics and Space Administration (NASA) also included quinoa within the Controlled Ecological Life Support System (CELSS) to augment the inadequate protein intake of astronauts in long-duration space travel (Schlick and Bubnehiem 1993).

BOTANICAL AND GENETIC CHARACTERISTICS OF THE QUINOA PLANT

Quinoa is an annual Amaranthaceae. This Andean grain is an important crop of the Andean region in South America from Colombia (2°N) to central Chile (40°S) (Risi and Galwey 1984; Jacobsen 2003). Despite its wide latitudinal distribution, quinoa also has a broad altitudinal distribution. Quinoa may be cultivated at sea level, middle mountain (between 2,000 and 3,000 masl), and high mountain (above 3,000 masl). In relation to this altitudinal and latitudinal distribution pattern, Tapia (2009) distinguished at least five ecotypes of quinoa: (i) *Valley* quinoa, which are late-ripening, with plant heights 150–200 cm or more, and growing at 2,000 and 3,000 masl; (ii) *Altiplano* quinoa, which can withstand severe frost and low precipitation, growing around Titicaca Lake in Bolivia and Perú; (iii) *Salar* quinoa, which can tolerate salty soils with high pH values, growing on the plains of the Bolivian Altiplano such as Uyuni and Coipasa; (iv) *Sea level* quinoa, generally small plants (near 100 cm) with a few stems and bitter grains, found in the south of Chile; and (v) *Subtropical* quinoa, which have small white or yellow grains, growing in the inter-Andean valleys of Bolivia. Royal Quinoa (Quinoa Real) is probably the most recognized quinoa cultivar in the international market. It is a bitter variety and is only produced in Bolivia, particularly in the districts of Oruro and Potosí, around the salt flats of Uyuni and Coipasa. The microclimatic conditions and physicochemical properties of the soil offer the appropriate habitat for the production of this type of quinoa (Rojas et al. 2010). Morphophenological characteristics of quinoa show that there is a huge diversity in varieties or local ecotypes (del Castillo et al. 2007). Therefore, available commercial quinoas exhibit wide genetic diversity, showing great variability in plant color, inflorescence and seeds, inflorescence type, protein, saponin and betacyanine contents, and calcium oxalate crystals in leaves. This extreme variability may reflect wide adaptation to different agroecological conditions such as soil, rainfall, nutrients, temperature, altitude, drought, salinity, and UV-B radiation.

Quinoa is a dicotyledonous annual herbaceous plant usually erect, with a height of about 100–300 cm, depending on environmental conditions and genotype. Leaves are generally lobed, pubescent, powdery, rarely smooth, and alternatively inserted on a woody central stem. The plant may be branched or unbranched, depending on variety and sowing density. Stem color may be green, red, or purple. The leafy flower cluster (a panicle with groups of flowers in glomerulus) arises predominantly from the top of the plant and may also arise from the leaf junction (axil) on the stem. Flowers are sessile, of the same color as the sepals, and may be hermaphrodite, pistillate, or male sterile. The stamens have short filaments bearing basifixed anthers; the style has two or three feathery stigma. The fruit occurs in an indehiscent achene, protected by the perigonium. The seeds are usually somewhat flat, measure 1–2.6 mm, and approximately 250–500 seeds comprise 1 g. The seeds also exhibit a great variety of colors – white, yellow, red, purple, brown, and black, among others. Seed embryo can be up to 60% of the seed weight and forms a ring around the endosperm. The taproot (20–50 cm long) is profusely branched and forms a dense web of rootlets that penetrate to about the same depth as the plant height (National Research Council 1989).

The vegetative period of quinoa is related to photoperiod sensitivity and varies between 120 and 240 days. Some varieties, such as CO-407 from Chile, have a vegetative period between 110 and 120 days, but others, such as the CICA variety, have more than 200 days. On the other hand, *C. quinoa* is a C_3 species confirmed by anatomical studies and carbon isotope discrimination (González et al. 2011). The $\delta^{13}C$ values of leaves of 10 varieties of quinoa ranged from a minimum of −27.3‰ to a maximum of −25.2‰ (Table 1.2). Typical values of $\delta^{13}C$ in C3 species can ranges from −35 to −20‰ (Ehleringer and Osmond 1989).

C. quinoa is an allotetraploid ($2n = 4x = 36$) and exhibits disomic inheritance for most qualitative traits (Simmonds 1971; Risi and Galwey 1989; Ward 2001; Maughan et al. 2004). The species closest to cultivated quinoa are *Chenopodium hircinum* and *Chenopodium berlandieri*, whose basic chromosome number ($2n = 4x = 36$) is the same as that of the cultivated types, and *Chenopodium petiolare* and *Chenopodium pallidicaule*, which have $2n = 2x = 18$ chromosomes (Fuentes et al. 2009). Quinoa species includes both domesticated cultivars (subsp. *quinoa*) and free-living, weedy forms (subsp. *milleanum* or *melanospermum*) (Wilson 1981, 1988). Domesticated and weedy quinoa populations are sympatric, and share a fundamentally autogamous reproductive system as well as a wide range of variation in leaf and grain size and color (del Castillo et al. 2007). Wild and domesticated populations of quinoa exist under cultivation, which indicates that domesticated quinoas are generally accompanied by wild populations in their various distribution areas. Thus, natural hybridization between wild and domesticated populations probably occurs easily (Fuentes et al. 2009). The highest variation in cultivated quinoa is found near Titicaca Lake, between Cuzco (Peru) and Lake Poopó (Bolivia), and this is where scientists believe the crop was first domesticated (Heiser and Nelson 1974). The main varieties known in this region are Kancolla, Cheweca, Witulla, Tahuaco, Camacani, Yocara, Wilacayuni, Blanca de Juli, Amarilla de Maranganí, Pacus, Rosada de Junín, Blanca de Junín, Hualhuas, Huancayo, Mantaro, Huacariz, Huacataz, Acostambo, Blanca Ayacuchana, and Nariño in Peru and Sajama,

Table 1.2 Carbon isotope composition $\delta^{13}C$ of 10 varieties of quinoa.

Cultivar	$\delta^{13}C$
Amilda	−25.6
Chucapaca	−26.3
CICA	−26.6
Kancolla	−27.3
Kamiri	−26.7
Ratuqui	−26.4
Sayaña	−26.3
Robura	−25.7
Sajama	−25.2
Samaranti	−25.6

Real Blanca, Chucapaca, Kamiri, Huaranga, Pasancalla, Pandela, Tupiza. Jachapucu, Wila Coymini, Kellu, Uthusaya, Chullpi, Kaslali, and Chillpi in Bolivia (Hernández Bermejo and León 1994). Throughout the Andean region, there are several genebanks where over 2,500 quinoa accessions are preserved in cold-storage rooms: in Peru, at the experimental stations of Camacani and Illpa (Puno), K'ayra and Andenes (Cuzco), Canaan (Ayacucho), Mantaro y Santa Ana (Huancayo), Baños del Inca (Cajamarca); in Bolivia, at the Patacamaya station of the IBTA; and in Ecuador, at the Santa Catalina station of INIAP.

QUINOA AND ENVIRONMENTAL STRESSES: DROUGHT AND SALINITY

Soil salinization is one of the major environmental issue affecting crop production, especially in marginal landscapes or areas with limited resources (Munns and Tester 2008; Rengasamy 2010; Munns 2011; Hussin et al. 2013). The intensive use of valuable natural resources such as land and water, along with high soil evapotranspiration and inefficient irrigation systems associated with poor water and soil management, inevitably accelerate secondary salinization that usually results in the loss of productive areas (Munns 2005; Hussin et al. 2013). Nearly 20% of the world's cultivated areas and about half of the world's irrigated lands are salt affected (FAO 2008). Out of the current 230 Mha of irrigated land, 45 Mha are salt-affected soils (19.5%), and of the almost 1,500 Mha dry agricultural land, 32 Mha are salt affected to varying degrees by human-induced processes (Munns and Tester 2008). Salinization of irrigated lands causes a loss of US$12 billion of the annual global income (Ghassemi et al. 1995).

In this context, enhancing salt tolerance of the conventional crops has proved to be somewhat elusive in terms of genetic manipulation to allow greater yields in salt-affected soils and marginal areas (Flowers 2004). The results, although promising, remain insignificant so far (Läuchli and Grattan 2007). An alternative approach is the use of naturally occurring xero-halophyte for crop production, "cash crop halophytes," as they already have the required level of salt tolerance (Lieth et al. 1999). The sustainable utilization of halophytes as cash crops may significantly contribute toward food, feed, fuel, wood, fiber, chemical production, and environmental quality (dune stabilization, combating desertification, bioremediation, or CO_2 sequestration) in many countries (Geissler et al. 2010; Hussin et al. 2013). Hence, research has focused more and more on the identification and selection of plant species such as *C. quinoa* that are naturally tolerant to drought and salinity.

Quinoa is one of the few crops, if not the only crop, able to grow in the most extreme environmental conditions (Jacobsen et al. 2003). In effect, quinoa can be cultivated from sea level to 4,000 masl, even in the Bolivian Altiplano with an extreme altitude of 4,200 masl. Quinoa is also remarkably adaptable to different agroecological zones. It adapts to hot, dry climates, can grow in areas of varying relative humidity, ranging from 40% to 88%, and can withstand temperatures from −4 to 38°C. Quinoa can grow in marginal soils lacking in nutrients, in soils with a wide range of pH from acid to basic (Boero et al. 1999), and even tolerates soil infertility (Sanchez et al. 2003). It also has excellent tolerance to extreme frost (Halloy and González 1993; Jacobsen et al. 2005, 2007), long drought periods (Vacher 1998; González et al. 2009a; Jacobsen et al. 2009), salinity (González and Prado 1992; Prado et al. 2000; Rosa et al. 2009; Ruffino et al. 2010; Hariadi et al. 2011), and high solar radiation (Palenque et al. 1997; Sircelj et al. 2002; Hilal et al. 2004; González et al. 2009b). It has high water use efficiency (WUE) shown by its tolerance or resistance to lack of soil moisture and produces acceptable yields with rainfall of 100–200 mm (Garcia et al. 2003, 2007; Bertero et al. 2004). Quinoa resists up to 3 months of drought at the beginning of its growth cycle. To make up for this part of its growth cycle, the stalk becomes fibrous and roots strengthen. When rains come, it recovers physiological activity (National Research Council 1989). Some varieties can grow in salt concentrations similar to those found in seawater

(40 dS/m) and even higher, well above the threshold for any known crop species (Hariadi et al. 2011; Razzaghi et al. 2011).

Salt tolerance is a complex trait and attributed to a plethora of interconnected morphological, physiological, biochemical, and molecular mechanisms. These mechanisms are linked to the major constraints of salinity on plant growth (i.e., osmotic effects, restriction of CO_2 gas exchange, ion toxicity, and nutritional imbalance) and operate in coordination to alleviate both the cellular hyperosmolarity and ion disequilibrium (Koyro 2006; Flowers and Colmer 2008; Geissler et al. 2009). The primary deleterious effect of soil salinity on plant growth is due to an osmotic effect, resulting from the lower soil water potential (Ψ), defined as the work water can do as it moves from its present state to the reference state. The reference state is the energy of a pool of pure water at an elevation defined to be zero (Munns 2002; Koyro et al. 2012). A low value of (Ψ) interferes with plant ability to take up water from the soil and, hence, causes a growth reduction, along with a range of physiological and biochemical changes similar to those caused by water deficit (Larcher 2001; Schulze et al. 2002; Munns 2005). To endure osmotic constraint, salt-tolerant plants are more restrictive with water loss via transpiration by a sensitive stomatal closure response. Inevitably, this leads to a decrease in the apparent photosynthetic rate due to a restricted availability of CO_2 for the carboxylation reaction (stomatal limitation of photosynthesis) (Huchzermeyer and Koyro 2005; Flexas et al. 2007; Dasgupta et al. 2011; Benzarti et al. 2012), thereby suppressing plant growth and productivity (D'Souza and Devaraj 2010; Gorai et al. 2011; Tarchoune et al. 2012; Yan et al. 2013).

According to several studies, quinoa tolerance to drought and salinity stresses is dependent on its vegetative stage (Bosque Sanchez et al. 2003; Garcia et al. 2003; Jacobsen et al. 2003). At the cotyledonary stage, the high adaptability of quinoa to soil salinity is related to metabolic adjustment. In studies carried out with seedlings of the Sajama cultivar, it was demonstrated that salinity tolerance depends on improved metabolic control of ion absorption and osmotic adjustment through osmolyte accumulation derived from a salt-induced altered carbohydrate metabolism (Rosa et al. 2009; Ruffino et al. 2010), whereas in early maturing stage, it is also related to structural and physiological adaptations. In this way, quinoa avoids the negative effects of drought through the development of a deep and dense root system, reduction of the leaf area, leaf dropping, special vesicular glands (salt bladders), small and thick-walled cells adapted to losses of water without loss of turgor even at severe water losses, and stomatal closure (Jensen et al. 2000; Adolf et al. 2013).

Although quinoa was classified as a highly salt-tolerant species (Jacobsen 2003; Hariadi et al. 2011; Razzaghi et al. 2011; Eisa et al. 2012; Adolf et al. 2013), many quinoa cultivars show distinct variability in their germination and growth responses to salinity. More than 200 quinoa accessions have been tested under saline conditions and found to be different in their responses to salinity. Differences were observed at germination stage and also later during the vegetative growth stage (Adolf et al. 2012). Moreover, salt tolerance at germination is not necessarily correlated with the degree of tolerance at later developmental stages. Eisa et al. (2012) found that the growth of the Peruvian quinoa cultivar "Hualhuas" was slightly stimulated in response to a low salinity level (20% seawater salinity). The same trend of salt-induced growth stimulation has been recently observed for the cultivar "CICA" (Fig. 1.1). The overall growth of CICA plants based on fresh weight (FW) gain was significantly increased ~85% compared with control plants grown under non-saline conditions. This increase was mainly a result of increased shoot FW rather than root FW (Fig. 1.1). Similar salt-induced stimulation of growth has also been reported for other Peruvian and Bolivian quinoa cultivars (Wilson et al. 2002; Koyro and Eisa 2008; Hariadi et al. 2011). Furthermore, the Andean hybrid grown at salinity level of 11 dS/m showed increases in both leaf area and dry mass when comparing with plants grow at control salinity level of 3 dS/m. As shown in Fig. 1.1, salinity tolerance threshold for CICA variety was at 200 mM NaCl, whereas

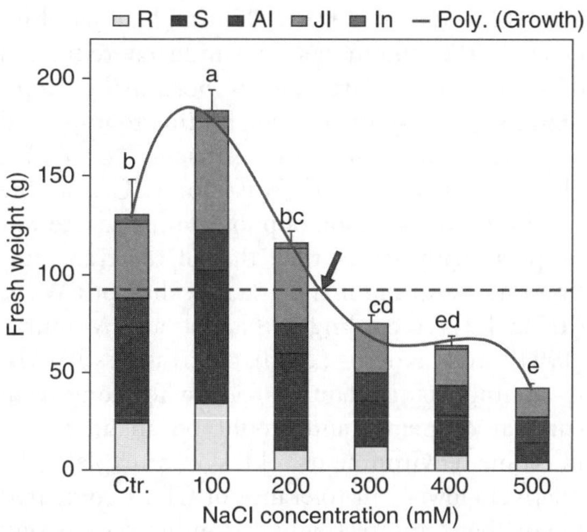

Fig. 1.1 Development and growth responses of different organs (expressed as fresh weights) of *C. quinoa* cv. CICA grown at different NaCl concentrations. The dotted line marks the C_{50} value. Each column represents the mean value of three replicates and the bars represent standard deviations. Columns with the same letter are not significantly different at $P \leq 0.05$, Duncan test. (R) root, (S) stem, (Al) adult leaf, (Jl) juvenile leaf, and (In) inflorescence.

C_{50} was slightly above 40% seawater salinity. The salinity tolerance threshold is the salt level that leads to the initial significant reduction in the maximum expected yield (Shannon and Grieve 1999), whereas C_{50} is the water salinity leading to 50% growth reduction in the maximum expected yield. In contrast, salinity levels above threshold value (supraoptimal condition) severely inhibit plant growth in many quinoa cultivars (Hariadi et al. 2011; Eisa et al. 2012). Seawater salinity level (500 mM NaCl) led to a significant reduction (~66%) in the FW of CICA plants relative to the control (Fig. 1.1). Inhibition of the initiation of new leaves and the formation of small leaves, some with symptoms of nutrient disorders, might contribute to the low FW observed at this salinity level. Interestingly, the plants displayed conspicuous growth and continued to grow even at seawater salinity levels (Fig. 1.1). Together, these results indicate that the CICA cultivar is highly salt tolerant and productive, capable of growing even under sea water salinity levels.

Salinity stress results in a decrease of photosynthesis in a wide variety of plant species (Sudhir and Murthy 2004). However, many halophyte species show higher level of photosynthesis under conditions of elevated salinity (Andersone et al. 2012), depending on the level of salt tolerance of the species and/or genotypes (Brock et al. 2007). Quinoa cultivars also show different photosynthetic responses, depending on parental origin. Recently, Adolf et al. (2012) found significant differences in both photosynthetic CO_2 assimilation and stomatal conductance when two varieties of quinoa when grown under saline conditions. "Utusaya," originating from the salar region of Bolivia, maintained a relatively high stomatal conductance, with only 25% reduction in net CO_2 assimilation when compared with the untreated control plants. In contrast, the cultivar "Titicaca" that has been bred in Denmark showed a higher decrease in stomatal conductance and also a 67% reduction in CO_2 assimilation. Interestingly, in the Utusaya variety, both the stomatal conductance and the photosynthesis rate were generally low under non-saline conditions, whereas these did not decrease in the Titicaca variety. Thus, it may be assumed that in saline environments, the Utusaya variety has a genetically improved osmoregulator mechanism to counteract the deleterious osmotic effects of salt and has less need to reduce water loss by transpiration (Adolf et al. 2013). A similar trait was observed between the CICA (less salt tolerant) and the Hualhuas (more salt tolerant) cultivars grown under increasing saline levels. The CO_2 assimilation (net photosynthetic rate, P_N) of the CICA cultivar steadily and significantly declined with increasing water salinity, reaching only 1.5% of the control values at seawater salinity treatment (Table 1.3). This result was consistent with observations on the effect of salinity on photosynthesis in many salt-tolerant species (Ashraf 1999; Bayuelo-Jiménez et al., 2003; Qiu et al. 2003; Koyro 2006). In a previous study, however, Eisa et al. (2012) showed that the photosynthetic activity of the Hualhuas cultivar was less affected with salt-induced reduction of about 72% at seawater salinity level. Furthermore, the photosynthetic responses of the cultivars CICA and Hualhuas correspond with the assumptions of Kao et al. (2006) and Moradi and Ismail (2007), who assume that relatively higher salt-tolerant species would have less reduced net photosynthesis. On the other hand, the reduction of P_N observed in CICA coincided with the progressive decrease of stomatal conductance (C_S), suggesting that salinity impacted the photosynthesis of CICA plants, at least partly, by an enhanced stomatal closure. Positive correlations between P_N and C_S have been found in *C. quinoa*, Hualhaus cultivar (Eisa et al. 2012), *Atriplex prostrata* (Wang et al. 1997), *Atriplex nummularia* and *Atriplex hastata* (Dunn and Neales 1993), *Atriplex centralasiatica* (Qiu et al. 2003), and *Avicennia marina* (Ball and Farquhar 1984).

According to Moradi and Ismail (2007) and Centritto et al. (2003), reduction of stomatal conductance is a significant way to decrease water loss from the leaves via transpiration and could be considered as an adaptive feature for salt tolerance. In CICA plants, the salt-induced reduction of C_S gives a strong inhibition of the transpiration rate (E), which reaches a minimum value at the highest salinity treatment (Table 1.3). This would contribute to conservation of water and also maintain a positive water balance. In fact, lower values of E represent an additional adaptive mechanism for coping with high salinity levels, as it could reduce salt loading into leaves and hence prolong the leaf lifespan by maintaining a subtoxic level of salt (Everard et al. 1994; Koyro 2006).

The coordinated regulation of CO_2/H_2O gas exchange is considered a key determinant for plant growth and biomass production under saline conditions (Romero-Aranda et al. 2001; Lu et al. 2002; Gulzar et al. 2003, 2005). In Hualhuas, Eisa et al. (2012) found that salt-induced reduction of transpiration rate was proportionally larger than the photosynthetic rate, leading to improved photosynthetic water use efficiency (PWUE). However, this is not the case for CICA, as the salt-induced reduction of photosynthetic rate was proportionally larger than that of the transpiration rate, resulting in a marked decline of PWUE (Table 1.3). According to Naidoo and Mundree (1993) and Koyro (2000), increasing PWUE is an important adaptive feature for long-term survival of plants and would be an advantage in saline environments. This may explain the relatively lower salt tolerance of CICA compared to Hualhaus. Interestingly, salt-induced reduction of P_N in CICA showed a positive correlation with C_S, but not with intercellular CO_2 concentration (C_i) (Table 1.3), suggesting that C_i is not the limiting factor for photosynthesis reduction in CICA under saline conditions.

Non-stomatal inhibition of photosynthesis in salt-stressed plants, particularly under severe stress conditions, has also been reported for several other crop species such as *Gossypium hirsutum* and *Phaseolus vulgaris* (Brugnoli and Lauteri 1991), *Oryza sativa* (Dionisio-Sese and Tobita 2000), *Helianthus annuus* (Steduto et al. 2000), and *Beta vulgaris* (Dadkhah 2011), among others. This inhibition of photosynthetic capacity has been attributed to an inhibited coupling factor activity (Tezara et al. 2008), reduced carboxylation efficiency (Wise et al. 1992; Jia and Gray 2004), reduced amount and/or activity of crucial photosynthetic enzymes such as Rubisco (Parry

Table 1.3 Effect of elevated water salinity on the net photosynthesis rate (P_N), transpiration rate (E), Stomatal conductance (C_s), ratio of the internal to the external CO_2 concentration (C_i/C_a), and photosynthetic water use efficiency (PWUE) of *C. quinoa* cv. CICA. All of these values are at the light saturation point of photosynthesis.

Treatments	P_N (μmol m^{-2} s^{-1})	E (mmol/m^{2}s)	C_s (mmol H_2O/m^{2}s)	C_i/C_a	PWUE (%)
Control	$16.615^a \pm 1.011$	$2.733^a \pm 0.234$	$0.164^a \pm 0.018$	$0.491^a \pm 0.022$	$0.625^a \pm 0.019$
100 mM	$12.310^b \pm 0.122$	$2.417^a \pm 0.045$	$0.140^b \pm 0.003$	$0.588^b \pm 0.006$	$0.510^{bc} \pm 0.006$
200 mM	$10.907^c \pm 0.119$	$1.998^b \pm 0.019$	$0.111^c \pm 0.002$	$0.550^b \pm 0.010$	$0.546^b \pm 0.007$
300 mM	$8.088^d \pm 0.398$	$1.232^c \pm 0.148$	$0.064^d \pm 0.008$	$0.577^b \pm 0.018$	$0.446^c \pm 0.034$
400 mM	$1.105^e \pm 0.240$	$0.357^d \pm 0.032$	$0.017^e \pm 0.002$	$0.747^c \pm 0.029$	$0.256^d \pm 0.034$
500 mM	$0.237^e \pm 0.048$	$0.280^d \pm 0.009$	$0.012^e \pm 0.000$	$0.882^e \pm 0.015$	$0.171^e \pm 0.018$

Means within a column followed by the same letter are not significantly different at $P \leq 0.05$, as determined by Duncan test. Each mean represents three replicates.

et al. 2002), reduced ribulose-1,5-bisphosphate (RuBP) regeneration (Giménez et al. 1992; Gunasekera and Berkowitz 1993), and reduction of the contents of photosynthetic pigments (Seemann and Critchley 1985; Hajar et al. 1996; Koyro 2006).

Salinity and drought may also impair photosynthesis by disturbing the photochemical reactions in the chloroplast (Tezara et al. 2005; Hura et al. 2007). Furthermore, as an indirect consequence of stomatal closure induced by salt and/or drought stress, restriction in intercellular CO_2 concentration should increase susceptibility to photochemical damages as excessive light energy at PSII level increases when CO_2 assimilation rates are low (Silva et al. 2010). This effect, however, seems to be species specific. For example, sorghum (*Sorghum bicolor*) plants subjected to salt stress showed a strong disturbance of photochemical activity (Netondo et al. 2004), whereas cowpea (*Vigna unguiculata*) plants subjected to progressive drought displayed slight changes in the PSII activity (Souza et al. 2004). Moreover, it has been demonstrated that stomatal closure reduces the CO_2/O_2 ratio in leaves and inhibits the fixation of CO_2, which induces an increased ROS generation via enhanced leakage of electrons to oxygen (Foyer and Noctor 2000). Therefore, in salt-treated plants, a low rate of CO_2 assimilation can result in oxidative stress.

Salt-induced leaf succulence and reduction in chlorophyll content has also been observed in quinoa plants in response to high water salinity (Eisa et al. unpublished results). With quinoa being a salt-tolerant species, it is conceivable that in salt-stressed plants the stomatal closure allows the leaves to either develop an additional scavenging mechanism in their light reaction centers or utilize the excessive energy for ion excretion or sequestration. This condition may lead to a reduction of the flow of electrons through the photosystems (reduction of the apparent quantum efficiency) (Table 1.3). Furthermore, the presence of a dense layer of bladder hairs filled with salt on the surface of leaves can form a strong reflective light (Fig. 1.2). Thus, this light-reflecting layer is thought to protect the photosystems from overreduction and photoinhibition under stress conditions (Freitas and Breckle 1992; Agarie et al. 2007; Orsini et al. 2011).

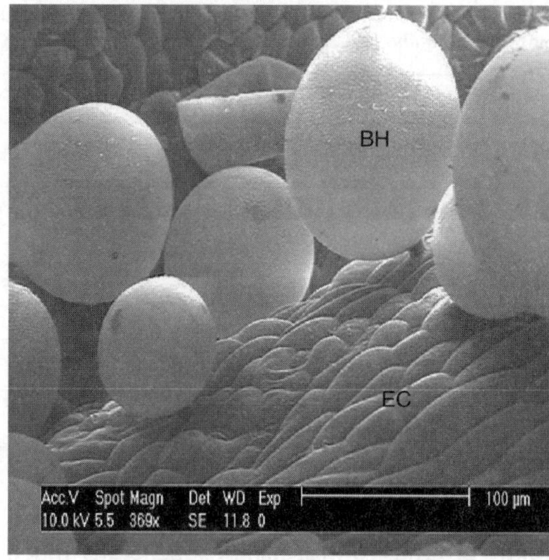

Fig. 1.2 Representative SEM micrographs of the juvenile leaf surface showing the various stages of bladder hairs development. BH, bladder hair and EC, epidermal cells.

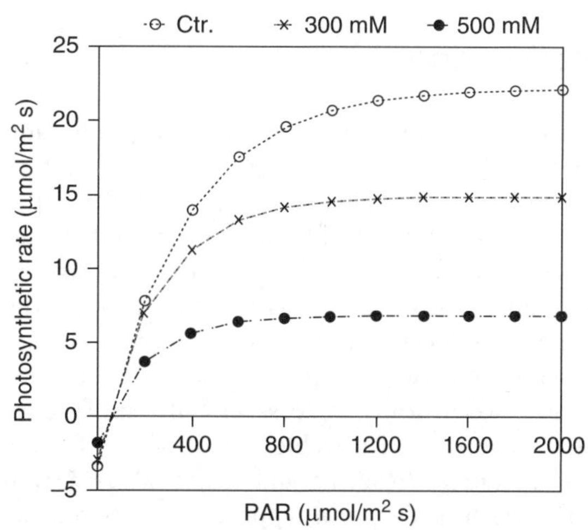

Fig. 1.3 Light response curves of *C. quinoa*, CICA cultivar, at different NaCl concentrations. Values are the mean of three independent measurements.

Light saturation point (L_s) gradually decreased with increasing water salinity, as shown in the cultivar CICA, commensurate with the reduction in photosynthetic capacity (Fig. 1.3). This might partially be due to salt-induced reduction in chlorophyll concentration per unit area

Table 1.4 Calculated photosynthetic efficiency (Φ_c), dark respiration (D_r), light compensation point (L_c), and light saturation point (L_s) of *C. quinoa* cv. CICA plants grown under various NaCl salinities.

Treatments	Φ_c[μmol CO_2 μmol^{-1} Quantum]	D_r[μmol m^{-2} s^{-1}]	L_c[μmol m^{-2} s^{-1}]	L_s[μmol m^{-2} s^{-1}]
Control 300 mM	0.062 0.052	−3.343 −2.627	49.945 46.172	872.297 652.115
500 mM	0.034	−1.756	45.722	506.239

The calculation was done using SigmaPlot software.

(Eisa et al. 2012). As a consequence, the calculated CO_2 compensation point (L_c) decreased in response to water salinity. Furthermore, the calculated dark respiration (D_r) decreased markedly with elevated water salinity, being minimal at 500 mM NaCl (Table 1.4). Salt-induced reduction in respiration rates might be due to the fact that the maintenance respiration of rapidly growing control plants is generally much higher than that of the more slowly growing plants grown under high saline stress (Koyro and Huchzermeyer 1999).

CONCLUSION

New goals and insights into food production and market development are needed in light of dwindling fresh water resources and the rapid loss of arable land due to soil salinization. Domestication of native halophytes and increasing the salt tolerance of glycophytic crops through the genetic engineering could achieve these goals, but research on these processes is still in the early stages. Realistically, success in both approaches will require considerable investment of time and resources (Rozema and Schat 2013). Given this scenario, *C. quinoa* appears to be a reliable new crop option to sustain the food supply for a rapidly growing world population. Its high tolerance to salinity and drought, together with its excellent nutritional quality, makes it an ideal crop to contribute to food security for the twenty-first century.

REFERENCES

Abugoch LE. 2009. Quinoa (*Chenopodium quinoa* Willd.): composition, chemistry, nutritional and functional properties. Adv Food Nutr Res 58:1–31.

Adolf VI, Shabala S, Andersen MN, Razzaghi F, Jacobsen SE. 2012. Varietal differences of quinoa's tolerance to saline conditions. Plant Soil 357:117–129.

Adolf VI, Jacobsen SE, Shabala S. 2013. Salt tolerance mechanisms in quinoa (*Chenopodium quinoa* Willd.). Environ Exp Bot 92:43–54.

Agarie S, Shimoda T, Shimizu Y, Baumann K, Sunagawa H, Kondo A, Ueno O, Nakahara T, Nose A, Cushman JC. 2007. Salt tolerance, salt accumulation, and ionic homeostasis in an epidermal bladder-cell-less mutant of the common ice plant *Mesembryanthemum crystallinum*. J Exp Bot 58:1957–1967.

Ahamed N, Singhal R, Kulkarni P, Pal M. 1996. Physicochemical and functional properties of *Chenopodium quinoa* starch. Carbohydr Polym 3:99–103.

Alexandratos N. 2005. Countries with rapid population growth and resource constraints: issues of food, agriculture, and development. Popul Dev Rev 31:237–258.

Alvarez-Jubete L, Arendt EK, Gallagher E. 2009. Nutritive value and chemical composition of pseudocereals as gluten-free ingredients. Int J Food Sci Nutr 60:240–257.

Andersone U, Samsone I, Ievinsh G. 2012. Protection of photosynthesis in coastal salt marsh plants *Aster tripolium* and *Hydrocotyle vulgaris* in conditions of increased soil salinity. Environ Exp Biol 10:89–97.

Ando H, Chen YC, Tang HJ, Shimizu M, Watanabe K, Mitsunaga T. 2002. Food components in fractions of quinoa seed. Food Sci Tech Res 8:80–84.

Ashraf M. 1999. Interactive effect of salt (NaCl) and nitrogen form on growth, water relations and photosynthetic capacity of sunflower (*Helianthus annuus* L.). Ann Appl Biol 135:509–513.

Atwell W, Patrick B, Johnson L, Glass R. 1983. Characterization of quinoa starch. Cereal Chem 60:9–11.

Ball MC, Farquhar GD. 1984. Photosynthetic and stomatal responses of the grey mangrove *Avicennia marina* to transient salinity conditions. Plant Physiol 74:7–11.

Bayuelo-Jiménez JS, Debouck GD, Lynch JP. 2003. Growth, gas exchange, water relations, and ion composition of *Phaseolus* species grown under saline conditions. Field Crop Res 80:207–222.

Benzarti M, Rejeb KB, Debez A, Messedi D, Abdelly C. 2012. Photosynthetic activity and leaf antioxidative responses of *Atriplex portulacoides* subjected to extreme salinity. Acta Physiol Plant 34:1679–1688.

Bertero HD, De la Vega AJ, Correa G, Jacobsen SE, Mujica A. 2004. Genotype and genotype-by-environment interaction effects for grain yield and grain size of quinoa (*Chenopodium quinoa* Willd.) as revealed by pattern analysis of multi-environment trials. Field Crop Res 89:299–318.

Bhargava A, Shukla S, Ohri D. 2006. *Chenopodium quinoa* - an Indian perspective. Ind Crop Prod 23:73–87.

Bhargava A, Shukla S, Ohri D. 2007. Genetic variability and interrelationship among various morphological and quality traits in quinoa (*Chenopodium quinoa* Willd.). Field Crop Res 101:104–116.

Boero C, González JA, Prado FE. 1999. Germination in different varieties of quinoa (*Chenopodium quinoa* Willd.) under different conditions of salinity and pH (In Spanish). Primer Taller Internacional sobre Quinua. Recursos Genéticos y Sistemas de producción. Lima, Perú. Roma: FAO. Available from: www.rlc.fao.org/prior/segalim/prodalim/prodveg/cdrom/contenido/libro05/cap2.htm 103k.

Bosque Sanchez H, Lemeur R, Van Damme P, Jacobsen SE. 2003. Ecophysiological analysis of drought and salinity stress in quinoa (*Chenopodium quinoa* Willd.). Food Rev Int 19:111–119.

Brinegar C, Goundan S. 1993. Isolation and characterization of chenopodin, the 11S seed storage protein of quinoa (*Chenopodium quinoa*). J Agric Food Chem 41:182–185.

Brock J, Aboling S, Stelzer R, Esch E, Papenbrock J. 2007. Genetic variation among different populations of *Aster tripolium* grown on naturally and anthropogenic salt-contaminated habitats: implications for conservation strategies. J Plant Res 120:99–112.

Brugnoli E, Lauteri M. 1991. Effects of salinity on stomatal conductance, photosynthetic capacity, and carbon isotope discrimination of salt tolerant (*Gossypium hirsutum* L.) and salt-sensitive (*Phaseolus vulgaris* L.) C3 non-halophytes. Plant Physiol 95:628–635.

del Castillo C, Winkel T, Mahy G, Bizoux JP. 2007. Genetic structure of quinoa (*Chenopodium quinoa* Willd.) from the Bolivian altiplano as revealed by RAPD markers. Genet Res Crop Evol 54:897–905.

Centritto M, Loreto F, Chartzoulakis K. 2003. The use of low CO_2 to estimate diffusional and non-diffusional limitations of photosynthetic capacity of salt stressed olive saplings. Plant Cell Environ 26:585–594.

Chakraborty S, Tiedemann AV, Teng PS. 2000. Climate change: potential impact on plant diseases. Environ Poll 108:317–326.

Comai S, Bertazzo A, Bailoni L, Zancato M, Costa CVL, Allegri G. 2007. The content of proteic and nonproteic (free and protein-bound) tryptophan in quinoa and cereal flours. Food Chem 100:1350–1355.

Cortés Hogea, 1558. 2009. In: Tapia M, editor. La quinua. Historia, distribución geográfica actual, producción y usos. Revista Ambienta 99:104–119.

D'Souza MR, Devaraj VR. 2010. Biochemical responses of hyacinth bean (*Lablab purpureus* L.) to salinity stress. Acta Physiol Plant 32:341–353.

Dadkhah A. 2011. Effect of salinity on growth and leaf photosynthesis of two sugar beet (*Beta vulgaris* L.) cultivars. J Agr Sci Tech 13:1001–1012.

Dasgupta N, Nandy P, Das S. 2011. Photosynthesis and antioxidative enzyme activities in five Indian mangroves with respect to their adaptability. Acta Physiol Plant 33:803–810.

Dini I, Tenore GC, Dini A. 2005. Nutritional and antinutritional composition of Kancolla seeds: an interesting and underexploited andine food plant. Food Chem 92:125–132.

Dionisio-Sese ML, Tobita S. 2000. Effects of salinity on sodium content and photosynthetic responses of rice seedlings differing in salt tolerance. J Plant Physiol 157:54–58.

Dunn GM, Neales TF. 1993. Are the effects of salinity on growth and leaf gas-exchange related. Photosynthetica 29:33–42.

Ehleringer JR, Osmond CB. 1989. Stable isotopes. In: Pearcy RW, Ehleringer JR, Rundel PW, editors. Plant physiological ecology. Field methods and instrumentation. UK: Chapman and Hall Ltd. pp. 281–300.

Eisa S, Hussin S, Geissler N, Koyro HW. 2012. Effect of NaCl salinity on water relations, photosynthesis and chemical composition of quinoa (*Chenopodium quinoa* Willd.) as a potential cash crop halophyte. Aust J Crop Sci 6:357–368.

Estrada A, Li B, Laarveld B. 1998. Adjuvant action of *Chenopodium quinoa* saponins on the induction of antibody responses to intragastric and intranasal administered antigens in mice. Comp Immunol Microb 21:225–236.

Everard JD, Gucci R, Kann SC, Flore JA, Loescher WH. 1994. Gas exchange and carbon partitioning in the leaves of celery (*Apium graveolens* L.) at various levels of root zone salinity. Plant Physiol 106:281–292.

Flexas J, Diaz-Espejo A, Galme's J, Kaldenhoff H, Medrano A, Ribas-Carbo M. 2007. Rapid variations of mesophyll conductance in response to changes in CO_2 concentration around leaves. Plant Cell Environ 30:1284–1298.

Flowers TJ. 2004. Improving crop salt tolerance. J Exp Bot 55:307–319.

Flowers TJ, Colmer TD. 2008. Salinity tolerance in halophytes. New Phytol 179:945–963.

Food and Agriculture Organization. 2006. Rural land sustainable management (in Spanish). FAO/Japan Regional Projects in Latin America 1988–2006. Tokyo, Japan.

Food and Agriculture Organization. 2008. Land and plant nutrition management service. Available from: www.fao.org/ag/agl/agll/spush.

[FAO/WHO] Food and Agriculture Organization/World Health Organization of United Nations. 1990. Protein quality evaluation. Rome, Italy: Report of a Joint FAO/WHO Expert Consultation.

[FAO/WHO] Food and Agriculture Organization/World Health Organization of United Nations. 2010. Fats and fatty acids in human nutrition: report of an expert consultation. Rome, Italy: Food and Nutrition Paper 91.

Foyer C, Noctor G. 2000. Tansley review 112. Oxygen processing in photosynthesis: regulation and signaling. New Phytol 146: 359–388.

Freitas H, Breckle SW. 1992. Importance of bladder hairs for salt tolerance of field-grown Atriplex species from a Portuguese salt marsh. Flora 187:283–297.

Friedman M. 1996. Nutritional value of proteins from different food sources. A review. J Agric Food Chem 44:6–29.

Fuentes FF, Maughan PJ, Jellen EN. 2009. Diversidad genética y recursos genéticos para el mejoramiento de la quinoa (*Chenopodium quinoa* Willd). Rev Geogr Valpso 42:20–33.

Galvez Ranilla L, Apostolidis E, Genovese MI, Lajolo FM, Shetty K. 2009. Evaluation of indigenous grains from the Peruvian Andean region for antidiabetes and antihypertension potential using in vitro methods. J Med Food 12:704–713.

Galwey NW, Leakey CLA, Price KR, Fenwick GR. 1990. Chemical composition and nutritional characteristics of quinoa (*Chenopodium quinoa* Willd.). Food Sci Nutr 42:245–261.

Gandarillas H. 1979. Genética y origen. In: Tapia ME, editor. Quinua y Kaniwa. Cultivos Andinos. Serie Libros y Materiales Educativos, vol. 49. Bogotá, Colombia: Instituto Interamericano de Ciencias Agrícolas. pp. 45–64.

Garcia M, Raes D, Jacobsen SE. 2003. Evapotranspiration analysis and irrigation requirements of quinoa (*Chenopodium quinoa*) in the Bolivian highlands. Agric Water Manag 60:119–134.

Garcia M, Raes D, Jacobsen SE, Michel T. 2007. Agroclimatic contraints for rainfed agriculture in the Bolivian Altiplano. J Arid Environ 71:109–121.

Geissler N, Hussin S, Koyro HW. 2009. Interactive effects of NaCl salinity, elevated atmospheric CO_2 concentration on growth, photosynthesis, water relations and chemical composition of the potential cash crop halophyte *Aster tripolium* L. Environ Exp Bot 65:220–231.

Geissler N, Hussin S, Koyro HW. 2010. Elevated atmospheric CO_2 concentration enhances salinity tolerance in *Aster tripolium* L. Planta 231:583–594.

Ghassemi F, Jakeman AJ, Nix HA. 1995. Salinisation of land and water resources: human causes, extent, management and case studies. Wallingford, England: CAB International. p. 544.

Giménez C, Mitchell VJ, Lawlor DW. 1992. Regulation of photosynthesis rate of two sunflower hybrids under water stress. Plant Physiol 98:516–524.

Gómez-Caravaca AM, Iafelice G, Lavini A, Pulvento C, Caboni MF, Marconi E. 2012. Phenolic compounds and saponins in quinoa samples (*Chenopodium quinoa* Willd.) grown under different saline and nonsaline irrigation regimens. J Agric Food Chem 60:4620–4627.

González JA, Prado FE. 1992. Germination in relation to salinity and temperature in *Chenopodium quinoa* Willd. Agrochimica 36:101–107.

González JA, Roldán A, Gallardo M, Escudero T, Prado FE. 1989. Quantitative determinations of chemical compounds with nutritional value from Inca crops: *Chenopodium quinoa* ("quinoa"). Plant Foods Hum Nutr 39:331–337.

González JA, Gallardo M, Hilal M, Rosa M, Prado FE. 2009a. Physiological responses of quinoa (*Chenopodium quinoa*) to drought and waterlogging stresses: dry matter partitioning. Bot Stud 50:35–42.

González JA, Rosa M, Parrado MF, Hilal M, Prado FE. 2009b. Morphological and physiological responses of two varieties of a highland species (*Chenopodium quinoa* Willd.) growing under near-ambient and strongly reduced solar UV-B in a lowland location. J Photochem Photobiol B 96:144–151.

González JA, Bruno M, Valoy M, Prado FE. 2011. Genotypic variation of gas exchange parameters and leaf stable carbon and nitrogen isotopes in ten quinoa cultivars grown under drought. J Agron Crop Sci 197:81–93.

González JA, Konishi Y, Bruno M, Valoy M, Prado FE. 2012. Interrelationships among seed yield, total protein and amino acid composition of ten quinoa (*Chenopodium quinoa*) cultivars from two different agroecological regions. J Sci Food Agric 92:1222–1229.

Gorai M, Ennajeh M, Khemira H, Neffati M. 2011. Influence of NaCl-salinity on growth, photosynthesis, water relations and solute accumulation in *Phragmites australis*. Acta Physiol Plant 33:963–971.

Gorinstein S, Pawelzik E, Delgado-Licon E, Haruenkit R, Weisz M, Trakhtenberg S. 2002. Characterisation of pseudocereal and cereal proteins by protein and amino acid analyses. J Sci Food Agric 82:886–891.

Gulzar S, Khan MA, Ungar IA. 2003. Salt tolerance of a coastal salt marsh grass. Comm Soil Sci Plant Anal 34:2595–2605.

Gulzar S, Khan MA, Ungar IA, Liu X. 2005. Influence of salinity on growth and osmotic relations of *Sporobolus ioclados*. Pak J Bot 37:119–129.

Gunasekera D, Berkowitz GA. 1993. Use of transgenic plants with Rubisco antisense DNA to evaluate the rate limitation of photosynthesis under water stress. Plant Physiol 103: 629–635.

Hajar AS, Zidan MA, Al-Zahrani HS. 1996. Effect of salinity stress on the germination, growth and some physiological activities of black cumin (*Nigella sativa* L.). Arab Gulf J Sci Res 14:445–454.

Halloy S, González JA. 1993. An inverse relation between frost survival and atmospheric pressure. Arct Alp Res 25:117–123.

Hariadi Y, Marandon K, Tian Y, Jacobsen SE, Shabala S. 2011. Ionic and osmotic relations in 10 quinoa (*Chenopodium quinoa* Willd.) plants grown at various salinity levels. J Exp Bot 62:185–193.

Heiser CB, Nelson CD. 1974. On the origin of cultivated Chenopods (*Chenopodium*). Genetics 78:503–505.

Hernández Bermejo JE, León J. 1994. Neglected crops: 1492 from a different perspective. FAO Plant Production and Protection Series. No. 26. p. 341.

Hevia F, Wilckens R, Berti M, Badilla R. 2001. Características del almidón y contenido de proteína de quinoa (*Chenopodium quinoa* W.) cultivada bajo diferentes niveles de nitrógeno en Chillán. Agro Sur 29:42–50.

Hilal M, Parrado MF, Rosa M, Gallardo M, Massa EM, González JA, Prado FE. 2004. Epidermal lignin deposition in quinoa cotyledons in response to UV-B radiation. Photochem Photobiol 79:205–210.

Huchzermeyer B, Koyro HW. 2005. Salt and drought stress effects on photosynthesis. In: Pessarakli M, editor. Handbook of photosynthesis. 2nd edition. Florida: CRC Press, Taylor and Francis Publishing Company. pp. 751–777.

Hunter, M.D. 2001. Effects of elevated atmospheric carbon dioxide on insect-plant interactions. Agr Forest Entomol 3: 153–159.

Hura T, Hura K, Grzesiak M, Rzepka A. 2007. Effect of long term drought stress on leaf gas exchange and fluorescence parameters in C3 and C4 plants. Acta Physiol Plant 29:103–113.

Hussin S, Geissler N, Koyro HW. 2013. Effect of NaCl salinity on *Atriplex nummularia* (L.) with special emphasis on carbon and nitrogen metabolism. Acta Physiol Plant 35:1025–1038.

Intergovernmental Panel on Climate Change. 2007. Climate Change 2007. Mitigation of Climate Change. In: Metz B, Davidson OR, Bosch PR, Dave R, Meyer LA, editors. Contribution of working group III to the fourth assessment report of the Intergovernmental Panel on Climate Change. Cambridge, MA: Cambridge University Press. p. 851.

Jacobsen SE. 1997. Adaptation of quinoa (*Chenopodium quinoa*) to Northern European agriculture: studies on developmental pattern. Euphytica 96:41–48.

Jacobsen SE. 2003. The worldwide potential of quinoa (*Chenopodium quinoa* Willd.). Food Rev Int 19:167–177.

Jacobsen SE, Hill J, Stolen O. 1996. Stability of quantitative traits in quinoa (*Chenopodium quinoa* Willd). Theo Appl Gen 93:110–116.

Jacobsen SE, Mujica A, Jensen CR. 2003. The resistance of quinoa (*Chenopodium quinoa* Willd.) to adverse abiotic factors. Food Rev Int 19:99–109.

Jacobsen SE, Monteros C, Christiansen JL, Bravo LA, Corcuera LJ, Mujica A. 2005. Plant responses of quinoa (*Chenopodium quinoa* Willd.) to frost at various phenological stages. Eur J Agron 22:131–139.

Jacobsen SE, Monteros C, Corcuera LJ, Bravo LA, Christiansen JL, Mujica A. 2007. Frost resistance mechanisms in quinoa (*Chenopodium quinoa* Willd.). Eur J Agron 26:471–475.

Jacobsen SE, Liu F, Jensen CR. 2009. Does root-sourced ABA play a role for regulation of stomata under drought in quinoa (*Chenopodium quinoa* Willd.). Sci Hort 122:281–287.

Jellen EN, Benlhabib O, Maughan PJ, Stevens MR, Sederberg MDM, Bonifacio A, Coleman CE, Fairbanks DJ, Jacobsen SE. 2005. Introduction of the Andean crop quinoa in Morocco. Soil and water management interaction on crop yields. The ASA-CSSA-SSSA International Annual Meetings, Salt Lake City, UT.

Jensen CR, Jacobsen SE, Andersen MN, Núñez N, Andersen SD, Rasmussen L, Mogensen VO. 2000. Leaf gas exchange and water relation characteristics of field quinoa (*Chenopodium quinoa* Willd.) during soil drying. Eur J Agron 13:11–25.

Jia YS, Gray VM. 2004. Interrelationship between nitrogen supply and photosynthetic parameters in *Vicia faba* L. Photosynthetica 41:605–610.

Jiang Q, Christen S, Shigenaga MK, Ames BN. 2001. γ-Tocopherol, the major form of vitamin E in the US diet, deserves more attention. Am J Clin Nutr 74:714–722.

Johnson DL, Ward SM. 1993. Quinoa. In: Janick J, Simon JE, editors. New Crops. New York, NY: Wiley. pp. 219–221.

Kao WY, Tsai TT, Tsai HC, Shih CN. 2006. Response of three glycine species to salt stress. Environ Exp Bot 56:120–125.

Konishi Y, Hirano S, Tsuboi H, Wada M. 2004. Distribution of minerals in quinoa (*Chenopodium quinoa* Willd) seeds. Biosci Biotechnol Biochem 68:231–234.

Koyro HW. 2000. Effect of high NaCl-salinity on plant growth, leaf morphology, and ion composition in leaf tissues of *Beta vulgaris* ssp. *maritima*. J Appl Bot-Angew Bot 74:67–73.

Koyro HW. 2006. Effect of salinity on growth, photosynthesis, water relations and solute composition of the potential cash crop halophyte *Plantago coronopus* (L.). Environ Exp Bot 56:136–146.

Koyro HW, Eisa SS. 2008. Effect of salinity on composition, viability and germination of seeds of *Chenopodium quinoa* Willd. Plant Soil 302:79–90.

Koyro HW, Huchzermeyer B. 1999. Salt and drought stress effects on metabolic regulation in maize. In: Pessarakli M, editor. Handbook of plant and crop stress. 2nd edition. New York, NY: Marcel Dekker Inc. pp. 843–878.

Koyro HW, Ahmad P, Nicole G. 2012. Abiotic stress responses in plants: an overview. In: Ahmad P, Prasad MNV, editors. Environmental adaptations and stress tolerance of plants in the era of climate change. New York, NY: Springer Science + Business Media. pp. 1–28.

Koziol MJ. 1992. Chemical composition and nutritional evaluation of quinoa. J Food Compos Anal 5:35–68.

Kuljanabhagavad T, Wink M. 2009. Biological activities and chemistry of saponins from *Chenopodium quinoa* Willd. Phytochem Rev 8:473–490.

Kuljanabhagavad T, Thongphasuk P, Chamulitrat W, Wink M. 2008. Triterpene saponins from *Chenopodium quinoa* Willd. Phytochem 69:1919–1926.

Larcher W. 2001. Physiological plant ecology: ecophysiology and stress physiology of functional. Heidelberg, Germany: Springer-Verlag. p. 513.

Läuchli A, Grattan SR. 2007. Plant growth and development under salinity stress. In: Jenks MA, Hasegawa PA, Jain SM, editors. Advances in molecular-breeding towards salinity and drought tolerance. Heidelberg, Germany: Springer-Verlag. p.1-31.

Cieza de León, 1560. 2009. In: Tapia M, editor. La quinua. Historia, distribución geográfica actual, producción y usos. Revista Ambienta 99:104–119.

Lieth H, Moschenko M, Lohmann M, Koyro HW, Hamdy A. 1999. Halophyte uses in different climates I. Ecological and ecophysiological studies. Progress in biometeorology. Volume 13. Leiden, The Netherlands: Backhuys Publishers. p. 258.

Lu C, Qiu N, Lu Q, Wang B, Luang T. 2002. Does salt stress lead to increased susceptibility of photosystem II to photoinhibition and changes in photosynthetic pigment composition in halophyte *Suaeda salsa* grown outdoors? Plant Sci163:1063–1068.

Mastebroek H, Limburg H, Gilles T, Marvin H. 2000. Occurrence of sapogenins in leaves and seeds of quinoa (*Chenopodium quinoa* Willd). J Sci Food Agric 80:152–156.

Maughan J, Bonifacio A, Jellen E, Stevens M, Coleman C, Ricks M, Mason S, Jarvis D, Gardunia B, Fairbanks D. 2004. A genetic linkage map of quinoa (*Chenopodium quinoa*) based on AFLP, RAPD, and SSR markers. Theor Appl Genet 109:1188–1195.

Montzka SA, Dlugokencky EJ, Butler JH. 2011. Non-CO_2 greenhouse gases and climate change. Nature 476:43–50.

Moradi F, Ismail AM. 2007. Responses of photosynthesis, chlorophyll fluorescence and ROS-scavenging systems to salt stress during seedling and reproductive stages in rice. Ann Bot 99:1161–1173.

Munns R. 2002. Comparative physiology of salt and water stress. Plant Cell Environ 25:239–250.

Munns R. 2005. Genes and salt tolerance: bringing them together. New Phytol 167: 645–663.

Munns R. 2011. Plant adaptations to salt and water stress: differences and commonalities. Adv Bot Res 57:1–32.

Munns R, Tester M. 2008. Mechanisms of salinity tolerance. Ann Rev Plant Biol 59: 51–681.

Naidoo G, Mundree SG. 1993. Relationship between morphological and physiological responses to water logging and salinity in *Sporobolus virginicus* (L). Kunth Oecologia 93:360–366.

National Research Council. 1989. Lost crops of the Incas: little known plants of the Andes with promise for worldwide cultivation. Washington, DC: National Academy Press. p. 415.

Netondo GW, Onyango JC, Beck E. 2004. Sorghum and salinity: II. Gas exchange and chlorophyll fluorescence of sorghum under salt stress. Crop Sci 44:806–811.

Nuñez LA. 1974. La agricultura prehistorica en los Andes Meridionales. Universidad del Norte, Editorial Orbe. Santiago de hile, Chile. p. 197.

Orsini F, Accorsi M, Gianquinto G, Dinelli G, Antognoni F, Carrasco KBR, Martinez EA, Alnayef M, Marotti I, Bosi S, Biondi S. 2011. Beyond the ionic and osmotic response to salinity in *Chenopodium quinoa*: functional elements of successful halophytism. Funct Plant Biol 38:818–831.

Palenque ER, Andrade M, González JA, Forno R, Lairana V, Prado FE, Salcedo JC, Urcullo S. 1997. Effect of UVB radiation on quinoa (*Chenopodium quinoa* Willd.). Rev Bol Fis 3:120–128.

Parry MA, Andralojc PJ, Khan S, Lea P, Keys AJ. 2002. Rubisco activity: effects of drought stress. Ann Bot 89:833–839.

Parry M, Rosenzweig C, Livermore M. 2005. Climate change, global food supply and risk of hunger. Philos Trans R Soc Lond B Biol Sci 360:2125–2138.

Prado FE, Gallardo M, González JA. 1996. Presence of saponin-bodies in pericarp cells of *Chenopodium quinoa* Willd. (quinoa). Biocell 20:261–266.

Prado FE, Boero C, Gallardo M, González JA. 2000. Effect of NaCl on growth germination and soluble sugars content in *Chenopodium quinoa* Willd. seeds. Bot Bull Acad Sinica 41:19–26.

Prado FE, Fernández-Turiel JL, Bruno M, Valoy M, Rosa M, González JA. 2010. Mineral content of seeds from quinoa varieties in Amaicha del Valle (Tucumán, Argentina). Biocell 34(2):157

Prakash D, Nath P, Pal M. 1993. Composition, variation of nutritional contents in leaves, seed protein, fat and fatty acid profile of Chenopodium species. J Sci Food Agric 62:203–205.

Pulvento C, Riccardi M, Lavini A, d'Andria R, Iafelice G, Marconi E. 2010. Field trial evaluation of two *Chenopodium quinoa* genotypes grown under rain-fed conditions in a typical Mediterranean environment in South Italy. J Agron Crop Sci 196:407–411.

Qiu QS, Barkla BJ, Vera-Estrella R, Zhu JK, Schumaker KS. 2003. Na+/H+ exchange activity in the plasma membrane of Arabidopsis. Plant Physiol 132:1041–1052.

Ranhotra GS, Gelroth JA, Glaser BK, Lorenz KJ, Johnson DL. 1993. Composition and protein nutritional quality of quinoa. Cereal Chem 70:303–305.

Razzaghi F, Ahmadi SH, Adolf VI, Jensen CR, Jacobsen SE, Andersen MN. 2011. Water relations and transpiration of quinoa (*Chenopodium quinoa* Willd.) under salinity and soil drying. J Agron Crop Sci 197:348–360.

Rengasamy P. 2010. Soil processes affecting crop production in salt-affected soils. Funct Plant Biol 37:613–620.

Repo-Carrasco R, Espinoza C, Jacobsen SE. 2003. Nutritional value and use of the Andean crops quinoa (*Chenopodium quinoa*) and Kañiwa (*Chenopodium pallidicaule*). Food Rev Intern 19:179–189.

Repo-Carrasco-Valencia RAM. 2011. Andean indigenous food crops: nutritional value and bioactive compounds. Ph.D. Thesis, University of Turku, Turku, Finland. p. 176.

Ridout CL, Price KR, DuPont MS, Parker ML, Fenwick GR. 1991. Quinoa saponins-analysis and preliminary investigations into the effects of reduction by processing. J Sci Food Agric 54:165–176.

Risi J, Galwey NW. 1984. The chenopodium grains of the Andes: Inca crops for modern agriculture. Adv Appl Bot 10:145–216.

Risi J, Galwey NW. 1989. The pattern of genetic diversity in the Andean grain crop quinoa (*Chenopodium* (*quinoa* Willd.). I. Associations between characteristics. Euphytica 41: 147–162.

Rojas W, Soto JL, Pinto M, Jäger M, Padulosi S. 2010. Granos Andinos. Avances, logros y experiencias desarrolladas en quinua, cañahua y amaranto en Bolivia. Roma, Italia: Bioversity International. p. 178.

Romero-Aranda R, Soria T, Cuartero J. 2001. Tomato plant-water uptake and plant-water relationships under saline growth conditions. Plant Sci 160:265–272.

Rosa M, Hilal M, González JA, Prado FE. 2009. Low-temperature effect on enzyme activities involved in sucrose-starch partitioning in salt-stressed and salt-acclimated cotyledons of quinoa (*Chenopodium quinoa* Willd.) seedlings. Plant Physiol Biochem 47:300–307.

Rozema J, Schat H. 2013. Salt tolerance of halophytes, research questions reviewed in the perspective of saline agriculture. Environ Exp Bot 92:83–95.

Ruales J, Nair BM. 1992. Nutritional quality of the protein in quinoa (*Chenopodium quinoa*, Willd.) seeds. Plant Foods Hum Nutr 42:1–11.

Ruales J, Nair BM. 1993. Content of fat, vitamins and minerals in quinoa (*Chenopodium quinoa* Willd.) seeds. Food Chem 48:131–136.

Ruffino AMC, Rosa M, Hilal M, González JA, Prado FE. 2010. The role of cotyledon metabolism in the establishment of quinoa (*Chenopodium quinoa*) seedlings growing under salinity. Plant Soil 326:213–224.

Sanchez HB, Lemeur R, Van Damme P, Jacobsen SE. 2003. Ecophysiological analysis of drought and salinity stress of quinoa (*Chenopodium quinoa* Willd.). Food Rev Intern 19:111–119.

Schlick D, Bubnehiem DL. 1993. Quinoa: an emerging "new" crop with potential for CELSS. Ames Research Center, Technical Paper 3422. Moffett Field California: National Aeronautics and Space Administration.

Schoenlechner R, Wendner M, Siebenhandl-Ehn S, Berghofer E. 2010. Pseudocereals as alternative sources for high folate content in staple foods. J Cereal Sci 52:475–479.

Schulze DH, Polumuri SK, Gille T, Ruknudin A. 2002. Functional regulation of alternatively spliced Na^+/Ca^{2+} exchanger (NCX1) isoforms. Ann NY Acad Sci 976:187–196.

Seemann JR, Critchley C. 1985. Effects of salt stress on the growth, ion content, stomatal behavior and photosynthetic capacity of salt-sensitive species, *Phaseolus vulgaris* L. Planta 164:151–162.

Shannon MC, Grieve CM. 1999. Tolerance of vegetable crops to salinity. Sci Hort 78:5–38.

Silva EN, Ribeiro RV, Ferreira-Silva SL, Viégas RA, Silveira JAG. 2010. Comparative effects of salinity and water stress on photosynthesis, water relations and growth of *Jatropha curcas* plants. J Arid Environ 74:1130–1137.

Simmonds NW. 1971. The breeding system of *Chenopodium quinoa*. I Male sterility Heredity 27:73–82.

Simopoulos AP. 2001. Evolutionary aspects of diet, the omega-6/omega-3 ratio and genetic variation: nutritional implications for chronic diseases. Biomed Pharmacother 60:502–507.

Sircelj MR, Rosa MD, Parrado MF, González JA, Hilal M, Prado FE. 2002. Ultrastructural and metabolic changes induced by UV-B radiation in cotyledons of quinoa (*Chenopodium quinoa* Willd.). Biocell 26:180.

Solíz-Guerrero JB, Jasso de Rodriguez D, Rodríguez-García R, Angulo-Sánchez JL, Méndez-Padilla G. 2002. Quinoa saponins: concentration and composition analysis. In: Janick J, Whipkey A, editors. Trends in new crops and new uses. Alexandria: ASHS Press. pp.110-114.

Solomon S, Plattner GK, Knott R, Friedlingstein P. 2009. Irreversible climate change due to carbon dioxide emissions. Proc Natl Acad Sci U S A 106:1704–1709.

de Sotelo, 1583. 2009. In: Tapia M, editor. La quinua. Historia, distribución geográfica actual, producción y usos. Revista Ambienta 99:104–119.

Souza RP, Machado EC, Silva JAB, Lagoa AMMA, Silveira JAG. 2004. Photosynthetic gas exchange, chlorophyll fluorescence and some associated metabolic changes in cowpea (*Vigna unguiculata*) during water stress and recovery. Environ Exp Bot 51:45–56.

Steduto P, Albrizio R, Giorio P, Sorrentino G. 2000. Gas-exchange response and stomatal and non-stomatal limitations to carbon assimilation of sunflower under salinity. Environ Exp Bot 44:243–255.

Sudhir P, Murthy SDS. 2004. Effects of salt stress on basic processes of photosynthesis. Photosynthetica 42:481–486.

Tapia M. 2009. La quinua. Historia, distribución geográfica actual, producción y usos. Revista Ambienta 99:104–119.

Tarchoune I, Degl'Innocenti E, Kaddour R, Guidi L, Lachaal M, Navari-Izzo F, Ouerghi Z. 2012. Effects of NaCl or Na_2SO_4 salinity on plant growth, ion content and photosynthetic activity in *Ocimum basilicum* L. Acta Physiol Plant 34:607–615.

Tezara W, Marín O, Rengifo E, Martínez D, Herrera A. 2005. Photosynthesis and photoinhibition in two xerophytic shrubs during drought. Photosynthetica 43:37–45.

Tezara W, Driscoll S, Lawlor DW. 2008. Partitioning of photosynthetic electron flow between CO_2 assimilation and O_2 reduction in sunflower plants under water deficit. Photosynthetica 46:127–134.

Thanapornpoonpong SN, Vearasilp S, Pawelzik E, Gorinstein S. 2008. Influence of various nitrogen applications on protein and amino acid profiles of amaranth and quinoa. J Agric Food Chem 56:11464–11470.

Triboi E, Martre P, Triboi-Blondel AM. 2003. Environmentally-induced changes in protein composition in developing grains of wheat are related to changes in total protein content. J Exp Bot 54:1731–1742.

Uhle M. 1919 La arqueología de Arica y Tacna. Boletín de la Sociedad Ecuatoriana de Estudios Históricos Americanos III (7–8):1–48.

Ulloa Mogollón J. 1586. Relación de la provincial de los Collaguas para la discripción de las Indias que su majestad manda hacer. En: Jimenez de la Espada M, editor. Relaciones geográficas de las Indias. Vol 1. Madrid, España.

Vacher JJ. 1998. Response of two main Andean crops, quinoa (*Chenopodium quinoa* Wild.) and papa amarga (*Solanum juzepczukii* B.) to drought on the Bolivian altiplano: significance of local adaptation. Agric Ecosyst Environ 68:99–108.

de la Vega G. 1966. El Inca. Royal commentaries of the Incas and general history of Peru. Austin: University of Texas Press.

Wallington TJ, Srinivasan J, Nielsen OJ, Highwood EJ. 2004. Greenhouse gases and global warming. In: Sabljic A, editor. Environmental and ecological chemistry. Encyclopedia of life support systems (EOLSS). Oxford, England: EOLSS Publishers. p. 27.

Wang LW, Showalter AM, Ungar IA. 1997. Effect of growth, ion content and cell wall chemistry in *Atriplex prostrata* (Chenopodiaceae). Am J Bot 84:1247–1255.

Ward SM. 2001. A recessive allele inhibiting saponin synthesis in two lines of Bolivian quinoa (*Chenopodium quinoa* Willd.). J Hered 92:83–86.

Wilson HD. 1981. Genetic variation among South American populations of tetraploid *Chenopodium* sect. *chenopodium* subsect. *cellulata*. Syst Bot 6:380–398.

Wilson HD. 1988. Quinoa biosystematics II: free living populations. Econ Bot 42:478–494.

Wilson C, Read JJ, Abo-Kassem E. 2002. Effect of mixed-salt salinity on growth and ion relations of a quinoa and a wheat variety. J Plant Nutr 25:2689–2704.

Wise RR, Frederick JR, Alm DM, Kramer DM, Hesketh JD, Crofts AR, Ort DR. 1992. Investigation of the limitations to photosynthesis induced by leaf water deficit in field-grown sunflower (*Helianthus annuus* L.). Plant Cell Environ 15: 755–756.

Woldemichael GM, Wink M. 2001 Identification and biological activities of triterpenoid saponins from *Chenopodium quinoa*. J Agric Food Chem 49:2327–2332.

Wood SG, Lawson LD, Fairbanks DJ, Robison LR, Andersen WR. 1993. Seed lipid content and fatty acid composition of three quinoa cultivars. J Food Compos Anal 6:41–44.

Yan K, Shao H, Shao C, Chen P, Zhao S, Brestic M, Chen X. 2013.Physiological adaptive mechanisms of plants grown in saline soil and implications for sustainable saline agriculture in coastal zone. Acta Physiol Plant DOI: 10.1007/s11738-013-1325-7.

Zhu N, Sheng S, Sang S, Jhoo JW, Bai N, Karwe M, Rosen R, Ho CT. 2002. Triterpene saponins from debittered quinoa (*Chenopodium quinoa*) seeds. J Agric Food Chem 50:865–867.

Zuidmeer L, Goldhahn K, Rona RJ, Gislason D, Madsen C, Summers C, Sodergren E, Dahlstrom J, Lindner T, Sigurdardottir ST, et al. 2008. The prevalence of plant food allergies: a systematic review. J Allergy Clin Immunol 121:1210–1218.

Chapter 2

History of Quinoa: Its Origin, Domestication, Diversification, and Cultivation with Particular Reference to the Chilean Context

Enrique A. Martínez[1], Francisco F. Fuentes[2], and Didier Bazile[3]

[1]*Centro de Estudios Avanzados en Zonas Áridas, La Serena and Facultad de Ciencias del Mar, Universidad Católica del Norte, Coquimbo, Chile*

[2]*Facultad de Agronomía e Ingeniería Forestal, Pontificia Universidad Católica de Chile, Casilla 306-22, Santiago, Chile*

[3]*UPR47, GREEN, CIRAD (Centre de Coopération Internationale en Recherche Agronomique pour le Développement), TA C-47/F, Campus International de Baillarguet, 34398 Montpellier Cedex 5 France*

QUINOA ORIGINS IN THE CENTRAL ANDES

Quinoa, a tetraploid crop plant, was described for the first time in 1797 by the German botanist and pharmacist Carl Ludwig Willdenow. It has been cultivated for the past 8,000 years in the South American Andes. It is hypothesized that the closest ancestors of quinoa could be the species *Chenopodium berlandieri* var. *nuttalliae*, distributed in North America, or a complex of species growing in the southern hemisphere, including *Chenopodium pallidicaule* Aellen (Kañahua), *Chenopodium petiolare* Kunth, *Chenopodium carnasolum* Moq., and the tetraploïd species, *Chenopodium hircinum* Schard or *Chenopodium quinoa* var. *melanospermum*. All these species are from the Andes (Wilson and Heiser 1979; Heiser and Nelson 1974; Mujica and Jacobsen 2000; Fuentes et al. 2009a). The areas cultivated with quinoa in South America goes from 2° North latitude in Colombia to 47° South latitude in Chile, and from 4,000 m in the high Andes to the sea level in southern latitudes. Particular adaptations of this species to certain geographical areas along the Andes gave rise to five major ecotypes associated with subcenters of diversity, differing in branching morphology and adaptations to rainfall regimes with precipitation of 2000 mm per year to strong drought stress of 150 mm per year. These ecotypes are the (i) Inter Andean valleys quinoa (in Colombia, Ecuador, and Peru); (ii) Highlands quinoa (in Peru and Bolivia); (iii) Yungas quinoa (in Bolivian subtropical forest); (iv) "Salares" quinoa in salt flats (in Bolivia, northern Chile, and Argentina); and (v) Coastal quinoa, from lowlands or from sea level (in central and southern Chile). The expansion routes from the Titicaca Lake were summarized by Fuentes et al. (2012) and supported with genetic data as revealed with the use of molecular markers.

The domestication process must have included all factors of the domestication syndrome, including larger fruit size, higher and uniform yields, reduced branching and bigger inflorescence,

Quinoa: Improvement and Sustainable Production, First Edition. Edited by Kevin Murphy and Janet Matanguihan.

reduced seed dormancy, and less auto dehiscence or seed fall. The quinoa landraces had also adapted to different soils, climates, and particularly day-lengths as day-lengths grow longer in the spring and summer seasons toward the southern latitudes.

ANCIENT EXPANSION TO SOUTHERN LATITUDES IN CHILE

Quinoa cultivation in Chile is centered primarily on two of the five main ecotypes, namely the salt flat ("salares") and the coastal ecotypes. The "salares" ecotype is distributed in the Tarapacá and Antofagasta regions (18–25°S) in northern Chile, with elevations over 3,000 m high (Fig. 2.1). In these regions, highland indigenous communities (Aymara and Quechua people) traditionally cultivate these quinoas in saline soils, with precipitation fluctuating between 100 and 200 mm per year falling during the southern hemisphere summer, from December to February (Lanino 2006). These ecotypes are closely related to the quinoa Bolivian varieties that are also of the "salares" ecotype, probably because these are also cultivated by the Aymara and Quechua communities on both sides of the current Chile–Bolivia border. On the other hand, there is evidence for the introduction of some quinoa genetic materials from the Peruvian Andean zone to the Antofagasta region. Despite this evidence, the dominant morphology of most of the quinoa studied so far in Chile is of the "salt flat" ecotype (Fuentes et al. 2009b).

In central Chile (Fig. 2.1) and even at the more southern latitudes (43°S) (O'Higgins to Lakes' political regions), the cultivated quinoa are different landraces of the coastal ecotype. Areas of quinoa cultivation are rain-fed and have variable altitudes between sea level and 1,000 m height. A remarkable difference is that compared to the extremely dry conditions where the "salares" quinoa is grown in northern Chile, rainfall in the central and southern zones of Chile occurs during the southern hemisphere winter (June–August), with rainfall fluctuations between 500 and 2,000 mm per year. This rainfall increases steadily across 34–40°S.

When the two ecotypes, "salares" and coastal ecotypes cultivated in Chile, are compared, there is a recognized and remarkable difference in terms of their adaptation to altitude, tolerance to drought and salinity, and day-length sensitivity (Bertero et al. 1999; Bertero 2001). In addition, the genetic backgrounds of the two main ecotypes cultivated in Chile are also very different. Interestingly, even within the southern coastal quinoa, the genetic backgrounds are extremely diverse (Fuentes et al. 2012).

Only one hybrid has been produced and recorded in the Chilean national system of protection for new varieties. The hybrid "La Regalona" (Von Baer et al. 2009) has been bred for higher yields and wide adaptation and is able to grow under day-lengths in latitudes close to the equator to latitudes as far as the polar latitudes up to 40° (S or N).

The expansion pattern of quinoa in Chile implies that it underwent a micro-evolutionary process, supported by a high genetic diversity that made it possible for ancient peoples to select quinoa adapted to contrasting and extended agroecological gradients (Fuentes et al. 2012). The quinoa adaptation process in Chile occurred at least since the past 3,000 years, as revealed by recent dating of seeds found in El Plomo hill in Santiago, at central Chile (Planella et al. 2011). Seed exchanges occurred throughout Chile even when ancient peoples from the north (Aymara, Quechua, Atacameños or Licanantay, Coyas, Diaguitas), from the center (Picunches, Pehuenches), and from the south (Mapuches, Huilliches) spoke different languages. The people from the south even gave quinoa another name, *dawue* (Sapúlveda et al. 2003).

REINTRODUCTION OF QUINOA IN ARID CHILE AFTER LOCAL EXTINCTION

Quinoa cultivation probably disappeared very early in the first two regions colonized by the Spanish conquerors some 400 years ago, in the Santiago and the Coquimbo regions, at 33 and

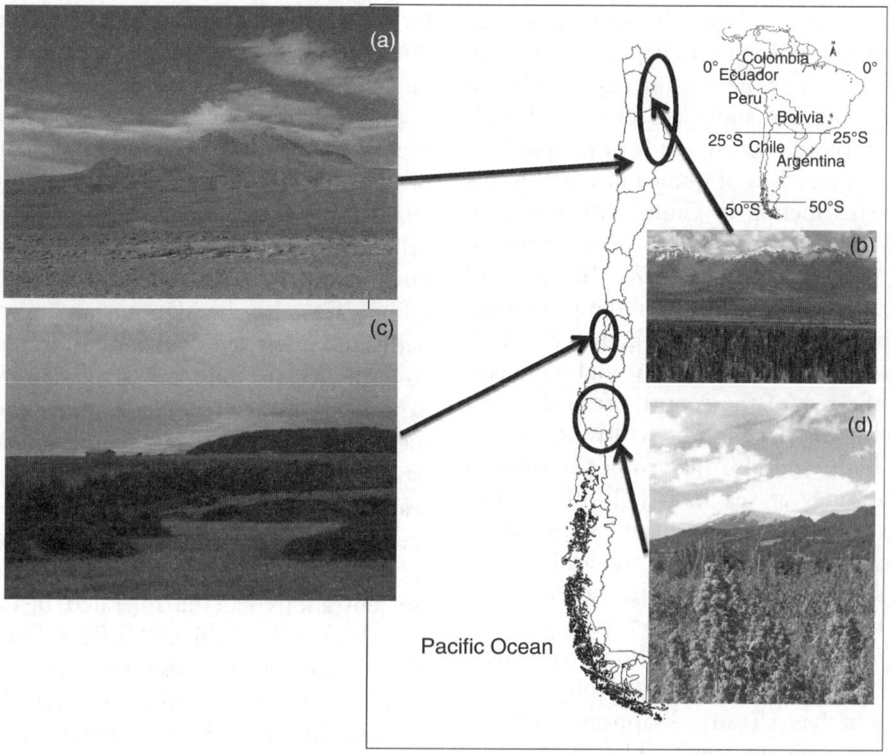

Fig. 2.1 Position of Chile in South America (right upper corner) where a long Atacama Desert (a) isolates the country from southern Peru and Bolivia. Quinoa is cultivated in places as the eastern Altiplano (b) at 4,000 m high ("salares" ecotypes), in the center (c) and south (d) of the country ("coastal" ecotypes) at sea level or piedmont (1,000 masl).

30°S, respectively. The recent ongoing efforts to reintroduce quinoa in the arid region of Coquimbo are supported by the scientific community. In 2003, a new research center, Centro de Estudios Avanzados en Zonas Áridas (CEAZA), started its activities in this northern zone of Chile. One of the objectives of the research conducted in this region is to relate climate change and natural and human-induced activities on natural and cultivated lands and coastal waters. In this region (29–32°S), the climate is mediterranean-desertic and semi-desertic with a marked seasonality, with rainfall occurring in winter and 8–10 dry months per year (Novoa and López 2001). The weather information available indicates that average rainfall in La Serena (30°S) has dropped about 100 mm (50%) in the past century (Martínez et al. 2009a), placing it among those regions with the greatest decrease in precipitation worldwide (http://www.ipcc.ch/pub/tpbiodiv_s.pdf). These changes in precipitation coincided with increases of 0.6°C in the earth's temperature during the past century (http://www.ipcc.ch/pub/un/ipccwg1s.pdf).

The environmental conditions in the Coquimbo region are transitional between the Mediterranean climate and the Atacama Desert. Its main transversal (East-West) valleys (Elqui, Limarí, and Choapa) present an increasing pluviometry from the North to the South, with approximately 60 mm per year rainfall in the Elqui Valley to 300 mm per year in the Choapa valley (Favier et al. 2009). Farmers lost income from grain crops such as wheat due to low yields. Farmers who had more income started farming fruit plantations for export and have been using drip irrigation (Jorquera 2001). Even though quinoa had been cultivated as early as 3000 B.P. in the arid region of Coquimbo (Planella et al. 2011), it probably disappeared very early during the Spanish conquest when the conquistadores introduced wheat and other European crops. The city of La Serena was the second city founded in Chile, after Santiago, and farmers around both cities have forgotten even the word "quinoa." The social memory loss of quinoa has driven this crop to near extinction in Chile, with less than 300 quinoa farmers represented in the national agronomic statistics (INE 2007). In fact, all ancestral seeds have been lost from the arid region of Coquimbo. When this fact was known, CEAZA started to acquire quinoa seeds through field collections in the rest of the country, from farmers in the Andes highlands, who cultivate quinoa at 4,000 m high, and also from other sources in southern latitudes at sea level (34–40°S). The first efforts focused on evaluation of the seeds collected for their adaptation to the arid region, particularly to determine if quinoa could be cultivated under the current low precipitation. The tests were also conducted to determine if quinoa could replace the wheat crop in areas from which it has already disappeared due to the increasing drought trend. The first results indicated that seeds from the center and south of Chile gave higher yields than those collected from the Andes highlands if sown in the spring season (Martínez et al. 2007, 2009b). Later on, results showed that extremely low irrigation was adequate for the quinoa to produce seeds but under experimental conditions, where all water applied was from artificial irrigation (Martínez et al. 2009a). Such experiments showed that quinoa could grow and produce seeds under extremely low levels of irrigation, equivalent to 50 mm of rainfall, but applied at very precise moments of the cycle (growth, flowering, and rain filling). However, precipitation at critical growth stages cannot be assured for farmers who depend solely on rainfall. Even though rainfall could reach 100 mm per year, there are years when majority of the rainfall comes in a span of a few days, as in the case in 2012, when 90% of the rainfall came in a single autumn day. Thus, experimental results can only predict the economic profit and yields from the arid region of Chile if artificial irrigation is provided. However, for other southern and more rainy regions of the country, quinoa undoubtedly can be a good rain-fed crop for the future.

Quinoa's tolerance to abiotic stresses is one of the reasons why the Food and Agriculture Organization (FAO) of the United Nations declared 2013 as the International Year of Quinoa and to promote it as one of the crops that alleviate world hunger and poverty. Other reasons to promote quinoa are related to its outstanding capacity to withstand other stressful conditions such as frost and salinity in soils and irrigation water. Our experience in reintroducing quinoa in the arid region of Chile confirmed that germinating seeds have a high tolerance to salinity. Experimental results also showed that genetic mechanisms are triggered in response to salinity: either salts are rejected from plant tissues or higher salt concentrations can be tolerated inside the cell vacuoles (Orsini et al. 2011; Ruiz-Carrasco et al. 2011). Both studies also revealed that quinoas from central Chile and those from the high Andes do have landraces highly tolerant to salt stress. However, those landraces from the more southern latitudes (39°S) are less tolerant to salt stress, as shown by the study of Delatorre-Herrera and Pinto (2009).

The nutritional value of quinoa is another aspect invoked by FAO to promote its world cultivation and consumption. The presence of the 20 amino acids in quinoa seed, and twice the quantity of proteins than that of many cereals, in addition to minerals, vitamins, good quality oils and antioxidants, and good quality starch, make quinoa seed of high nutritional and functional value (Galwey 1992; Schlick and Bubenheim 1996; Vega-Gálvez et al. 2010). All these nutritional properties have been confirmed for landraces of quinoa from the three ancestral production zones in Chile (Miranda et al. 2011, 2012a, 2012b). Isoflavones, important for improving milk during breast milk production, have also been found in Chilean quinoa (Lutz et al. 2013). Other flavonoids, probably from its seed coat saponins, seem to be involved in antibacterial activity (Miranda et al. 2013).

FINAL REMARKS

The ancestral progenitors of quinoa might have originated in North America, but quinoa production and the culture that developed alongside its cultivation and consumption is known to be shared among the ancient peoples of South America. In ancient times, quinoa was grown from the southern part of South America from the highlands of what is now Bolivia, to the furthest austral latitudes and lowlands of Chile and Argentina, where the agrocultural tradition of quinoa almost disappeared. This long latitudinal gradient implies at least 3,000 years of quinoa acclimation to new lands, new climates, and longer day-lengths. Interestingly, the nutritional quality and stress-tolerant properties of quinoa did not change. Efforts to reintroduce quinoa to the arid regions of Chile have shown that certain quinoa landraces could produce seed even under extremely low irrigation levels, but it has to be applied at precise points during the growth cycle. At present, there is a wide range of quinoa landraces that can be adapted to new areas in the world. Stress-tolerant quinoa can be grown in marginal areas or under harsher environments where the more traditional crops cannot be grown. Moreover, quinoa seed is extremely nutritious and can fulfill the need and demands of a growing world population for high-quality food.

REFERENCES

Bertero HD. 2001. Effects of photoperiod, temperature and radiation on the rate of leaf appearance in quinoa (*Chenopodium quinoa* Willd.) under field conditions. Ann Bot 87:495–502.

Bertero HD, King RW, Hall AJ. 1999. Modelling photoperiod and temperature responses of flowering in quinoa (*Chenopodium quinoa* Willd.). Field Crops Res 63:19–34.

Delatorre-Herrera J, Pinto M. 2009. Importance of ionic and osmotic components of salt stress on the germination of four quinua (*Chenopodium quinoa* Willd.) selections. Chilean J Agr Res 69:477–485.

Favier V, Falvey M, Rabatel A, Praderio E, López D. 2009. Interpreting discrepancies between discharge and precipitation in high-altitude area of Chile's Norte chico region (26–32 S). Water Res Res 45:W02424 10.1029/2008WR006802.

Fuentes FF, Espinoza PA, Von Baer I, Jellen EN, Maughan PJ. 2009a. Determinación de relaciones genéticas entre *Chenopodium quinoa* Willd. del sur de Chile y parientes silvestres del género *Chenopodium*. Anales del XVII Congreso Nacional de Biología del Perú, Tacna, Perú, p. 45.

Fuentes FF, Martínez EA, Hinrichsen PV, Jellen EN, Maughan PJ. 2009b. Assessment of genetic diversity patterns in Chilean quinoa (*Chenopodium quinoa* Willd.) germplasm using multiplex fluorescent microsatellite markers. Conserv Genet 10:369–377.

Fuentes FF, Bazile D, Bhargava A, Martínez EA. 2012. Implications of farmers' seed exchanges for on-farm conservation of quinoa, as revealed by its genetic diversity in Chile. J Agric Sci 150:702–716.

Galwey NW. 1992. The potential of quinoa as a multi-purpose crop for agricultural diversification: a review. Ind Crops Prod 1:101–106.

Heiser CB, Nelson CD. 1974. On the origin of cultivated Chenopods (*Chenopodium*). Genetics 78:503–505.

INE-Instituto Nacional de Estadísticas. 2007. VII Censo Nacional Agropecuario y Forestal (Internet) (cited June 24, 2013). Available at: http://www.ine.cl/canales/base_datos/otras_bases_datos.php.

Jorquera C. 2001. Evolución Agropecuaria de la Región de Coquimbo: Análisis contextual para la conservación de la vegetación nativa. Squeo FA, Arancio G, Gutierrez JR. Libro rojo de la flora de la región de Coquimbo, y de los sitios prioritarios para su conservación. La Serena, Chile: Ediciones Universidad de La Serena. 386.

Lanino M. 2006. Características climáticas de ancovinto durante 2005 a 2006. Iquique, Chile: Boletín Tecnico FIA-UNAP-CODECITE. 1–3.

Lutz M, Martínez EA, Martínez A. 2013. Daidzein and genistein contents in seeds of quinoa (*Chenopodium quinoa* Willd) from local ecotypes grown in arid Chile. Ind Crops Prod 49:117–121.

Martínez EA, Delatorre J, Von Baer I. 2007. Quínoa: las potencialidades de un cultivo sub-utilizado en Chile. Tierra Adentro (INIA) 75:24–27.

Martínez EA, Veas E, Jorquera C, San Martín R, Jara P. 2009a. Re-introduction of *Chenopodium quinoa* Willd. into arid Chile: cultivation of two lowland races under extremely low irrigation. J Agron Crop Sci 195:1–10.

Martínez EA, Jorquera-Jaramillo C, Veas E, Chía E. 2009b. El futuro de la quínoa en la región árida de Coquimbo: lecciones y escenarios a partir de una investigación sobre su biodiversidad en chile para la acción con agricultores locales. Revista de Geografía de Valparaíso 42:95–111.

Miranda M, Bazile D, Fuentes FF, Vega-Gálvez A, Uribe E, Quispe I, Lemus R, Martínez EA. 2011. Quinoa crop biodiversity in Chile: an ancient plant cultivated with sustainable agricultural practices and producing grains of outstanding and diverse nutritional values. In: 6th International CIGR Technical Symposium – Section 6: "Towards a Sustainable Food Chain" Food Process, Bioprocessing and Food Quality Management, Nantes, France.

Miranda M, Vega-Gálvez A, Quispe-Fuentes I, Rodríguez MJ, Maureira H, Martínez EA. 2012a. Nutritional aspects of six quinoa (*Chenopodium quinoa* Willd.) ecotypes from three geographical areas of Chile. Chilean J Agric Res 72:175–181.

Miranda M, Vega-Gálvez A, Martinez EA, López J, Rodríguez MJ, Henríquez K, Fuentes FF. 2012b. Genetic diversity and comparison of physicochemical and nutritional characteristics of six quinoa (*Chenopodium quinoa* Willd.) genotypes cultivated in Chile. Food Sci Tech 32:835–843.

Miranda M, Vega-Gálvez A, Jorquera E, López J, Martínez EA. 2013. Antioxidant and antimicrobial activity of quinoa seeds (*Chenopodium quinoa* Willd.) from three geographical zones of Chile. Méndez-Vilas A. Worldwide research efforts in the fight against microbial pathogens: from basic research to technological development. Boca Raton, FL: Brown Walker Press. 83–86.

Mujica A, Jacobsen SE. 2000. Agrobiodiversidad de las aynokas de quinua (*Chenopodium quinoa* Willd.) y la seguridad alimentaria. Seminario Agrobiodiversidad en la Región Andina y Amazónica, pp. 151–156.

Novoa JE, López D. 2001. IV Región: El escenario geográfico físico. En Libro rojo de la flora nativa y de los sitios prioritarios para su conservación: región de Coquimbo. Squeo FA, Arancio G, Gutierrez JR, Libro rojo de la flora de la región de Coquimbo, y de los sitios prioritarios para su conservación. La Serena, Chile: Ediciones Universidad de La Serena. 13–28.

Orsini F, Accorsi M, Gianquinto G, Dinelli G, Antognoni F, Ruiz-Carrasco KB, Martínez EA, Alnayef M, Marotti I, Bosi S, Biondi S. 2011. Beyond the ionic and osmotic response to salinity in *Chenopodium quinoa*: functional elements of successful halophytism. Funct Plant Biol 38:818–831.

Planella MT, Scherson R, McRostie V. 2011. Sitio El Plomo y nuevos registros de cultígenos iniciales en cazadores del Arcaico IV en alto Maipo, Chile central. Chungara, Revista de Antropología Chilena 43:189–202.

Ruiz-Carrasco KB, Antognoni F, Coulibaly AK, Lizardi S, Covarrubias A, Martínez EA, Molina-Montenegro MA, Biondi S, Zurita-Silva A. 2011. Variation in salinity tolerance of four lowland genotypes of quinoa (*Chenopodium quinoa* Willd.) as assessed by growth, physiological traits, and sodium transporter gene expression. Plant Physiol Biochem 49:1333–1341.

Schlick G, Bubenheim DL. 1996. Quinoa: candidate crop for NASA's Controlled Ecological Life Support Systems. Janick J. Progress in new crops. Arlington, TX: ASHS Press. 632–640.

Sepúlveda J, Thomet M, Palazuelos P, Mujica MA. 2003. La Kinwa Mapuche, recuperación de un cultivo para la alimentación. Chile: CET-Sur, Fundación para la Innovación Agraria, Ministerio de Agricultura.

Vega-Gálvez, A., M. Miranda, J. Vergara, E. Uribe, L. Puente, E. A. Martínez. 2010. Nutrition facts and functional potential of quinoa (*Chenopodium quinoa* Willd.), an ancient Andean grain: a review. J Sci Food Agr 90:2541–2547.

Von Baer I, Bazile D, Martínez EA. 2009. Cuarenta años de mejoramiento de la quínoa (*Chenopodium quinoa* Willd.) en la Araucanía: origen de "La Regalona-B". Revista Geográfica de Valparaíso 42:34–44.

Wilson HW, Heiser CB. 1979. The origin and evolutionary relationships of 'huauzontle' (*Chenopodium nuttalliae* Safford), domesticated chenopod of Mexico. Am J Bot 66:198–206.

Chapter 3
Agroecological and Agronomic Cultural Practices of Quinoa in South America

Magali Garcia[1], Bruno Condori[2], and Carmen Del Castillo[1]
[1]*Faculty of Agronomy, Universidad Mayor de San Andres, La Paz, Bolivia*
[2]*Consultative Group on International Agricultural Research – International Potato Center, La Paz, Bolivia*

INTRODUCTION

The Food and Agriculture Organization (FAO) of the United Nations identified quinoa as a potential crop to combat global malnutrition. Quinua or quinoa (*Chenopodium quinoa* Willd.) is an underutilized crop with enormous potential, mainly due to its nutritional and physiological properties. The crop produces small grains and has several intrinsic qualities for developing superior varieties, including a major gene pool, that is, more than 6,000 varieties and several wild relatives with outstanding characteristics such as earliness, color, and grain size. Quinoa also has a large genetic variability and plasticity, resistance to biotic and abiotic factors, and ability to adapt to adverse soil and climatic conditions in places where most agriculture is marginal. Moreover, quinoa is efficient in its use of production inputs, and thus can be grown through a large range of production environments, from the equator to high latitudes, and from sea level to 4,000 m. It has the ability to produce grains even under highly saline soils. It is a functional and ideal food for human nutrition due to its nutritional profile, as represented by the quality and quantity of its essential amino acid composition. Unlike many other regional products, quinoa grain can be stored in natural conditions, constituting a reserve that can be consumed even after several months or in seasons with food scarcity. This property increases the value of quinoa in support of human nutrition. The diversity of its uses and its importance, both cultural and nutritional, make it a principal crop in South America.

Together with potato, quinoa was the staple food of the indigenous population (Hellin and Higman 2003). Among the food grains cultivated in the South American highlands, quinoa ranks as a very significant crop in terms of quality and nutritional value. At the time of the Spanish conquest of the Inca Empire in 1532, quinoa, potatoes, and maize were the principal staple foods in Andean South America, with quinoa cultivation extending slightly beyond the region occupied by the Incas (Galwey 1993; Cusack 1984; Risi and Galwey 1984). Following the conquest, quinoa cultivation declined drastically and was displaced by crops preferred by the "conquistadores." Furthermore, quinoa was not adopted as a crop by European settlers in South America or in Europe, as were other New World crops such as maize and potatoes. It was not until the last decades of the twentieth century that interest in quinoa rose again. The crisis in food security worldwide made it imperative to look

Quinoa: Improvement and Sustainable Production, First Edition. Edited by Kevin Murphy and Janet Matanguihan.

for crop alternatives. Quinoa, with its excellent nutritional profile and genetic diversity, makes it an ideal crop to be developed further.

ANDEAN DOMESTICATION

Quinoa has been cultivated in the Bolivian and Peruvian Andean region for ~7,000 years by indigenous population groups (e.g., the Aymara and the Quechua in Bolivia) (Garcia 2003). Several names were given to the crop, such as "mother cereal" in Quechua, or the internationally known names "Inca-rice," "Inca-wheat," or "Han" (China). People of the Chibcha (Bogota) culture called quinoa "suba" (Valencia-Chamorro 2004), the Tiahuanacotas (Bolivia) called it "jupha," the inhabitants of the Atacama desert (currently in Chile) knew it by the name "dahue," and in Ecuador, it was called "quimián" (Pulgar Vidal 1954). Tapia (1979) wrote that the names "quinua" and "quinoa" were used in Bolivia, Peru, Ecuador, Argentina, and Chile.

The monikers "Grain of the Incas" or "Inca-rice," though internationally known, is misleading since archaeological evidence shows that the Andean quinoa was domesticated before the Incas (Heisser and Nelson 1974; Jacobsen 2003). Through migration and growing trade, the crop spread throughout the Andean regions. Pre-Inca native populations practiced intensive agriculture, using irrigation, composting, rotations, and construction of terraces to conserve soil fertility in the mountains and increase agricultural production, especially in the areas surrounding Lake Titicaca. When the Incas established their kingdom in Cuzco (1100–1533 AC), they quickly recognized the extraordinary agricultural and nutritional qualities of quinoa and gave the quechua name "chisiya mama" or "mother grain" and introduced the grain in religious activities. Quinoa served very well as concentrated food for the Incan army during the marches of conquest. Thus, through the extension of the Inca kingdom, quinoa was also systematically distributed from Chile to Colombia. The variation in quinoa seed size and the wide range of seed color, from black to yellow, pink, and white, is a clear indication that the ancient Andean farmer successfully bred quinoa.

The botany of quinoa was first described by Willdenow in 1778. It was recognized as native to South America, with the center of origin located in the Andes of Bolivia and Peru (Cardenas 1944). The distribution range is fairly wide geographically but restricted to the Andes. The greatest diversity of ecotypes can be found in the Andes, both cultivated and wild relatives. Social, cultural, and economic importance was also given to quinoa by peoples of the Andes region (Gandarillas 1979b).

Bonifacio (2001), Del Castillo et al. (2008), and the FAO (2011) identify four main groups of quinoa according to the agroecological conditions of the areas where it is grown: valleys, high plains, salt flats, and sea level. These ecotypes have different and specific botanical agronomic and adaptive characteristics. Most probably, it took decades for the actual varieties to be developed from the wild forms, though certain wild forms of quinoa are still consumed as vegetable in local communities in the Andes (Mujica et al. 2001a). Although its production declined significantly during the Spanish conquest, the popularity of quinoa rose again in the past century and is widely consumed at present.

The area where quinoa was domesticated is hardly suitable for agriculture. The Andes, and more the higher planes, are exposed to harsh climatic and soil conditions. Water shortage is a major problem due to the combined effect of low rainfall, high rate of evapotranspiration, and the low water retention capacity of the soils. Plants are exposed to extremely high rates of solar radiation and large daily thermal amplitudes. During spells of dry periods, farmers frequently deal with yield reductions that lead to cumulative shortages of human food and animal feed (Garcia et al. 2003; Jensen et al. 2000). In addition, plant growth is exceedingly hampered by frost in the southern part of Peru and Bolivia and by high soil salinity, especially in the salt deserts of the southern part of Bolivia (Jacobsen et al. 2003).

As quinoa and other robust agricultural crops have been domesticated in the harsh environment of the Andes, with its highly variable conditions,

Fig. 3.1 (a–c) Examples of quinoa plants in farmer's field showing the vast diversity of colors and the forms of panicles (Province of Omasuyos, bordered to the south and west by Lake Titicaca, Bolivia) (Del Castillo and Winkel, IRD – CLIFA, 2002–2008). (*See color insert for representation of this figure.*)

these crops are characterized by exceptional environmental adaptation, specifically to mountain and arid conditions with tolerance to drought (Jensen et al. 2000; Garcia et al. 2003, 2007; Jacobsen et al. 2001, 2003, 2003b, 2003c; Bois et al. 2006;Geerts et al. 2008b), frost (Jacobsen et al. 2005, 2007), saline soils (Koyro and Eisa 2008; Rosa et al. 2009; Ruffino et al. 2010; Jacobsen and Mujica 2003; Hariadi et al. 2011; Bosque 1998), large ranges in daily temperature, and other abiotic and biotic factors (Jacobsen et al. 2003, 2003b; Bertero et al. 2004). These Andean crops have yielded products of high nutritional value for millennia (Gonzalez et al. 1989, 2009, 2010; Grau 1997; Gross et al. 1989; Hermann and Heller 1997; Repo-Carrasco et al. 2003; Jacobsen and Mujica 2003) and are consumed by the indigenous population.

BOTANICAL AND TAXONOMICAL DESCRIPTION

Quinoa is commonly known as a pseudo-cereal, as its characteristics are very different from those of the true cereals. First, it is dicotyledonous and not a member of the *Poaceae* (grasses) family as are the monocotyledonous cereals. Moreover, quinoa and other specimens such as the amaranth do not contain gluten, which most typical cereals do. The crop has enormous intraspecies variability and plasticity, enabling it to grow under extremely diverse agronomic conditions. Owing to its physiological efficiency, it is classified as a C_3 plant. The Andean quinoa (Fig. 3.1) has a height of 0.5–2 m, with grains of approximately 2-mm diameter. Its roots, often with numerous ramifications, can reach up to 1.80 m in depth in times of severe drought in sandy soils. Different genotypes exist with and without shoot ramifications and intermediate forms. The crop has efficient mechanisms against various abiotic stresses such as hail, frost, and drought. Most phenotypic differences are reported in relation with variety and agroclimatic conditions.

Leaves are polymorphic on the same plant and also differ much in form and color (green, red, purple) among varieties (Mujica et al. 2001a). They are amphistomatal, with the younger leaves often covered by Ca-oxalate glands on the upper leaf surface. Flower buds are organized in typical inflorescences with a central axis and secondary and tertiary axes. They can be of the lax form (amaranthiform) or the compact form (glomerous). Flowers are incomplete and autogamous or allogamous. Both hermaphrodite and unisexual female flowers are present (Bhargava et al. 2006). The percentage of allogamy differs between varieties and goes from nearly from 0% to 80%. Asynchronous flowering is typical of the crop. Flowering occurs over a total of 12–15 days, although individual flowers remain open only during 5–7 days, with maximal opening between 10 AM and 2 PM (Erquinigo 1970).

Fruits are grain shaped and contain approximately15% moisture at harvest (Gallardo 1997). The pseudo-grains (dicotyledonous embryo, episperm and perisperm) are embedded in a perigonium (modified leaf structure) that has to be removed during postharvest processing. In general, large and sweet grains that are white in color are preferred in the international market, although recently, colored grains are becoming popular in exotic markets (Geerts et al. 2008b).

Initially, the crop was classified within the *Chenopodiaceae* family (Cronquist 1981), but phylogenic revision has merged the *Amaranthaceae* and *Chenopodiaceae* in the family of the *Amaranthaceae* (Angiosperm Phylogeny Group 2003), and the crop is now a member of the subfamily *Chenopodioidae*. It is a member of the family of the Amaranths (*Amaranthus* spp.) and of the Himalayan grain chenopods such as the domesticated forms of *Chenopodium album* L. (Partap and Kapoor 1985a, 1985b). Several crops such as sugar beet, beetroot, mangold, spinach and other goosefoot weeds also belong to the family. The genus *Chenopodium* also includes several grain crops in South America such as *C. quinoa* Willd. and *Chenopodium pallidicaule* Aellen, vegetables in Mexico such as *Chenopodium nuttalliae* Safford and *Chenopodium ambrosioides* L., and vegetables or medicinal plants in South America such as *Chenopodium carnosolum* Moq. and *C. ambrosioides* (FAO 2011).

GENETIC BACKGROUND AND RESEARCH ON QUINOA GENETICS

The Andean region is the center of origin of quinoa, more specifically the Bolivian and Peruvian Andes (Bonifacio 2003). In terms of its genetic variability, quinoa can be considered as an oligocentric species with its center of origin around Lake Titicaca (Mujica et al. 2001a). In comparison with the wild types (*C. album*, *C. carnosolum*, *Chenopodium hircinum*, *Chenopodium murale*, *Chenopodium petiolare*), the flower buds of quinoa are more condensed, plant and grain size larger, and the level of pigmentation higher (Mujica et al. 2001a). Ancient peoples have selected genotypes on the basis of their use and resistance to adverse biotic and abiotic factors. Over the decades, these genotypes have been developed to the currently known ecotypes (Table 3.1 and Fig. 3.2). Nowadays, traditional wild types are still locally conserved for medicinal purposes or as security crop in case of natural disasters (Geerts et al. 2008b). In addition, wild quinoa plants, widely distributed in the South American Andes, have valuable genes that can be

Table 3.1 Different ecotypes of quinoa, their local names, and their property or principal use.

Ecotype	Property or principal use
Chullpi	Soup
Pasankalla	Roasted (toasted)
Coytos	Flower
Reales	Grain or "pissara"
Utusaya	Good performance under saline conditions
Witullas and Achachinos	Good performance under cold temperatures and frost
Kcancollas	Good performance under drought conditions
Quellus (yellow forms)	High yield
Chewecas	Good performance under excess water
Ayaras	High nutritional value
Ratuquis	Short cycle

Source: Mujica et al. 2001a.

Fig. 3.2 (a–c) Close-up view of quinoa grains (Del Castillo and Winkel, IRD – CLIFA, 2002–2008).

exploited in the future to increase crop resistance to climate hazards and adaptation, thus maintaining high production rates. Some populations are characterized by tolerance and resistance to insects and diseases, frost, and drought. They also possess favorable traits in terms of nutritional value and the duration of the productive cycle, which also add commercial value to the crop (Rojas et al. 2008; Del Castillo et al. 2007).

At present, the large genetic diversity of quinoa is conserved in germplasm banks (Bonifacio 2003) such as the Bolivian quinoa germplasm bank (PROINPA Foundation 2003a), which is currently under the management of the National Institute for Agricultural and Forestry Research (INIAF). There are several other quinoa germplasm banks, such as those managed by the Centro Internacional de la Papa (CIP), the Institute of Plant Genetics and Crop Plant Research (IPK, Germany), the US Department of Agriculture (USDA), and the Faculty of Agronomy of UMSA in La Paz. As reported in 2010 (FAO 2011), the Bolivian germplasm bank holds 3,121 accessions, including wild-type variations, though this number continues to increase due to on-going germplasm collection trips. In Peru, the Germplasm Bank at the Illpa Experimental Station in Puno holds 536 quinoa accessions, whereas the gene bank at the University of La Molina holds about 2,000 accessions (Bravo and Catacora 2010). In Ecuador, the National Institute of Agricultural Research has 608 accessions (Peralta 2009).

In general, ecotypes of the north and central Andes have small to medium grains compared to the larger grains of the southern ecotypes of the salt flats of Bolivia (PROINPA Foundation 2003d). A large genetic diversity is also traditionally preserved for auto-consumption in local cultivation systems called "aynokas" (Mujica et al. 2001c). "Aynokas" are agricultural production units formed by a division of communal areas to ensure agricultural and ecological sustainability and necessary crop rotation (Aguilar and Jacobsen 2003). Unfortunately, due to market pressure, several farming systems in the Altiplano have shifted to more commercial quinoa varieties and even different crops altogether, thus reducing the use of local varieties (Geerts et al. 2008b).

Owing to the large differences in agroecological environments and the wide distribution of quinoa in the Andes, three main classes of quinoa ecotypes can be distinguished. The valley ecotypes have small grain size, tall plants, and high resistance to mildew. The ecotypes around Lake Titicaca also have small grains, medium resistance to mildew, intermediate growth cycles, and low saponin content. In the Southern Altiplano, the ecotypes have higher saponin content, large grains, and low resistance to mildew. The various ecotypes with local names and the properties for which they are cultivated are presented in Table 3.1.

Quinoa is a diploid allotetraploid ($2n = 4x = 36$), with 36 somatic chromosomes. The basic number of chromosomes of the *Chenopodium* genus is nine (Mujica et al. 2001a). There has been considerable research on the floral biology of quinoa, such as the proportion of auto-pollination and cross-pollination. This type of research has brought about important advances in hybridization, selection, and genetic improvement of quinoa varieties (Mujica et al. 2001a). Breeding programs are generally directed toward developing cultivars with higher yields, high protein, and low saponin contents (Bhargava et al. 2006), while increasing or maintaining resistance against biotic and abiotic factors, such as downy mildew and drought (Bonifacio 2003).

ECOLOGY AND PHYTOGEOGRAPHY

The geographical distribution of quinoa extends from 5° North Latitude in southern Colombia to 43° South Latitude in the Tenth Region of Chile and the Argentinean Andes. The altitudinal distribution ranges from sea level to 4,000 masl. Quinoa is planted mainly in the highlands shared by Chile, Peru, and Bolivia. There are coastal quinoa, valley quinoa, and highlands quinoa. The Bolivian Altiplano, with a cultivated area of more than 100,000 ha, has the world's largest concentrated area of quinoa production. A large part of this extension is located in the south of the country, close to the salt flats. The second largest producer is Peru with a cultivated area of around 55,000 ha, concentrated mainly in southern Puno producing more than 41,000 tons per year. In Ecuador, 1,700 ha are dedicated to quinoa production in the provinces of Carchi, Imbabura, Pichincha, Cotopaxi, Chimborazo, Loja, Latacunga, Ambato, and Cuenca. In Colombia, some 700 ha are grown with quinoa, nearly all in the south of Nariño, in the municipality of Sapuyes. The north of Chile and the Argentine highlands are also working to increase quinoa production.

In South America, different quinoa varieties are distributed according to eco-geographic zones. For example, one of the most popular varieties is the variety "Real" (PROINPA Foundation 2003a) cultivated mainly in the largest production area in the Andes, at the south of the Bolivian Altiplano. It is grown under conditions of intensive drought and frost to which it is extremely resistant. In this area, annual rainfall varies from exceptionally dry in the extreme south to dry in the intersalar region, from 150 to 340 mm precipitation, respectively (Geerts et al. 2006b). Farm households rely mostly on quinoa production (Laguna 2000). The "Real" variety is in demand due to its large white grains. Nevertheless, it has not been grown successfully outside its center of origin apparently due to the fact that its flowering is closely linked with light hours at high altitude and high solar radiation intensity that is absent in other latitudes and altitudes. To the north of the Bolivian Altiplano, specific varieties more adapted to their local environments are grown, generally with small and medium grains (PROINPA Foundation 2002), but which are completely unable to thrive in the salt flats of Bolivia. In general, native varieties have a moderate yield but are more resistant to abiotic factors and have a higher-than-average nutritional quality if cultivated close to their center of origin.

The crop cycle varies largely from 120 to 240 days, depending on the varieties and on the zone of production and the attendant environmental conditions. In general, varieties grown in colder environments have longer crop cycles, whereas shorter cycle varieties are grown in the valleys and lowlands. Sanchez (2012), using a base temperature of 1°C for the crop, found in the Bolivian Highlands at a thermal time of 1,500 heat units, which was fairly constant from North to South. Apart from this study, there have been no more studies on the quantity of heat that the crop needs to complete its growing cycle. There are indications that climate change impacts are somewhat shortening the crops cycles, especially in the highlands. Despite this possible climate effect, research institutions at present are not working toward the release of early varieties.

The most important phenological stages that are morphoanatomically significant according to Espindola (1980, 1992) and Mujica et al. (2001b) are presented in Table 3.2 with their key properties (Geerts et al. 2008b).

CULTIVATION AND AGRONOMIC PRACTICES IN SOUTH AMERICA

In general, the production systems of quinoa in South America can be grouped into the systems of the southern highlands, the systems

Table 3.2 Phenological stages of quinoa.

	Phenological stage	Characteristics	Start (between *x*–*x* days after sowing)
0	Germination	Seed swelling and bursting	3–5
1	Cotyledonous phase	Plant emergence	3–10 days after sowing
2	Two real leaves	Initiation of the vegetative period; a rapid root development is noticed	10–20
3	Five alternate leaves	Early vegetative stage; sensitive to competition by weeds	35–45
4	Thirteen alternate leaves	Important root ramifications present	45–50
5	Preflowering (flower bud development)	Varieties with lax and compact flower buds can be distinguished	55–70
6	Flowering	Flowering starts on top of the flower bud and continues to the bottom; sensitive stage for hail, frost, drought, and diseases	90–130 (50% of flowers)
7	Early grain filling (milky)	Grains are still malleable and moist (50% moisture content or MC); sensitive stage for hail, frost, drought, and diseases	100–130
8	Late grain filling (pasty)	The specific color of the variety is obtained and grains are much drier (25% MC)	130–160
9	Physiological maturity	Hard and dry grains are obtained (15% MC)	160–180

Source: Adapted from Espindola 1992; Mujica et al. 2001b.

of the central and northern Altiplano, and those of the valleys in Ecuador and Colombia. Roughly, the northern and central Altiplano includes the North Altiplano of Bolivia and the Puno region in Peru. The so-called intersalar region is agroclimatologically classified as Southern Altiplano (Geerts et al. 2006b).

In the Southern Altiplano, quinoa is commonly cultivated mainly as a monoculture. In this system, fields are left fallow for one to three cropping cycles or even longer, before being cultivated again. In the rest of the Altiplano or Highlands, the crop is in rotation with potato or beans (Mujica et al. 2001b). In the Andean valleys of Ecuador and Colombia, quinoa is always cultivated as part of a rotation system after maize and potato (Peralta 2009). Nutritional conditions are generally favorable for subsequent quinoa cultivation after a potato crop (Mujica et al. 2001b).

Traditional cultivation includes the incorporation of some organic fertilizer, if available, such as sheep and/or llama dung. The dung is applied at sowing, after the first weeding and thinning, and at flowering. Mycorrhizal association has been reported in quinoa, which suggests possible maximization of nutrients (Mujica 1994).

QUINOA PRODUCTION

Soil conditions

Quinoa can grow in soils with a textural class ranging from sandy to clay, although soils with good drainage are better for the crop. Quinoa can also tolerate a wide range of soil pH (Mujica et al. 2001b), from acid soils (pH 4.5; e.g., Cajamarca region in Peru) to more alkaline soils (pH 9; e.g., salt depressions of Bolivia) with optimum growth in soils with almost neutral pH. The quinoa plant has high requirements for nitrogen (N) and calcium (Ca), moderate for phosphorous (P), and minimal for potassium (K) (Mujica et al. 2001a).

Quinoa is a facultative halophyte (Bosque et al. 2006; Jacobsen and Mujica 2003) and can grow in extreme saline conditions, such as soils with electrical conductivity that can be as high as 52 dS/m (Jacobsen et al. 2001). Adjustment of leaf water potential by the accumulation of salt ions in tissues enables the plant to maintain cell turgor and transpiration under saline conditions in the salt flats of the Altiplano. Potentially, quinoa could play an important role in cleaning salt-contaminated soils (Jacobsen et al. 2003). Gonzalez and Prado (1992) reported that germination is retarded due to increased soil salinity, but even at very high salinity, seeds can remain dormant and viable. Nevertheless, tolerance or sensitivity to soil salinity is dependent on the variety (Schabes and Sigstad 2005).

CLIMATE

Drought resistance

Owing to its phenological plasticity and resistance to climate constraints, quinoa is exceptionally adapted to the different arid climates of the Andean region (Mujica et al. 2001b). In the Andean Altiplano, droughts can occur at the end and beginning of the rainy season due to a decrease in the length of the rainy seasons. Droughts can also occur within the growing period because rainfall in the Altiplano is likely to occur in delimited episodes of rain, separated by dry periods (Garreaud et al. 2003; Garcia et al. 2007). Intraseasonal dry spells are by far as important to crop yield as interseasonal differences, especially when the dry spells occur during critical growth stages such as anthesis (Fox and Rockström 2000). Generally, quinoa is a remarkable crop in that it can still produce grains even under drought conditions (Garcia et al. 2003). Quinoa is very efficient in water use, despite being a C_3 plant species. Several mechanisms related to drought resistance are present in quinoa, including drought escape, tolerance, and avoidance (Jensen et al. 2000). Many of these mechanisms also serve to make quinoa tolerant or escape the effects of other abiotic stresses such as frost. However, when drought occurs during sensitive phenological stages, such as emergence, flowering, and milky grain, yields can be severely reduced.

Drought escape appears as a lengthening of the growth cycle in response to droughts during early vegetative stages and as an early maturing process in response to drought stress during the later growth stages (Jacobsen and Mujica 1999; Garcia 2003;Geerts et al. 2006a). Early maturing is an important escape mechanism in the Andean areas with frequent droughts at the beginning and end of the growing season (Jacobsen et al. 2003). Quinoa tolerates drought mainly through tissue elasticity and low osmotic potential, which is a measure of solutes in the plant. Proline might act as a main regulating agent for the osmotic balance in quinoa as it does in cotton (Parida et al. 2007). In turgid tissues, proline is known to oxidate quickly, whereas under water-deficit conditions, proline oxidation is inhibited. Indeed, Aguilar et al. (2003) reported that proline content was highest in varieties from eco-geographic locations with distinctly unfavorable drought conditions and large differences in day and night temperatures.

The quinoa plant also avoids the negative effects of drought through a high root/shoot ratio (Bosque et al. 2003; Sanchez et al. 2003), reduction in the leaf area by leaf dropping, dynamic stomatal behavior, and the presence of special vesicular glands of Ca-oxalate, which are small and thick-walled cells that preserve cell turgor even during severe water losses (Jensen et al., 2000). The Ca-oxalate glands are a very specific drought resistance mechanism for quinoa. Ca-oxalate crystals are hygroscopic and have a two-way function for drought stress mitigation. First, they are assumed to increase albedo and decrease the direct radiation on the leaves. Second, they are assumed to control excessive transpiration by humidification of the stomatal guard cells (Mujica et al. 2001a). Asynchronous flowering within the flower bud is another mechanism to spread the risk and is induced by drought and several other abiotic stresses.

Although quinoa possesses a wide variety of drought resistance mechanisms, water stress still decreases grain yields quite often (Bilbao et al. 2006; Bosque et al. 2003), except when droughts occur during the initial growth phases after successful emergence that induce a certain hardening (Huiza 1994; Bosque et al. 2000; Garcia 2003; Geerts et al. 2006a).

Temperature and photoperiod

In general, the adequate mean temperature for quinoa growth is 15–20°C, but it can grow at mean temperatures ranging from 10 to 25°C. Extremely high temperatures can cause flower abortion (Jacobsen et al. 2003). Apart from drought, frost is one of the major growth limiting factors in the Altiplano (Carrasco et al. 1997; Hijmans 1999; François et al. 1999). Frost can reduce yield due to cell destruction and even plant death (Fig. 3.3). Quinoa is one of the few crops that can tolerate frost to a certain extent, but this depends largely on the duration of the frost, the quinoa variety, the phenological stage of the plant when frost occurs, the relative humidity, and the micro-location of the fields (e.g., hill slopes have lower risk of frost compared to valleys).

Jacobsen et al. (2005) studied the influence of frost events of different duration and intensity in various phenological stages and for different cultivars of quinoa. They found that quinoa is most susceptible to frost from the flower bud formation stage onward but less susceptible during the vegetative stages (Bois et al. 2006). A temperature of −4°C lasting for four hours during the flowering stage caused seed reduction up to 66%, while quinoa in the vegetative period was considerably damaged when exposed to temperatures of −8°C which last for 2–4 h.

Recently, Jacobsen et al. (2007) reported that the main survival mechanism against frost in quinoa is a moderate supercooling that avoids ice formation. Supercooling is the cooling of liquid below its normal freezing point without crystallization of the liquid. The high level of soluble sugars in quinoa reported in their study may cause a reduction in freezing temperature and the mean lethal temperature. Proline content and the content of soluble sugars such as sucrose may serve as indicators of frost tolerance in different varieties.

Quinoa also tolerates a broad range of radiation intensities, from radiation at sea-level up to intensive radiation at high altitudes. Genotypes have been classified as short day, long day, and indifferent, in relation to photoperiod sensitivity (Bertero 2003). Bertero (2001, 2003) and Bertero et al. (1999a, 1999b, 2000) conducted extensive research on photoperiod and temperature sensitivity of quinoa development in South America. Varieties originating from Colombia to Southern Chile were used in the study, and all cultivars had a facultative short day response in terms of duration of emergence to flowering. It was demonstrated that the duration of all phases of

(a) (b)

Fig. 3.3 (a,b) Frost damage in quinoa (−5°C at 60 days after sowing).

development is sensitive to photoperiod (Bertero 2003). In the same way that water stress during certain phenological stages influences the duration of subsequent phenological stages (Garcia 2003), the duration of the photoperiod during certain phases also influences subsequent phases and is called delayed response (Bertero 2003). The mean incident radiation affects the phyllochron in quinoa, with higher incident radiation causing larger phyllochrons. However, cultivars with the highest photoperiod sensitivity and largest phyllochron were insensitive to radiation (Bertero 2001).

Temperature sensitivity was highest in quinoa cultivars originating from cold and dry climates but lower in cultivars from warmer and humid climates (Bertero et al. 2000), an observation that can possibly explain why varieties from the highlands are more affected by global warming. The higher sensitivity of cultivars from regions with frequent late season drought and frost predisposes them for a faster seed filling when photoperiods are shortening and unfavorable conditions are approaching (Bertero 2003).

Fig. 3.4 Quinoa after hail damage (Del Castillo and Winkel, IRD – CLIFA, 2002–2008).

Hail

Hail, and sometimes snow, are sporadic and quite localized in the Andes (Jacobsen et al. 2003) but may still cause substantial yield losses in quinoa, especially when flower buds are already present (Fig. 3.4, unpublished data). As is the case for other abiotic stresses, susceptibility to hail is largely dependent on the variety.

CULTIVATION

In the Andes, quinoa is mainly grown during the austral summer months (September to May), although the sowing times markedly differ and in some equatorial areas cultivation may go beyond that period. Similarly, the production techniques are in accordance to the local ecosystem, and so differ from one system to another. In the areas surrounding the southern salt flats, traditional cultivation includes field preparation carried out from March to May on virgin fields or fields that are fallow to capture moisture from the late rains and some winter snows and store it for the following season. Much of the intensity of field cultivation (plowing, breaking, harrowing, and leveling) depends on the years a field has been fallow, the soil texture, and the availability of a tractor.

In the northern Altiplano of Bolivia and Peru, field preparation is done either shortly after harvest of the previous crop or before sowing after the first rains. In Ecuador and Colombia, the cultivation period could range from November to February, after the onset of the rainy season.

Sowing

Quinoa production in areas close to its center of origin goes from fully traditional to semi-mechanized to fully mechanized. Field preparation and sowing is carried out either by hand or tractor (Fig. 3.5). In general, it has to be stressed that mechanization often causes excessive

Fig. 3.5 (a) Traditional quinoa farming with little soil surface opening and large spaces between plants (SUMAMAD-UMSA team). (b) Mechanized fields (Winkel 2013 ©IRD).

eolic erosion and fertility loss due to increased soil denitrification, especially in the southern highlands.

The average sowing date of quinoa in the Andes lies mostly between September and November, although in some minor areas close to the equator, the sowing date could extend until February. The differences in sowing dates are due to crop cycle lengths of individual varieties and local climatic conditions. The sowing date also depends mainly on availability of initial soil moisture. Sowing is one of the most important activities because the emergence of seedlings, which impacts plant density and final yields, depends on this stage. Sowing is a key practice for the success or failure of the quinoa crop and requires quite some experience. Quinoa seeds are sown at different times, depending on the place to be sown, the varietal traits, and soil moisture. These are also important factors in determining the sowing method to be used, whether manual or mechanical. Superficial sowing poses the risk of seed dehydration or burning by solar radiation, whereas deep sowing can prevent germination due to restricted growth (Rodriguez and Raffaillac 2003).

The sowing method most often used in the southern highlands near the salt flats of Bolivia is the manual method, which is to dig sowing pits at an average distance of 1 m. Using a "taquiza" (a local specialized tool), holes are dug until moist soil is reached. Several seeds are then deposited in the hole and immediately covered with soil to a depth ranging from 4 to 10 cm. Sowing density is generally from 8 to 15 kg/ha, with lower seed densities when high-quality, certified seeds are used (Mujica et al. 2001b; Aguilar and Jacobsen 2003; Yucra and Garcia 2007). Shortly after sowing, pits are often covered with grasses or small shrubs to prevent burning of seeds due to extreme solar radiation.

Previously, the cultivation of quinoa in the southern highlands near the salt flats of Bolivia relied exclusively on the manual sowing method as described. The production system also includes a strong linkage with the llama husbandry because of the positive effect of the animal dung to crop productivity. At present, though, the production system is shifting toward a more mechanized method (Cossio 2008). Owing to the higher demand for quinoa, the agricultural frontier has been extended. More virgin land on the flats is being plowed, drastically reducing the natural vegetation and giving rise to some environmental problems (Jacobsen 2011).

In contrast, sowing in the north Altiplano of Bolivia and south of Peru is generally by furrow, with a distance of 0.4–0.8 m in the furrows or on the ridges. Less common sowing methods are by transplanting (inter-Andean valleys) to avoid intensive weeding later on or by broadcasting the seeds. In the inter-Andean valleys of Ecuador and Colombia, the traditional cultivation technique consists of sowing under dry conditions in rotation with potato or on strips in maize crops, with little soil preparation and using only the residual organic fertilizers from the preceding crop. Sowing density varies between 15 and 20 kg seeds/ha. Regardless of the area where quinoa is cultivated, traditional Andean growers always look to reduce the risk, and sow several ecotypes at different times and in different locations.

Fertilization of quinoa

One of the problems in quinoa cultivation is the local belief that as it is an indigenous crop, it is unnecessary to provide a large quantity of external inputs. Actual field application by farmers is not a priority, although animal dung is applied to quinoa fields or pits, when available.

Research on quinoa fertilization has been limited. Schulte auf'm Erley et al. (2005) reported that quinoa responds strongly to nitrogen fertilization, producing up to 3.5 t/ha with a fertilization rate of 120 kg N/ha without reduction in nitrogen use efficiency. Harvest indexes were not affected by nitrogen application but grain nitrogen content increased significantly. Murillo (1995) investigated the application of 80 kg N/ha for quinoa in Bolivia at sowing, without additional P or K application. Yield and grain quality increased significantly. Moreover, they reported the necessity of additional irrigation to facilitate nitrogen absorption when rainfall is inadequate. Berti et al. (1997) applied 0–225 kg N/ha to quinoa sown at sea level in Chile and found the highest yields (3,555 kg/ha) at the highest fertilizer levels. The yield response to increasing N fertilization was quadratic in shape, and a slight decrease in nitrogen use efficiency was noted up to 225 kg N/ha. Moreover, the harvest index was significantly lower for treatments with high N fertilizer application. In other Chilean experiments, yields up to 5 t/ha were reported for quinoa fertilized with N combined with drip irrigation (De la Torre-Herrera 2003).

To ensure an adequate nutrient supply for crops, growers must strive to maintain a good balance of nutrients in the soil (Miranda 2012). The loss of nutrients has to be minimized, while maximizing efficiency in their addition as their removal during harvest is unavoidable. The higher and better the quality of yield is, the greater the removal of soil nutrients. Therefore, to increase quinoa yields in a sustainable way, it is extremely important to consider application of fertilizers, particularly as quinoa is well known for the high quality of its protein, suggesting a strong nitrogen demand from the soil. Without fertilizers, there could be permanent soil nitrogen and nutrient depletion, rapidly degrading an already poor soil.

As the organic market is the largest niche market for quinoa, most fertilization research had been conducted on the application and use efficiency of manure in fields. The application of organic fertilizers, especially manure, at inappropriate times and quantities will not necessarily improve soil characteristics but may even have negative effects on the soils, especially over the short-term period. The actual incorporation of manure into the soil in farmers' fields ranges only from 4 to 10 t/ha. Small quantities of manure incorporated within the production systems will increase field labor but with negligible effect on yield. With such results, farmers get discouraged and forgo manure application. To have a significant effect, manure should be applied at higher rates, up to 20–30 t/ha, to guarantee adequate revenue from crop harvest and commercialization, in addition to maintaining good soil structure (Miranda 2012). An adequate soil and water balance, as well as water availability during critical periods, is essential to reap benefits from soil fertilization, especially manure application.

The timing of fertilization is also important when working solely with organic fertilizers. There is limited effect on crop yield when manure is applied during sowing because the nitrogen will be liberated approximately 50–60 days after application, especially in the highlands of the Andes. The low air and soil temperatures prevent the rapid decomposition of manure and result in a slow release of nutrients. The best time to apply manure is around 2 months before sowing, or at least 45 days, to adequately age the manure. Working with compost, Miranda et al. (2012) demonstrated a reduction in the time needed for manure to be decomposed and showed the benefits to the crop because of the accelerated ageing of manure.

In general, in the northern Andes, where quinoa is sown after potatoes, organic matter content and nutrients is favorable for quinoa. In some cases, the slow decomposition of manure, coupled with previous application of nutrients for the potato crop, almost fulfill the nutrient requirements of quinoa. Afterward, the crop only needs supplemental fertilization.

However, when planted after a grain crop (corn or wheat on the coast and barley or oats in the mountains), it is necessary to use organic matter in a ratio of at least 3 t/ha. The average fertilization required is in the formula 80-40-00, equivalent to 174 kg/ha of 46% urea, 88 kg/ha of 46% triple calcium, and zero potassium. Potassium is freely available in the soils of the Andes and in South America in general, as there are large quantities of potassium-retaining clays in the soils.

On the coast, soil nutrients are scarce as the amount of organic matter is extremely low and soils are very sandy. In general, the formula recommended for fertilization is 240-200-80, equivalent to 523 kg/ha of 46% urea, 435 kg/ha of superphosphate, 46% calcium triple, and 134 kg/ha of 60% potassium chloride. In addition, application of manure, compost, humus, or organic matter is recommended, when available.

Cultural practices

Cultivation is limited to one or two hoeings, with an occasional hilling-up, particularly in the inter-Andean valleys. Hoeing is done during the first phenological phases, whereas hilling-up is specific for furrow-sown quinoa and is also used for additional fertilization just before bud formation to optimize yields. When there is excessive vegetative growth, hilling-up can be necessary to avoid plant lodging due to heavy canopy. Also, thinning of plants is necessary to prevent intraspecific competition. If sown in rows, quinoa plants should be spaced around 10 cm apart (Aguilar and Jacobsen 2003).

Early weeding (±30 days after sowing) is important in quinoa fields in the Altiplano as plants compete with each other for scarce nutrients (De Barros Santos et al. 2003; Bhargava et al. 2006). Additional weeding is generally carried out around flowering (±90 days after sowing). Chemical weed control is not recommended so as to preserve field biota. In the case of organic quinoa, chemical weed control is prohibited.

Crop water requirements and irrigation

Quinoa is known largely for its drought resistance (Jensen et al. 2000; Geerts et al. 2008a). Therefore, it is traditionally cultivated under rain-fed conditions, even in semi-arid and very arid locations. However, this high drought resistance is often translated in low yields, because quinoa sacrifices yield for survival and adaptation. Under this consideration, Geerts et al. (2008a, 2008b) studied the impact of additional water on quinoa production and found deficit irrigation (DI) to be highly beneficial in various experimental locations. DI is already practiced in the re-introduction of quinoa in the arid regions of Chile (Martínez et al. 2009). On the other hand, quinoa is rarely grown under full irrigation, probably because it is not traditional and because it does not respond well to high irrigation due to the increased risk of downy mildew.

Analysis of water requirements showed how much water quinoa requires for "optimal" production (Garcia et al. 2003). The analysis also determined which stages in quinoa production could be more efficient regarding water consumption and transformation into yield. The idea was to apply DI with the aim of concentrating irrigation during the most sensitive crop stages, while maximizing water use efficiency (WUE). The results were quite significant, showing that regardless of the cropping location, the most sensitive stages for successful or at least medium yields were crop establishment (emergence) and flowering. Moreover, an adequate balance between evapotranspiration before and after anthesis, with a larger proportion of water consumed after

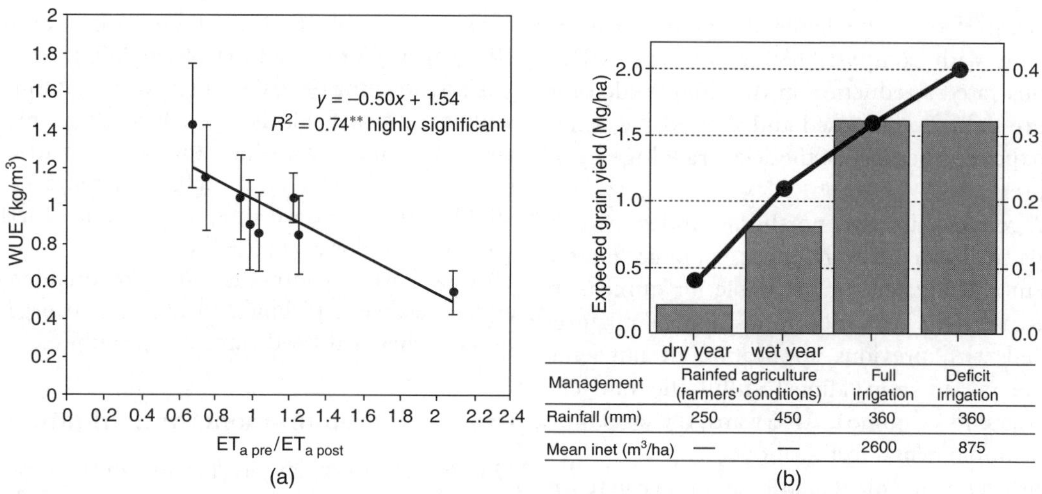

Fig. 3.6 (a) Relationship between water use efficiency (WUE) and pre- and postanthesis evapotranspiration rate. (b) Relationship among WUE, yield, and (full and deficit) irrigation requirements. (Source: Geerts et al. 2008b).

anthesis, will have a large positive effect on WUE (Fig. 3.6). Therefore, the general advice is to consider irrigation in years when the onset of the rainy season might be delayed, and during flowering and grain filling when there are frequent dry spells.

Biotic threats: pests and diseases

Although quinoa can tolerate extremely unfavorable growing conditions, pests such as birds, insects, and rodents, and several diseases caused by fungi, bacteria, nematodes, and viruses can cause significant yield losses (Danielsen et al. 2003; Rasmussen et al. 2003). Disease and pest occurrence in quinoa often depends on plant density, presence of weeds, relative humidity, the nutrient state of the field, and on the crop rotation used (Nieto et al. 1998; Mujica et al. 2001b; García et al. 2001; Danielsen et al. 2003). Preventive actions against pests and diseases in quinoa fields are of primary importance, and an essential component of integrated pest management (IPM). Because a complete overview of all biotic threats against quinoa is out of the scope of this chapter, only a few important pest and diseases of quinoa in the Andean highlands are described. Export organizations and companies, as well as researchers, have stressed the need for research on host resistance and pest population carry-over from one to another growing season to another. At present, there is a dearth in such research.

An important fungal disease in quinoa is downy mildew (*Peronospora farinosa* Fr.) (Danielsen et al. 2003; Butron et al. 2006). To prevent this disease, local extension services are encouraging farmers to use healthy, certified seed (Mujica et al. 2001b). Early application of fungicide or organic remedies is extremely important for an effective mildew control. Danielsen and Munk (2004) reported spectral reflectance measurements as a method for yield loss prediction because of downy mildew. The critical period attaining peak disease severity is around flower bud initiation, occurring around 60 days after sowing (Kumar et al. 2006). Siñani et al. (2006) stressed the relationship between relative humidity and mildew occurrence. In germplasm banks, several studies have been conducted to identify quinoa accessions resistant to mildew (Gamarra et al. 2001). Although resistance to mildew depends largely on the variety, Danielsen et al. (2001) reported severe yield losses up to 33% even in the most resistant varieties. Highly susceptible varieties had almost 100% yield losses.

Another important pest, occurring throughout the cropping cycle, is the moth "ticonas," which is actually a complex group of at least four genera.

Larvae of these insects are very harmful. It only takes one larva per plant to cause serious damage (PROINPA Foundation 2003b). In the central and southern Bolivian Altiplano, the Lepidopteran pest *Eurysacca melanocamta* Meyrick (K'cona K'cona) is very destructive, mostly during the larval stage (Avalos and Saravia 2006). This species can survive all year round as it has several possible hosts (PROINPA Foundation 2003c). For all pests and diseases, pesticides (either chemical or organic mixtures) have to be applied at the right stage during the generative cycle of the insect or fungus.

In addition to insects and fungi, bird and rodent pests are an emerging problem and a primary cause of yield loss in quinoa. Even though Bhargava et al. (2006) reported that losses due to birds and rodents are minor because of the defense accorded by saponins (which could be true for varieties with high saponin content), our own field experience and reports from other researchers estimate that yield losses could go as high as 60% (Rasmussen et al. 2003). Unfortunately, little research has been performed on this area to obtain a more generalized yield loss due to bird and rodent feeding.

Seed harvest and postharvest technology

The harvest time in the South American Andes may extend from February to May, but April is the principal harvest month. The optimum time for cutting plants depends on several factors, such as quinoa variety, soil type, humidity, and prevailing temperature. It is crucial to know when the plants are ready for harvest. Usually, the leaves turn yellow or red, depending on the variety. The grains can be seen in the panicle through the opening of the perigonium, also indicative of physiological maturity (Aroni 2005). Another way to test if the plant is ready for harvest is to tap the panicle with the hand. If the grains fall out, harvesting can begin.

Quinoa is harvested mainly by pulling out the plants and leaving them in stacks on the field to dry (Fig. 3.7). The disadvantage of this method is that it removes the roots from the soil instead of being left as organic matter. This method also contributes to soil erosion and lowers soil fertility. Another disadvantage of this method is that soil particles can be mixed with the grain.

Harvesting can also be done manually by cutting the plant with a sickle 10–15 cm above the soil and leaving the stubble in the soil, thus helping in soil conservation. The plants must be cut at the right time, that is, when the panicles still retain grain upon handling, because when the plants exceed maturity, grains drop from the panicle. The disadvantage of this method is that it cannot be used on very sandy soils because these soils do not have enough mechanical resistance to permit these actions. Also, it is quite difficult to cut large plants with thick stems and therefore the method of pulling out plants is preferred and commonly used by farmers.

Semi-mechanical harvesting involves cutting the plants with a mechanical mower, which is easier to use when the plants are arranged in holes or furrows. Use of a mechanical mower speeds up the harvest, which is a distinct advantage, and leaves the stem and roots in the soil for incorporation later (Aroni 2005). The first two methods described require more hand labor and more harvest losses are expected, as compared to semi-mechanical harvesting. The crop can also be harvested using either combine harvesters or stationary harvesters.

Actual yields range from 500 to 1,000 kg ha^{-1} of grain at farmers' level, even though yields of up to 5,000 kg/ha can be achieved under suitable climatic conditions. Adequate rainfall and favorable temperatures do not always prevail in the different agroecological areas of the Andes, thus the discrepancy between potential and actual yields. Approximately 5–10 tons of chaff per hectare is gained as a by-product and can be used as livestock feed.

Harvest can take up to 45 days due to asynchronous flowering and ripening, which is a natural defense mechanism of quinoa to the adverse climatic events common in the area. Threshing is carried out after additional drying of the quinoa and can vary from completely manual to mechanized (Salas 2003). Natural or mechanized ventilation is used to remove impurities and dust. Additional ventilation is

Fig. 3.7 Two methods of manual harvesting of quinoa: (a,b) pulling out the plants and leaving them in stacks on the field to dry; (c) plants are piled above ground and left to dry (Del Castillo and Winkel, IRD – CLIFA, 2002–2008).

also used when the percent grain moisture is too high from a commercial requirement (which should be <10%). Once the grain is cleaned, it is generally washed or toasted to remove saponin, a bitter substance in the pericarp that constitutes a chemical defense against the feeding activity of insects and animals. The use of chemicals (e.g., sodium hydroxide) to make saponins soluble is effective but is not widely used. Saponin content is higher in varieties classified as "bitter varieties" and lower in "sweet varieties." Processing costs are related to the saponin content of the variety. Another complementary postharvest process is grain size and color selection corresponding to different commercial uses such as direct grain consumption and flour.

Improper postharvest practices are a major cause of low yields in the South American Andes (up to 40% yield loss) and stagnant export (Salas 2003). Specifically, improper postharvest practices cause poor grain germination, fungi and mold damage, and color and odor change. To prevent postharvest losses and refine postharvest technology, fundamental researches on the physical properties of quinoa seeds are to be conducted (Prego et al. 1998; Sigstad and Prado 1999; Sigstad and Garcia 2001; Vilche et al. 2003; Tolaba et al. 2004; Gely and Santalla 2007).

Germination studies are often carried out together with seed and postharvest studies. Jacobsen and Bach (1998) and Jacobsen et al. (1999) report that light had no influence on germination rate or on final germination. On the other hand, harvest time, moisture content of seed at harvest, and germination temperature seemed to have significant effects. Seeds that were harvested later, and thus had lower moisture content at harvest, had higher germination percentage. Furthermore, the thermal time to germination, defined as root protrusion, was 30° days (Jacobsen and Bach 1998).

Many of the quinoa commercial chain in Andean countries go through traditional intermediaries who have ways of transporting the crop produce. The major portion of conventionally produced quinoa is processed traditionally in plants owned by intermediaries or wholesalers. Bio-quinoa for export is usually processed in semi-industrial plants (Brenes et al. 2001).

Within the quinoa cluster, organizations such as the National Association of Quinoa Producers (ANAPQUI) and the Central de Cooperativas Agropecuarias Operacion Tierra (CECAOT) are classified as supporting organizations (Brenes et al. 2001). ANAPQUI represents about 5,000 of the approximately 20,000 quinoa producers in Bolivia. Its activities are concentrated in the Southern Bolivian Altiplano and mostly on organic quinoa. The quinoa produce, which is mostly organic, is destined in part for the export market. Even though the international prices of organic quinoa are 10–15% higher than for conventional quinoa, the demand for organic quinoa is increasing worldwide (Hellin and Higman 2003).

ANAPQUI is an association well known for its positive support of quinoa producers due to guaranteed prices, extension, and certain mechanization (Magliocca 2002). The impact of such an organization is proof of the positive effect of integrating activities at various levels of the market and processing chain. There is also increased jurisdictional security by having an association of farmers. However, restrictions on the use of agricultural inputs, in some cases not adapted to local family production, make it extremely difficult for farmers to practice organic agriculture. In fact, farmers in other areas of the Andes do not produce much organic quinoa because they prefer to spend their labor on crops that they perceive to be less labor intensive and more economically rewarding.

Hellin and Higman (2003) investigated the stability and viability of potato and quinoa production systems in Bolivia, Ecuador, and Peru under current market pressure. They report that quinoa consumption has been improved by government-sponsored initiatives that use quinoa in food support programs (e.g., free school meals). Initiatives by private or nongovernment organizations also strengthen quality production, processing, public image, and market access among smallholder producers. Private as well as public interventions are essential and are to be supported in remote areas where quinoa is often the only subsistence crop (Hellin and Higman 2003).

REFERENCES

Aguilar PC, Cutipa Z, Machaca E, López M, Jacobsen, SE. 2003. Variation of proline content of quinoa (*Chenopodium quinoa* Willd.) in high beds (waru waru). Food Rev Int 19:121–127.

Aguilar PC, Jacobsen SE. 2003. Cultivation of quinoa on the Peruvian Altiplano. Food Rev Int 19:31–41.

Angiosperm Phylogeny Group. 2003. An update of the Angiosperm Phyllogeny Group classification for the orders and families of flowering plants: APG II. Bot J Linnean Soc 141:399–436.

Aroni JC. 2005. Fascículo 5 – Cosecha y poscosecha. In: PROINPA y FAUTAPO, editors. Serie de Módulos Publicados en Sistemas de Producción Sostenible en el Cultivo de la Quinoa: Módulo 2. Manejo agronómico de la Quinoa Orgánica. Fundación PROINPA, Fundación AUTAPO, Embajada Real de los Países Bajos. La Paz, Bolivia. Octubre de 2005. pp. 87–102.

Avalos FL, Saravia R. 2006. Ciclo biológico, fluctuación poblacional e identificación de la Kcona Kcona, plaga del cultivo de la quinoa. In: Del Castillo C, Garcia M, Vacher JJ, Winkel T, editors. Compendio de resúmenes de tesis de grado e investigaciones realizadas en quinoa, cañahua, amaranto y papa en la facultad de agronomía de la UMSA entre 1991 y 2006. Facultad de Agronomía, UMSA - IRD, Francia - QuinAgua, VLIR & K.U. Leuven, Bélgica, La Paz, Bolivia. p. 21.

Bertero HD. 2001. Effects of photoperiod, temperature and radiation on the rate of leaf appearance in quinoa (*Chenopodium quinoa* Willd.) under field conditions. Ann Bot 87:495–502.

Bertero HD. 2003. Response of developmental processes to temperature and photoperiod in quinoa (*Chenopodium quinoa* Willd.). Food Rev Int 19:87–97.

Bertero HD, King RW, Hall AJ. 1999a. Photoperiod-sensitive development phases in quinoa (*Chenopodium quinoa* Willd.). Field Crops Res 60:231–243.

Bertero HD, King RW, Hall AJ. 1999b. Modeling photoperiod and temperature responses of flowering in quinoa (*Chenopodium quinoa* Willd.). Field Crops Res 63:19–34.

Bertero HD, King RW, Hall AJ. 2000. Photoperiod and temperature effects on the rate of leaf appearance in quinoa (*Chenopodium quinoa* Willd.). Aust J Plant Physiol 27:349–356.

Bertero HD, De la Vega AJ, Correa G, Jacobsen SE, Mujica A. 2004. Genotype and genotype-by-environment interaction effects for grain yield and grain size of quinoa (*Chenopodium quinoa* Willd.) as revealed by pattern analysis of international multi-environment trials. Field Crops Res 89:299–318.

Berti M, Wilckens R, Hevia F, Serri H, Vidal I, Mendez C. 1997. Fertilizacion nitrogenada en Quinoa (*Chenopodium quinoa Willd.*). Ciencia de Investigacion Agraria 27:81–90.

Bhargava A, Shukla S, Ohri D. 2006. *Chenopodium quinoa*- An Indian perspective. Ind Crops Prod 23:73–87.

Bilbao M, Espíndola G, Peric Y. 2006. Semilla basica por selección masal estratificada en ocho variedades de quinoa (*Chenopodium quinoa* Willd.). In: Del Castillo C, Garcia M, Vacher JJ, Winkel T, editors. Compendio de resúmenes de tesis de grado e investigaciones realizadas en quinoa, cañahua, amaranto y papa en la facultad de agronomía de la UMSA entre 1991 y 2006. Facultad de Agronomía, UMSA - IRD, Francia - QuinAgua, VLIR & K.U. Leuven, Bélgica, La Paz, Bolivia. p. 22.

Bois JF, Winkel T, Lhomme JP, Raffaillac JP, Rocheteau A. 2006. Response of some Andean cultivars of quinoa (*Chenopodium quinoa* Willd.) to temperature: effects on germination, phenology, growth and freezing. Eur J Agron 25:299–308.

Bonifacio A. 2001. Recursos genéticos, etnobotánica y distribución geográfica. In: Mujica A, Jacobsen SE, Izquierdo J, Marathee JP, editors. Primer taller internacional sobre quinoa. Cultivos Andinos. [CD-ROM]. Santiago: FAO, UNA-Puno, CIP.

Bonifacio A. 2003. *Chenopodium sp.*: genetic resources, ethnobotany, and geographic distribution. Food Rev Int 19:1–7.

Bosque H, Lemeur R, Van Damme P. 2006. La fluorescencia de la clorofila, una herramienta para estudios de la fisiologia del estres: experiencia con la quinoa (*Chenopodium quinoa* Willd.). Revista Ciencia y Tecnología Agropecuaria 1:38–43.

Bosque HD. 1998. Ecophysiological analysis of drought and salinity stress of quinoa (*Chenopodium quinoa* Willd.). MSc thesis in Soil Science and Eremology. Faculty of Science, Faculty of Agricultural and Applied Biological Sciences, University of Gent. International Center for Eremology, Gent, Belgium. 122 p.

Bosque H, Lemeur R, Van Damme P. 2000. Análisis ecofisiológico del cultivo de la quínua (*Chenopodium quinoa* Willd.) en condiciones de estrés de la sequía y la salinidad. Tropicultura 18:198–202.

Bosque H, Lemeur R, Van Damme P, Jacobsen SE. 2003. Ecophysiological analysis of drought and salinity stress of quinoa (*Chenopodium quinoa* Willd.). Food Rev Int 19:111–119.

Bravo R, Catacora P. 2010. Situación actual de los bancos nacionales de germoplasma. In: Bravo R, Valdivia R, Andrade K, Padulosi S, Jagger M, editors. Granos Andinos: avances, logros y experiencias desarrolladas en quinoa, cañihua y kiwicha en Perú. Roma, Italia: Biodiversity International. pp. 15–18.

Brenes ER, Madrigal K, Condo A. 2001. El cluster de quinoa en Bolivia: diagnostico competitivo y recomendaciones estratégicas (Proyecto Andino de Competitividad).

Butron R, Mamani F, Coca Moranti M. 2006. Estudio de la incidencia y severidad del mildiu (*Peronospora farinosa* Fr.) en nueve variedades de quinoa en la estación experimental Belén. In: Del Castillo C, Garcia M, Vacher JJ, Winkel T, editors. Compendio de resúmenes de tesis de grado e investigaciones realizadas en quinoa, cañahua, amaranto y papa en la facultad de agronomía de la UMSA entre 1991 y 2006. Facultad de Agronomía, UMSA - IRD, Francia - QuinAgua, VLIR & KULeuven, Bélgica, La Paz, Bolivia. p. 19.

Cardenas M. 1944. Descripción preliminar de las variedades de *Chenopodium quinoa* de Bolivia. Revista de Agricultura. Universidad Mayor San Simón de Cochabamba 2(2):13–26.

Carrasco E, Devaux A, Garcia W, Esprella R. 1997. Frost-tolerant potato varieties for the Andean Highlands. In: CIP Program Report 1995–1996. Lima, Perú: Centro Internacional de la Papa (CIP). pp. 227–232.

Cossio J. 2008: Agricultura de conservación con un enfoque de manejo sostenible en el Altiplano sur. Revista Habitat 75:44–49.

Cronquist A. 1981. An integrated system of classification of flowering plants. The New York Botanical Garden, New York: Colombia University Press. 1262 pp.

Cusack DF. 1984. Quinoa: grain of the Incas. The Ecologist 14:21–30.

Danielsen S, Bonifacio A, Ames T. 2003.Diseases of quinoa (*Chenopodium quinoa*). Food Rev Int 19:43–59.

Danielsen S, Jacobsen SE, Echegaray J, Ames T. 2001. Impact of downy mildew on the yield of quinoa. Lima, Perú: Centro Internacional de la Papa (CIP). pp. 397–401.

Danielsen S, Munk L. 2004. Evaluation of disease assessment methods in quinoa for their ability to predict yield loss caused by downy mildew. Crop Prot 23:219–228.

De Barros Santos RL, Spehar, CR, Vivaldi L. 2003. Quinoa (*Chenopodium quinoa*) reaction to herbicide residue in a Brazilian Savannah soil. Pesquisa Agropecuária Brasileira, Brasília 38:771–776.

Del Castillo C, Winkel T, Mahy G, Bizoux JP. 2007. Genetic structure of quinoa (*Chenopodium quinoa* Willd.) from the Bolivian altiplano as revealed by RAPD markers. Genet Res Crop Evol 54:897–905.

Del Castillo C, Mahy G, Winkel T. 2008. La quinoa en Bolivie: une culture ancestrale devenue culture de rente "bio-équitable." Biotechnol Agron Soc Environ 12(4): 421–435.

De la Torre-Herrera J. 2003. Current use of quinoa in Chile. Food Rev Int 19:155–165.

Erquinigo F. 1970. Biología floral de la quinoa (*Chenopodium quinoa* Willd.). Tesis Ing. Agro. Facultad de Agronomía. Puno, Perú: Universidad Técnica del Altiplano. p. 89.

Espindola G. 1980. Estudio de componentes directos e indirectos del rendimiento en quinoa (*Chenopodium quinoa* Willd.). Tesis de Grado, Facultad de Ciencias Agrícolas y Pecuarias 'Martin Cardenas', Cochabamba, Bolivia.

Espindola G. 1992. V curso de producción de quinoa. Centro experimental para la industrialización de la quinoa. La Paz, Bolivia: Proyecto PNUD, FAO, MACA-IBTA. p. 70.

FAO. 2011. Quinoa: An ancient crop to contribute to world food security Regional office for Latin America and the Caribbean, 60 pp.

Fox P, Rockström J. 2000. Water-harvesting for supplemental irrigation of cereal crops to overcome intra-seasonal dry-spells in the Sahel. Phys Chem Earth 25:289–296.

François C, Bosseno R, Vacher JJ, Seguin B. 1999. Frost risk mapping derived from satellite and surface data over the Bolivian Altiplano. Agr Forest Meteorol 95:113–137.

Gallardo M, Gonzalez J, Ponessa G. 1997. Morfologia del fruto y semilla de *Chenopodium quinoa* Willd ("quinoa") Chenopodiaceae. (Fruit and seed morphology of *Chenopodium quinoa* Willd ("quinoa") Chenopodiaceae.) Lilloa 39(1):71–80.

Galwey NW. 1993. The potential of quinoa as a multi-purpose crop for agricultural diversification: a review. Ind Crops Prod 1:101–106.

Gamarra M, Bonifacio A, Peralta E. 2001. Mejoramiento genetico y participativo en quinoa al mildeu en Peru, Bolivia y Ecuador. Program text of the "Programa Nacional de Investigación en Cultivos Andinos."

Gandarillas H. 1979. Genética y origen. In: Tapia M (ed). Quinua y Kañiwa, cultivos andinos. Bogota, Colombia, CIID, Oficina Regional para América Latina. pp. 45–64.

Garcia M. 2003. Agroclimatic study and drought resistance analysis of quinoa for an irrigation strategy in the Bolivian Altiplano. Dissertationes de Agricultura, Faculty of Applied Biological Sciences, K.U. Leuven, Belgium. 556. p. 184.

Garcia M, Raes D, Jacobsen SE. 2003. Evapotranspiration analysis and irrigation requirements of quinoa (*Chenopodium quinoa*) in the Bolivian highlands. Agr Water Manage 60:119–134.

Garcia M, Raes D, Jacobsen SE, Michel T. 2007. Agroclimatic constraints for rainfed agriculture in the Bolivian Altiplano. J Arid Env 71(1):109–121.

García S, Marcos JF, Pallás V, Sánchez-Pina MA. 2001. Influence of the plant growing conditions on the translocation routes and systemic infection of carnation mottle virus in *Chenopodium quinoa* plants. Physiol Mol Plant Pathol 58:229–238.

Garreaud R, Vuille M, Clement AC. 2003.The climate of the Altiplano: observed current conditions and mechanisms of past changes. Palaeogeogr Palaeoclimatol Palaeoecol 194:5–22.

Geerts S, Mamani RS, Garcia M, Raes D. 2006a. Response of quinoa (*Chenopodium quinoa* Willd.) to differential drought stress in the Bolivian Altiplano: towards a deficit irrigation strategy within a water scarce region. Proceedings of the 1st International Symposium on Land and Water Management for Sustainable Irrigated Agriculture. CD-rom and book of abstracts. p. 200.

Geerts S, Raes D, Garcia M, Del Castillo C, Buytaert W. 2006b. Agro-climatic suitability mapping for crop production in the Bolivian Altiplano: a case study for quinoa. Agr Forest Meteorol 139:399–412.

Geerts S, Raes D, Garcia M, Condori O, Mamani J, Miranda R, Cusicanqui J, Taboada C, Vacher J. 2008a. Could deficit irrigation be a sustainable practice for quinoa (*Chenopodium quinoa* Willd.) in the Southern Bolivian Altiplano? Agric Water Manage 95:909–917.

Geerts S, Raes D, Garcia M, Vacher J, Mamani R, Mendoza J, Huanca R, Morales R, Miranda B, Cusicanqui J, Taboada C. 2008b. Introducing deficit irrigation to stabilize yields of quinoa (*Chenopodium quinoa* Willd.). Eur J Agron 28:427–436.

Gely MC, Santalla EM. 2007. Moisture diffusivity in quinoa (*Chenopodium quinoa* Willd.) seeds: effect of air temperature and initial moisture content of seeds. J Food Eng 78 (3): 1029–1033.

Gonzalez JA, Prado FE. 1992. Germination in relation to salinity and temperature in *Chenopodium quinoa* Willd. Agrochimica 36:101–107.

Hariadi Y, Marandon K, Tian Y, Jacobsen S-E, Shabala S. 2011. Ionic and osmotic relations in quinoa (*Chenopodium quinoa* Willd.) plants grown at various salinity levels. J Exp Bot 62:185–193.

Hellin J, Higman S. 2003. Quinoa and food security. In: Hellin J, Higman S, editors. Feeding the market: South American farmers, trade and globalization. UK: University of Manchester. pp. 131-169.

Hermann M, Heller J. (eds). 1997. Andean Roots and Tubers. Promoting the Conservation and Use of Underutilized and Neglected Crops. 21. Institute of Plant Genetics and Crop Plant Research, Gatersleben/International Plant Genetic Resources Institute, Rome, 256 p.

Hijmans RJ. 1999. Estimating frost risk in potato production on the Altiplano using interpolated climate data. In: Internacional de la Papa (CIP) Program Report 1997–1998. Lima, Peru: CIP. pp. 373–380.

Huiza LZ. 1994. Efecto del déficit hídrico a marchitez intensa sobre el ritmo de crecimiento de la quinoa (*Chenopodium quinoa* Willd.). Tesis de grado de la Facultad de Agronomía, La Paz, Bolivia: UMSA.

Jacobsen SE. 2003. Quinoa - research and development at the International Potato Center (CIP). Lima, Perú: CIP. p. 5.

Jacobsen SE, Mujica A. 1999. Fisiología de la resistencia a sequía en quinoa (*Chenopodium quinoa* Willd.). Lima, Peru: Centro Internacional de la Papa (CIP).

Jacobsen SE, Bach AP. 1998. The influence of temperature on seed germination rate in quinoa (*Chenopodium quinoa* Willd.). Seed Sci Tech 26:515–523.

Jacobsen SE, Mujica A. 2003. Quinoa: an alternative crop for saline soils. J Exp Bot 54 (Suppl. 1), i25.

Jacobsen SE, Jornsgard B, Christiansen JL, Stolen O. 1999. Effect of harvest time, drying technique, temperature and light on the germination of quinoa (*Chenopodium quinoa)*. Seed Sci Tech 27:937–944.

Jacobsen SE, Quispe H, Mujica A. 2001. Quinoa: an alternative crop for saline soils in the Andes. Lima, Peru: Centro Internacional de la Papa (CIP). pp. 403–408.

Jacobsen SE, Mujica A, Jensen CR. 2003a. The resistance of quinoa (*Chenopodium quinoa* Willd.) to adverse abiotic factors. Food Rev Int 19:99–109.

Jacobsen SE, Mujica A, Ortiz R. 2003b. The global potential for quinoa and other Andean crops. Food Rev Int 19:139–148.

Jacobsen S-E, Mujica A, Jensen CR. 2003c. Resistance of quinoa (*Chenopodium quinoa* Willd.) to adverse abiotic factors. J Exp Bot 54 (Suppl. 1): i21.

Jacobsen SE, Monteros C, Christiansen JL, Bravo LA, Corcuera LJ, Mujica A. 2005. Plant responses of quinoa (*Chenopodium quinoa* Willd.) to frost at various phenological stages. Eur J Agron 22:131–139.

Jacobsen SE, Monteros C, Corcuera LJ, Bravo LA, Christiansen JL, Mujica A. 2007. Frost resistance mechanisms in quinoa (*Chenopodium quinoa* Willd.). Eur J Agron 26:471–475.

Jacobsen SE. 2011. The situation for quinoa and its production in Southern Bolivia: from economic success to environmental disaster. J Agron Crop Sci 197 (5):390–399.

Jensen CR, Jacobsen SE, Andersen MN, Núñez N, Andersen SD, Rasmussen L, Mogensen VO. 2000. Leaf gas exchange and water relation characteristics of field quinoa (*Chenopodium quinoa* Willd.) during soil drying. Eur J Agron 13:11–25.

Koyro HW, Eisa SS. 2008. Effect of salinity on composition, viability and germination of seeds of *Chenopodium quinoa* Willd. Plant Soil 302:79–90.

Kumar A, Bhargava A, Shukla S, Singh HB, Ohri D. 2006. Screening of exotic *Chenopodium quinoa* accessions for downy mildew resistance under mid-eastern conditions of India. Crop Prot 25:879–889.

Laguna PF. 2000. The impact of quinoa export on peasants' livelihoods of Bolivian highlands. In: Sustainable Rural Livelihoods: Building Communities, Protecting Resources, Fostering Human Development, Xth World Congress of the International Rural Sociology Association, Rio de Janeiro, Brazil, August, 2000.

Magliocca F. 2002. Etude de marché: La Quinoa. Maîtrise IMTL 2001–2002.

Martínez E, San Martín R, Jorquera C, Veas E, Jara P. 2009. Reintroduction of quinoa into arid Chile: cultivation of two lowland races under extremely low irrigation. J Agron Crop Sci 195:1–10.

Miranda R. 2012. Organic matter requirements for a sustainable production of quinoa in the Central and Southern Bolivian Altiplano. Doctoral research. Universidad Federal de Santa Maria, Brazil.

Miranda R, Carlesso R, Huanca M, Mamani P, Borda A. 2012. Rendimiento y acumulación de nitrógeno en la quinua (*Chenopodium quinoa* Willd) producida con estiércol y riego suplementario. Revista Venesuelos 20(1).

Mujica A. 1994. Andean grains and legumes. In: Hernando Bermujo, JELJ, editor. Neglected Crops: 1492 from a different perspective. Rome, Italy: FAO. pp. 131–148.

Mujica A, Izquierdo J, Marathee JP. 2001a. Capítulo I. Origen y descripción de la quinoa. In: Mujica A, Jacobsen SE, Izquierdo J, Marathee JP, editors. Quinoa (*Chenopodium quinoa* Willd.): ancestral cultivo andino, alimento del presente y futuro. Santiago de Chile: FAO, UNA, Puno. pp. 9–29.

Mujica A, Canahua A, Saravia R. 2001b. Capítulo II. Agronomía del cultivo de la quinoa. In: Mujica A, Jacobsen SE, Izquierdo J, Marathee JP, editors. Quinoa (*Chenopodium quinoa* Willd.): Ancestral cultivo andino, alimento del presente y futuro. Santiago de Chile: FAO, UNA, Puno. pp. 20–48.

Mujica A, Jacobsen SE, Canahua A. 2001c. Los aynokas de quinoa (*Chenopodium quinoa* Willd.) - un sistema de conservación de recursos genéticos estratégicos del altiplano peruano. Conferencia Electrónica 2001 del Consorcio para el Desarrollo Sostenible de la Ecorregion Andina (http://www.condesan.org/infoandina/Foros/insitu2001/S.Jacobsen-spanish.htm).

Murillo R. 1995. Comportamiento del nitrógeno proveniente de fertilizantes minerales en el cultivo de la quinoa (*Chenopodium quinoa* Willd) bajo condiciones de riego y secano. Thesis dissertation for Agronomist. UMSA.

Nieto CC, Vimos CN, Caicedo CV, Monteros CJ, Rivera MM. 1998. Respuesta de la quinoa a diferentes tipos de rotación de cultivos en dos localidades de la sierra, durante cinco años. Centro Internacional de Investigaciones para el Desarrollo, Programa Cultivos Andinos (INIAP), working paper.

Parida AK, Dagaonkar VS, Phalak MS, Umalkar GV, Aurangabadka LP. 2007. Alterations in photosynthetic pigments, protein and osmotic components in cotton genotypes subjected to short-term drought stress followed by recovery. Plant Biotech Rep 1 (1): 37–48.

Partap T, Kapoor P. 1985a. The Himalayan grain chenopods. I. Distribution and ethnobotany. Agr Ecosys Env 14:185–199.

Partap T, Kapoor P. 1985b. The Himalayan grain chenopods. II. Comparative morphology. Agr Ecosys Env 14:201–220.

Peralta E. 2009. La quinoa en Ecuador. "Estado del Arte." Quito, Ecuador: INIAP.

Prego I, Maldonado S, Otegui M. 1998. Seed structure and localization of reserves in *Chenopodium quinoa*. Ann Bot 82:481–488.

PROINPA Foundation. 2002. Variedades de quinoa recomendadas para el Altiplano Norte y Central. La Paz, Bolivia: PROINPA.

PROINPA Foundation. 2003a. Catálogo de Quinoa Real. La Paz, Bolivia: PROINPA. p. 52.

PROINPA Foundation. 2003b. Biología y comportamiento de las ticonas. Ficha Técnica n°4. La Paz, Bolivia: PROINPA.

PROINPA Foundation. 2003c. Biología y comportamiento de la polilla de la quinoa. Ficha Técnica n°5. La Paz, Bolivia: PROINPA.

PROINPA Foundation. 2003d. Variedad "Quinoa Jacha Grano." Ficha Técnica n°6. La Paz, Bolivia: PROINPA.

Pulgar Vidal J. 1954. La quinoa o suba en Colombia. Publ. No. 3. Fichero Científico Agropecuario. Bogotá, Colombia: Ministerio de Agricultura. pp. 73–76.

Rasmussen C, Lagnaoui A, Esbjerg P. 2003. Advances in the knowledge of quinoa pests. Food Rev Int 19:61–75.

Repo-Carrasco R, Espinoza C, Jacobsen S-E. 2003. Nutritional value and use of the Andean crops quinoa (*Chenopodium quinoa*) and kañiwa (*Chenopodium pallidicaule*). Food Rev Int 19:179–189.

Risi J, Galwey NW. 1984. The Chenopodium grains of the Andes: Inca crops for modern agriculture. Adv Appl Biol 10:145–216.

Rodriguez JP, Raffaillac JP. 2003. El papel de tamaño de quinoa (*Chenopodium quinoa* Willd.) en el crecimiento y desarrollo de las plantas frente a diferentes profundidades de siembra. La Paz, Bolivia: Facultad de Agronomia, UMSA, internal working paper. p. 5.

Rojas W, Mamani E, Pinto M, Alanoca C, Ortuño T. 2008. Identificación taxonómica de parientes silvestres de quinua del Banco de Germoplasma de Granos Altoandinos. Revista de Agricultura–Año 60, Nro. 44. Cochabamba, Bolivia, Diciembre 2008. pp. 56–65.

Rosa M, Hilal M, Gonzalez JA, Prado FE. 2009. Low-temperature effect on enzyme activities involved in sucrose-starch partitioning in salt-stressed and salt-acclimated cotyledons of quinoa (*Chenopodium quinoa* Willd.) seedlings. Plant Physiol Biochem 47: 300–307.

Salas S. 2003. Quinoa - postharvest and commercialization. Food Rev Int 19:191–201.

Sanchez V. 2011. Influencia de la plasticidad de la quinua (*Chenopodium quínoa* Willd) sobre su constante térmica en el Altiplano Boliviano. Trabajo dirigido Lic. Ing. Agr. La Paz-Bolivia. Universidad Mayor de San Andrés. Facultad de Agronomía. 71 p.

Sanchez HB, Lemeur R, Van Damme P, Jacobsen S-E. 2003. Ecophysiological analysis of drought and salinity stress of quinoa (*Chenopodium quinoa* Willd.). Food Rev Int 1–2: pp. 111–119.

Schabes FI, Sigstad EE. 2005. Calorimetric studies of quinoa (*Chenopodium quinoa* Willd.) seed germination under saline stress conditions. Thermochimica Acta 428:71–75.

Schulte auf'm Erley G, Kaul HP, Kruse M, Aufhammer W. 2005. Yield and nitrogen utilization efficiency of the pseudocereals amaranth, quinoa, and buckwheat under differing nitrogen fertilization. Eur J Agron 22:95–100.

Sigstad EE, Prado FE. 1999. A microcalorimetric study of *Chenopodium quinoa* Willd. seed germination. Thermochimica Acta 326:159–164.

Sigstad EE, Garcia CI. 2001. A microcalorimetric analysis of quinoa seeds with different initial water content during germination at 25°C. Thermochimica Acta 366:149–155.

Siñani E, Espíndola G, Morales DM. 2006. Evaluación de la eficiencia de fungicidas en el control de mildiu (*Peronospora farinosa* Fr.) de la quinoa en el Altiplano central y norte. In: Del Castillo C, Garcia M, Vacher JJ, Winkel T, editors. Compendio de resúmenes de tesis de grado e investigaciones realizadas en quinoa, canahua, amaranto y papa en la Facultad de Agronomía de la UMSA entre 1991 y 2006. Facultad de Agronomía, UMSA - IRD, Francia - QuinAgua, VLIR & KU Leuven, Belgica, La Paz, Bolivia, p 24.

Tapia ME. 1979. Historia y distribución geográfica.. In: Tapia ME, editor. Quinoa y kaniwa: cultivos andinos. Bogotá, Colombia: CIID. pp. 11–19.

Tolaba MP, Peltzer M, Enriquez N, Lucia Pollio M. 2004. Grain sorption equilibria of quinoa grains. J Food Eng 61:365–371.

Valencia-Chamorro SA. 2004. Quinoa. In: Encyclopedia of grain science. Australia: Elsevier/CRC, pp. 4885–4892.

Vilche C, Gely M, Santalla E. 2003. Physical properties of quinoa seeds. Biosystems Eng 86:59–65.

Willdenow CL. 1798. Species plantarum, ed. 4, 1(2). – Berolini: G. C. Nauk.

Winkel T. 2013. Quinua y quinueros. Ed. IRD France. ISBN 978-2-7099-1749-0. 176 pp.

Yucra E, Garcia M. 2007. Producción y sistemas de cultivo de quinoa en zonas del altiplano boliviano. In: Annals of the International Congress of Quinoa. Iquique. Chile.

Chapter 4
Trends in Quinoa Yield over the Southern Bolivian Altiplano: Lessons from Climate and Land-Use Projections

Serge Rambal[1,3], Jean-Pierre Ratte[1], Florent Mouillot[2] and Thierry Winkel[2]

[1] *CNRS, UMR 5175 CEFE, F-34293 Montpellier cedex 5 France*
[2] *IRD UMR 5175 CEFE F-34293 Montpellier cedex 5 France*
[3] *Universidade Federal de Lavras Departamento de Biologia CP 3037, CEP 37200-000 Lavras MG Brazil*

SUMMARY

The Southern Altiplano of Bolivia is the world leader in quinoa export. In these arid highlands, quinoa yield is largely influenced by on-going changes affecting both climate drivers and land-use practices. We developed a simple "what-if" simulation model for anticipating such changes in quinoa yields over three time-slices: a recent past (1961–2000), a near future (2046–2065), and finally a remote one (2081–2100). Outputs from a global climate model (GCM) were used as drivers for projections of regional climate patterns and for providing risk information. For tackling further problems arising from our limited understanding of what the local-level impacts of climate change are likely to be in this arid mountain area, we decided to consider a grid-level area centered on the salar of Uyuni, a region experiencing marked changes in climate and land use for the past four decades. Local land-use change scenarios were derived from aerial photographs and data from high-resolution remote-sensing validated by ground truth observations on sample areas representative of the region. Climate change scenarios were linked to our yield model composed of a soil water balance model coupled with a simple crop growth component. The daily climate from GCM runs has been used for driving the soil water balance model and calculating the drought severity that the crop might undergo. At the field plot level, the yield index (defined as the simulated yield relative to a fixed yield potential) depends on soil drought, mean air temperature that largely affects carbon assimilation through photosynthesis, and minimal temperature that impairs the yield during frost nights. At the landscape scale, a mosaic of three land units is considered: *cerros* (steep slopes), *faldas* (gentle slopes in piedmonts), and *pampas* (flat plains). We simulated 2-year crop cycles taking into account complementarities among these landscape units. In each land unit, growth efficiency parameters are affected by the decline in the biological potential of the land. These parameters parallel the rate of encroachment of the quinoa crop, the subsequent decline of pasture lands in the

Quinoa: Improvement and Sustainable Production, First Edition. Edited by Kevin Murphy and Janet Matanguihan.

pampas and the soil aeolian erosion caused by such intensification. Results show an on-going decrease in the occurrence of frost that will be followed by an increasing drought frequency in the forthcoming decades. Long-term increase in soil dryness associated with a decline in the biological potential of the land may largely impair the sustainability of this area. This type of analysis, although speculative, provides insights that can assist research and development organizations target options for sustainability in crop production.

INTRODUCTION

The Andes are the most densely populated highlands in the world, with human settlements commonly found above 3800 masl and agricultural activities up to 4200 masl (Little 1982). GCMs suggest that climatic changes in the Andean Altiplano might have serious consequences for water management and local agriculture (Thibeault et al. 2010; Seth et al. 2010). Some climate scenarios even project that the montane climate currently known in the Andean region would disappear by the end of the twenty-first century (Williams et al. 2007). Specifically, climate analyses suggest that the warming observed in the Andean highlands is likely to accelerate in the coming decades at a rate greatly exceeding that of the surrounding lowlands (Vergara et al. 2007). Paradoxically, this climate warming would not necessarily lead to a decrease in the number of frost days (Thibeault et al. 2010). There is more uncertainty with projections on precipitation, but trends suggest an evolution toward a shorter, more intense wet season with less precipitation in spring (September–November) and more in summer (January–March) (Seth et al. 2010).

While our understanding of glacier retreat and its consequences has significantly improved (Vuille et al. 2008), assessing the vulnerability of Andean agricultural lands to the continuing climate change requires more specific studies, in particular the interactions between climate and land-use changes and their combined impacts on the economic yield of indigenous crop species (Valdivia et al. 2010).

Among the Andean crops, quinoa (*Chenopodium quinoa* Willd.) is an interesting case due to the huge expansion of its cultivation in the southern Altiplano of Bolivia, at the same time that the impacts of climate warming are observed in the region. Over the past 40 years, the southern Altiplano of Bolivia has become the first region to export quinoa worldwide. Quinoa has transitioned from an auto subsistence grain crop, limited mostly to small fields on slopes (a traditional frost avoiding strategy), to a commercial product cultivated over extended areas in the flat plains amenable to mechanization. In this arid region, quinoa is cultivated under rain-fed conditions and low plant density in fields that, since the 1970s, have rapidly encroached on pasture lands of native herb and shrub vegetation in the plains. It is estimated that the land surface dedicated to quinoa crop in Bolivia has increased more than 230% between 1980 and 1990 (FAO-Stat), most of this increase corresponding to newly cultivated plains in the southern Altiplano. Quinoa grain yield, with an average value of 600 kg/ha in the region, is notably unstable, due to harsh cropping conditions (Garcia et al. 2004; Geerts et al. 2009). What would the quinoa grain yield become through the twenty-first century under the projected trends of increasing climate instability and expanding crop land area?

Regional climate scenarios and crop modeling are useful tools to answer such a question, helping anticipate agroclimatic risks and prepare adaptation strategies that could reduce agricultural vulnerability under climate and land-use changes (Hertel et al. 2010; Parry and Livermore 2002). After a brief presentation of the southern Altiplano region, this chapter describes the methods used to analyze the climatic risks affecting the study area as well as the climatic trends derived from the current observations and GCM projections. The subsequent section introduces the yield index approach, the water balance model, and the land-use scenarios applied to scale quinoa productivity up to the landscape level. Results summarize the current and projected drought and frost limitations over the study area and show the subsequent effects on quinoa yield at the plot and the landscape levels at varying time points. From this, a picture of future quinoa crop production in the southern Altiplano can be seen, and adaptation options for a sustainable use of local land resources are discussed.

MATERIALS AND METHODS

The study area

The study area consists of highlands (Altiplano) located at the southwestern part of the Bolivian territory, at the borders with Argentina and Chile. This region, bordered on its eastern and western edges by the Andean cordilleras, is characterized by the presence at its center of a ca. 100 × 100 km dry salt expanse, the salar of Uyuni, whereas a smaller salt lake, the salar of Coipasa, lies at its northern edge. Except for the vast salt lake areas, the landscapes in this region show a mosaic of three types of land units: relatively extended flat plains at an altitude of ca. 3650 masl, and an alternation of volcanic slopes and valleys in the hinterland. The native vegetation of this tropical highland ecosystem, also known as *puna*, consists of a mountain steppe of herbaceous and shrub species traditionally used for extensive grazing of llama and sheep herds. The southern Altiplano has an arid climate, with an annual mean precipitation varying between 100 and 350 mm from the south to the north of the salar of Uyuni. Most of the precipitation (>70%) occurs during austral summer (December–January–February), when mid- and upper-level easterly winds bring moist air that feed convective storms over the plateau (Garreaud 2000; Falvey and Garreaud 2005). Summer precipitations exhibit significant synoptic variations, largely explained by both local boundary layer moisture and zonal wind aloft (Garreaud and Aceituno 2001). The rest of the year, the Altiplano is influenced by a mid-level westerly flow that brings very dry air from the Pacific and rare precipitation events (Vuille and Ammann 1997). Between-year variations in rainfall over this area may be explained by a tendency for more precipitation during La Niña and less during El Niño events (Ronchail and Gallaire 2006), although the connection between the rainfall regime and the El Niño Southern Oscillation (ENSO) still remains unclear (Garreaud et al. 2009; Thibeault et al. 2012; Seiler et al. 2013).

Mean annual air temperature is about +9°C, with daily thermal amplitudes much higher than seasonal ones and radiative night frost occurring virtually in any season. Frost risks for a standard temperature threshold of 0°C have been detailed by Pouteau et al. (2011). These authors demonstrated the nested effects of physiographic factors across scales in this mountain area. Results showed prominent effects of elevation, latitude, and salt lake proximity at regional scale, whereas slope, topographic convergence, and insulation had more influence at local scales. As a result, the shores of the salt lakes appear less prone to frost, whereas highlands at the west and south of the region are continuously exposed.

Recent past and present climate

We used the Standardized Precipitation Evapotranspiration Index (SPEI), which is a site-specific drought metric that quantifies deviations from the average water balance. The SPEI is calculated by means of a climate water balance, that is, the difference between precipitation and potential evapotranspiration, at monthly time intervals. Different SPEIs are obtained for different time windows representing the cumulative water balance over the previous months (Vicente-Serrano et al. 2013). Monthly climate data at high-resolution 0.5 × 0.5 degree grids for the study area are provided by the Climate Research Unit (CRU) of the University of East Anglia (United Kingdom). We used the CRU TS3.1 datasets for the period 1901–2009. Three time-windows of 3, 6, and 24 months were retained for the analysis. The first time-window (SPEI3) is related to the top soil water limitation that could impair the seedling stage, the second (SPEI6) is related to the whole productivity of the crop, and the last and longer one (SPEI24) is associated with the availability of rain from the previous year in the subsoil. Within a 2-year rotation, quinoa crop uses water stored in the subsoil during the previous fallow year. If this additional soil water is insufficient, quinoa productivity may be drastically impacted. We retain a value of SPEI less than −1 lasting more than 3 months as a drought threshold for any time-window (3, 6, or 24 months). For the frequency analysis of drought events, we cut the whole sequence into three successive 36-year periods: 1901–1936, 1937–1972, and 1973–2009.

Source of climate scenarios

We used climate scenarios from the Centre de Recherches Météorologiques Coupled Global Climate Model version 3 (CNRM-CM3). It is based on the coupling of the ARPEGE-Climate Atmospheric Global Climate Model, an oceanic General Circulation Model OPA8.1, a sea ice model GELATO2, and TRIP, a river routing scheme. CNRM-CM3 has been applied in the framework of the fourth assessment report of the Intergovernmental Panel on Climate Change (IPCC). We retained the IPCC-SRES A1B simulations for the twenty-first century that provided daily precipitation, minimal and maximal temperatures, and global radiation for three time-slices: 1960–2000, 2046–2065, and 2081–2100. We used a coarse resolution of 2.5 × 2.5° grid, which is a good compromise as it does not affect the large-scale patterns and allows a reasonable level of regional description over South America (Boulanger et al. 2006, 2007).

Simulating the yield index at local or plot scale

We defined three abiotic limitation factors that may impair crop productivity and yield at a daily time step: a soil water limitation factor, f_1, which will be discussed in more detail further on; a limitation factor for carbon assimilation, f_2, related to the within-day variations of air temperature; and a frost limitation factor, f_3, exponentially related to the minimal temperature. All these factors have values ranging between 0 and 1, and their product calculated over the entire growth period (from early November to late March) defines the potential yield index:

$$\mathrm{YI}_{\mathrm{plot}} = \sum f_3 \sum_{\text{growth period}} f_1\, f_2 \Delta t \tag{4.1}$$

$\mathrm{YI}_{\mathrm{plot}} = 1$ means that no limitation has been observed during the whole growth period.

For calculating f_2, it was considered that, within a given day, the pattern of change of air temperature follows two sinusoidal segments: first, a decline from the maximal value that occurred at 2 pm to the next minimal value that was observed at dawn; second, the increase of air temperature from its minimal value to the maximal 1. The theoretical duration of the daylight period, d, was calculated for the latitude of the study area using Milankovitch's equation. Relative instantaneous carbon assimilation, A_r, follows a parabolic pattern with current time air temperature, T_a, reaching a maximum ($A_r = 1$) at an optimal temperature of T_{opt}:

$$A_r = 1 - \alpha(T_a - T_{opt})^2 \tag{4.2}$$

with α being a thermal efficiency coefficient.

The f_2 factor was then calculated as

$$f_2 = {}^1\!/_d \sum_d A_r\, \Delta t \tag{4.3}$$

The f_3 factor affects sequentially the yield (Eq. 4.1), quantifying the percent yield loss due to frost. This factor is exponentially related to the minimal daily temperature and scaled by a vulnerability parameter. The vulnerability scaling parameter was fitted to produce realistic losses in the recent period and was applied further to the whole period.

The soil water balance model

The water balance of the top soil layer was simulated at a daily time step using a bucket-type model with a limited storage capacity and mathematically represented by a difference equation in which the daily change in soil water storage, ΔS, equals rain input, R, minus outputs that are deep drainage, D, and actual evaporation, AET:

$$\Delta S = R - \mathrm{AET} - D \tag{4.4}$$

Maximum stored water in the soil, S, is limited by the field capacity, FC. When the soil water storage exceeds the field capacity threshold, extra water flows down as deep drainage. Actual evaporation AET extracts water from the soil compartment. AET is a fraction of potential evaporation PET, related to both the amount of stored water in the soil and to PET itself. PET was calculated using Priestley–Taylor's equation (1972). Daily PET values were calibrated for our regional applications with a standard Priestley–Taylor parameter of 1.26 that corresponds to the original equation (Geerts et al. 2007, 2009). The net all-wave radiation term was related to global incoming short-wave radiation using Kalma et al.'s equation (2000) calibrated with unpublished measurements over quinoa field. For reducing PET to AET, Linacre (1973) proposed a simplified algorithm: AET rate equals whichever is the less of (i) the PET rate or (ii) $\beta(S/\mathrm{FC})^2$. Finally, the model may be summarized by two coupled equations:

$$\mathrm{AET} = \min\left[\beta(S/\mathrm{FC})^2, \mathrm{PET}\right] \tag{4.5}$$

$$S(t+1) = \min[S(t) + R - \mathrm{AET}, \mathrm{FC}] \tag{4.6}$$

β is fixed to 4.5 mm, a value slightly lower than the one retained in this area by Geerts et al. (2006). Retention properties of the sandy soil that covers the study area have been derived from Saxton et al. (1986). Percent sand ranges from 76% to 81% and clay fraction from 6% to 7%. Depending on the soil texture, field capacity may be estimated to extend from 0.153 to 0.167 cm^3 H_2O/cm^3 soil, corresponding to the water content reached at a soil water potential of −0.02 MPa. The water content at wilting point ranged from 0.067 to 0.074 cm^3 H_2O/cm^3 soil. For modeling purposes, we retained a soil layer of 0.50 m depth with a 75 mm field capacity and 37.5 mm storage at wilting point. For well-watered conditions, when S ranges from field capacity to a relative soil water content of 0.8, $f_1 = 1$ and then declines linearly down to 0 at wilting point.

Scenarios of land-use/land-cover changes

The whole land area potentially suitable for quinoa crops can be classified into three neighboring land units: *cerros*, *faldas*, and *pampas*. The hilly *cerros* correspond to the higher altitude areas cultivated manually and hardly affected by night frosts. The *pampas*, previously used for livestock grazing, are now undergoing rapid cropping intensification. The *faldas* correspond to the transition area between *cerros* and *pampas*. The area of fallow lands equal that of crop lands in this arid region, as a result of 2-year cropping cycles. This "dry farming" practice, as well as the clump (pocket) sowing, is typical of local quinoa fields. With this practice, for a given plot, the soil is maintained bare the first year so rain water can infiltrate the soil and will be used for the subsequent cultivation year. We considered that, on average, the partitioning of the whole area suitable for cultivation over the study area would be 10% *cerros*, 20% *faldas*, and 70% *pampas*.

Scenarios of land-use/land-cover (LULC) changes came from maps derived from aerial photographs taken in 1963, high-resolution scans of digital photographic data from the US geological survey of 1972, and high-resolution remotely sensed data from 1990 and 2005. Interpretations were validated by ground truth observations on sampled areas. Five communities representative of the study area were mapped since 1963: Chacoma, Chilalo, Kapura, Palaya, and Otuyo. For the more recent period, we used SPOT satellite data recorded in 1998, 1999, 2005, and 2007 over 14 communities: Alianza, Cacohota, Cacoma, Chilaco, Chusiquiri, Colcaya, Irpani, Jirira, Lia, Otuyo, Pastos Lobo, Salinas, Tahua, and Vituyo. Results obtained at the community-level were aggregated in three trajectories for each land unit expressed in percent changes of the suitable area and then extrapolated toward 2040.

Scaling local yield index up to the region

For all the simulations at plot scale, we assumed that land use did not change over the considered period.

The yield index at regional scale YI_{region} depends on both LULC and "growth efficiency" parameters related to the biological potential of the land (detailed further):

$$YI_{region} = \sum_{i=1,3} \varepsilon_i k_i YI_{plot} \quad (4.7)$$

This regional yield index thus includes the time changes of percent area k_i and the changes in the biological potential ε_i in the three land units considered over the entire region. We further applied the same changes for *faldas* and *pampas* and assumed no change in growth efficiency for *cerro*. For *faldas* and *pampas*, the change was related to the ratio of areas devoted to cultivation and those supporting pastures. We assumed that the growth efficiency was not affected and maintained equal to 1 until the pasture/crop ratio is lower than 10, after that it declined slowly to a new equilibrium. It reached 0.4 for pasture/crop ratios lower than 1. The fallow areas are considered independently in our simulation.

RESULTS

Drought history over the area

The time course of SPEI values over time is presented in Fig. 4.1. Main results are summarized in Table 4.1, and highlight (i) the frequency of drought events, (ii) their month of onset within the year, and (iii) their duration.

Over the whole 1901–2009 period, we observed an increase in the frequency of short droughts quantified by SPEI3 with return periods of 4 years in the last two time-slices. The overall return period is 4.9 years because we observed only four events in the first time-slice with a corresponding return period of 9 years. The most frequent period when drought begins to occur was December–January in the first time-slice. Progressively, this drought onset occurred later in the wet season. This pattern of change was observed whatever the time-window (3, 6, or 24 months) used for the SPEI calculation. An interesting result concerns the drought duration, which increased slightly with time from 3 to 6 months with SPEI3. This pattern of change was maintained for SPEI6 with a significant linear trend of 0.89 month per decade that increased the drought duration from 4 months at the beginning of the series to 13 months at its end. For SPEI24, the whole return period was 6.4 years with slightly higher frequency within 1936–1972 that had seven drought events. The drought duration increased with time with 7.0 ± 3.4, 8.1 ± 5.4, and 17.2 ± 13.6 years for the three time-slices, respectively. The most severe and long droughts were observed in 1983 and 1992 with durations of 28 and 33 months, respectively.

Climate projections and soil drought limitation

In our climate scenario, if we compare 1981–2000 to 1961–1980 used as baseline, the on-going warming trends were 0.24 and 0.35°C per decade for minimal and maximal air temperatures, respectively. For the near future, we projected an increase of 2.1 and 4.0°C in the 2046–2065 time-slice and 3.0 and 5.1°C in the 2081–2100 time-slice (Table 4.2) for minimal and maximal air temperatures, respectively. Direct results on the amount of rain are not presented because we prefer to highlight their effects on the soil water regime, thus answering the question: "how do the change of the rain regime and the concomitant increase of potential evaporation modify the availability of soil water for quinoa crops?"

The seasonal pattern of the soil water storage displays three distinct phases (Fig. 4.2). The first phase corresponds to a well-watered period with soil water storage S greater than 40 mm and no significant negative impact on the crop yield. It includes the period from March to early May. Furthermore, the dry period begins in late May and ends mid-November with soil water storage lower than the wilting point. Finally, we could consider the period from late November to January as a transition period. This period is characterized by both a reduced amount of stored water and a large between-year variation. This transition period corresponds to the onset of the rainy season and the establishment of the crop, a growth stage crucial for the final crop yield.

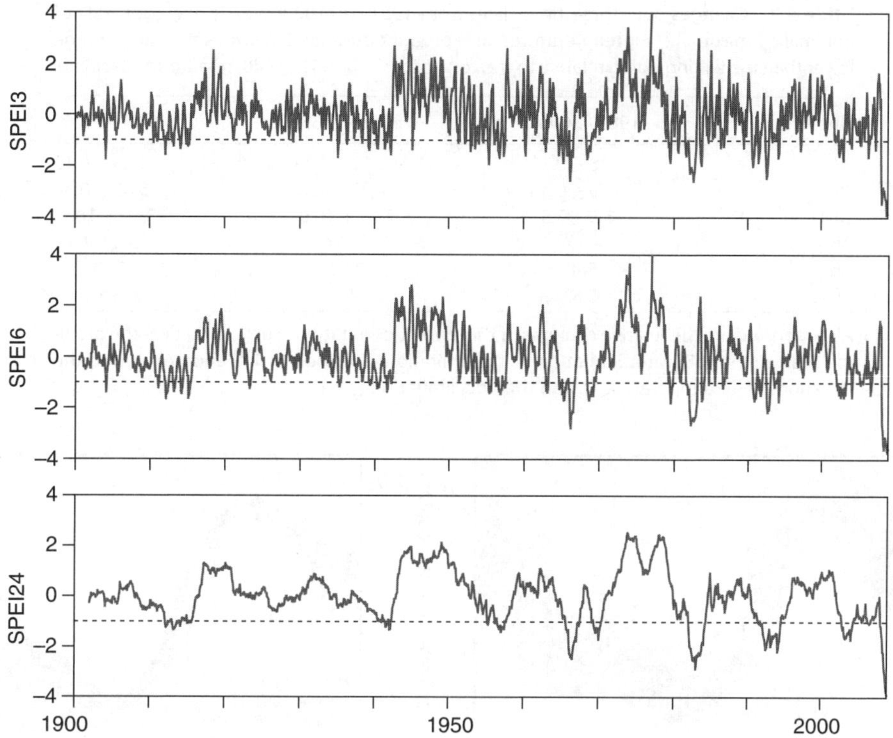

Fig. 4.1 Drought evolution from 1901 to 2009 on different timescales as assessed by the SPEIs. The series represents the evolution of the SPEIs with time-windows of 3, 6, and 24 months. Dry periods display negative SPEI values and humid ones have positive SPEI values. The dotted lines at SPEI = 1 show the threshold value defining a moderate drought.

Table 4.1 Summary of results obtained by analyzing the Standardized Precipitation Evapotranspiration Index (SPEI) with three time-windows of 3, 6, and 24 months.

SPEI3	F	Return period (RP) = 4.9 years, larger RP for 1901–1936 with only four dry events, against nine events in the second and third time-slices
	O	The drought onset was observed earlier (December–January) at the beginning of the series and progressively occurred later
	D	No significant increase of drought duration (D) but longer events in the second and third time-slices (3; 4.2 ± 1.7; 6 ± 5.3 months)
SPEI6	F	RP = 6.4 years; lower RP (5.1 years) from 1936 to 1972 with seven events
	O	Same pattern of change as for SPEI3
	D	Significant increase of D with time (0.89 month each 10 years)
SPEI24	F	RP = 6.4 years; shorter RP (4.5 years) from 1936 to 1972 with seven events
	O	Same pattern of change as for SPEI3 and SPEI6
	D	Significant increase of D in the third time-slice (7.0 ± 3.4; 8.1 ± 5.4; 17.2 ± 13.6 years); very severe drought events in 1983 and 1992, when D = 28 and 33 months, respectively

The SPEI threshold value retained for defining a moderate drought is −1. F refers to the frequency of drought events in year, O corresponds to the month of the onset of the drought period, and D is the drought duration in months.

For the three time-slices that follow 1961–1980, we did not observe any significant change in the soil water storage pattern (Fig. 4.3). However, the transition period showed a decline in S for values lower than 40 mm. At this level of soil water storage, corresponding to relative water contents ranging from 0.6 to 0.8, the impact of soil drought on carbon

Table 4.2 Changes over three time-slices from 1981 to 2100 in yearly averages ±SD of minimal t_n, maximal t_x air temperatures, and drought duration DD with soil water storage lower than the wilting point, relative to the control period (1961–1980) retained as baseline.

Time-slice	1981–2000	2046–2065	2081–2100
Δt_n	+0.48 ± 0.11	+2.11 ± 0.11	+3.03 ± 0.13
Δt_x	+0.69 ± 0.30	+4.03 ± 0.36	+5.05 ± 0.38
ΔDD	−1.5 ± 7.5	39.1 ± 10.9	34.9 ± 10.1
$\Delta f_1/f_1$	−1.7	−26.6	−24.4
$\Delta f_2/f_2$	5.8	23.3	27.1
$\Delta f_3/f_3$	2.5	7.3	8.1

Also reported are the relative changes of three abiotic limitations cumulated over the growing season: the soil water limitation f_1, the limitation of carbon assimilation related to the variations of air temperature f_2, and the frost limitation f_3.

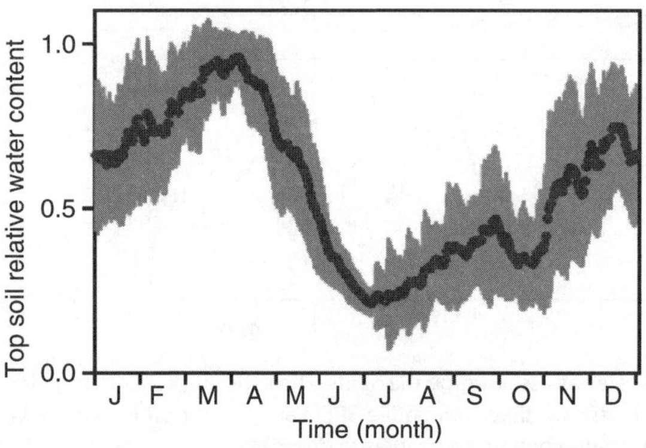

Fig. 4.2 Seasonal course of the daily relative soil water content averaged over the second time-slice (1981–2000). The grey area shows the standard error.

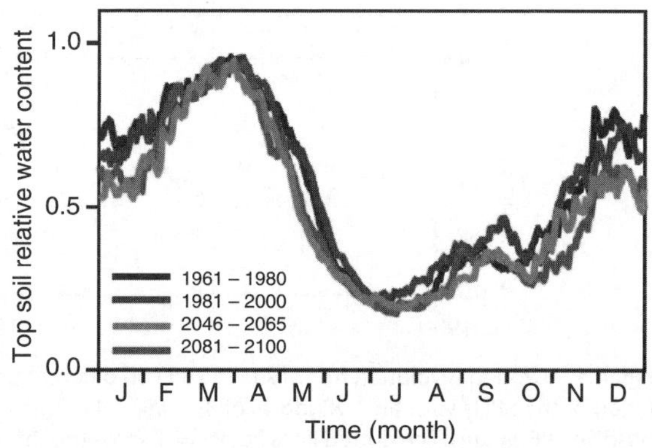

Fig. 4.3 Seasonal time courses of the daily relative soil water content averaged over each of the four time-slices from 1961 to 2100.

assimilation and growth is already significant. The number of days within the growth period in which S is lower than the wilting point did not increase significantly in the 1981–2000 time-slice (Table 4.2). Moreover, the number of dry days increased largely to 39.1 ± 10.9 and 34.9 ± 10 in the last two time-slices, respectively. This direct effect of climate change was associated with a drastic increase in the between-year variation.

For any given day of the year, we calculated an unpredictability index based on the coefficient of variation of the soil water storage S (Fig. 4.4). In the first two time-slices, the unpredictability index reached 40% at the beginning of the crop season, in early November (Fig. 4.4). It increased sharply to 60% in the projected time-slices (2046–2065 and 2081–2100). Thus, the expected consequence of this higher climate unpredictability is an extremely large between-year variation in crop yields at both field plot and landscape levels in the near future.

Time variation of yield at local or plot scale

Initially, we analyzed how the three abiotic limitations changed with time. For that, we calculated the relative variation for each time-slice and each limitation factor. The value observed in the 1961–1980 time-slice was used as a baseline for comparison.

The relative drought limitation quantified here by $\Delta f_1/f_1$ increased with time. This increase (resulting in negative values in Table 4.2) was

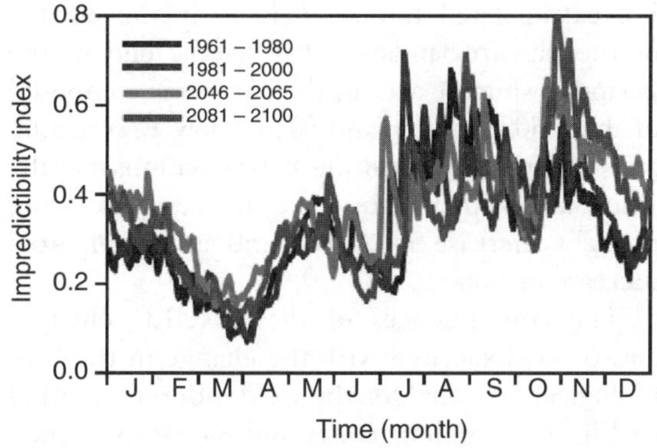

Fig. 4.4 Seasonal time course of the daily unpredictability index calculated over four time-slices from 1961 to 2100.

slight in the 1981–2000 time-slice and reached larger values in the last two time-slices, −26.6% and −24.4%. The positive effects of climate warming appeared in both the carbon assimilation factor, f_2, and the frost factor, f_3. The first effect considerably reduced the negative drought effect in the 1981–2000 time-slice. Furthermore, in 2046–2065 and then in 2081–2100, this positive effect increased largely with values of 23.3% and 27.1%, respectively. The cost of the frost declined with the increase of $\Delta f_3/f_3$.

The consequences of these changes in the climate parameters on the yield index are summarized in Table 4.3 and Fig. 4.5. The yield index was calculated on a unit area basis. Our calculations did not include the fact that one unit area of crop is always associated with one unit area of fallow (this effect will be integrated in the simulation at landscape scale presented later). Neither did the yield simulations at plot scale consider any change in the land biological potential with time. Over the recent periods, a positive effect of the warming trend facilitated photosynthesis and reduced the impact of frost events. This positive linear trend was associated with a reduced between-year variation. Water limitations contributed significantly in impairing the yield but these effects were largely counterbalanced by the increase in air temperature. In the first two time-slices pooled from 1961 to 2000, the positive effect of warming had a significant effect on the quinoa yield that increased linearly by 5% per decade (Fig. 4.5). In the third time-slice, from

Table 4.3 Changes of the mean yield index at plot scale YI_{plot} and its standard deviation over different time-slices from 1961 to 2100.

Time-slice	$\bar{YI}_{plot} \pm$ SD	$\Delta YI_{plot}/\Delta t$ (% per year)
1961–2000	0.31 ± 0.08	+0.5
1961–1980	0.30 ± 0.07	–
1981–2000	0.33 ± 0.09	–
2046–2065	0.32 ± 0.16	Negligible
2081–2100	0.35 ± 0.13	−1.7

Linear trends in yield increase or decrease with time are reported for 1961–2000 and 2081–2100, respectively.

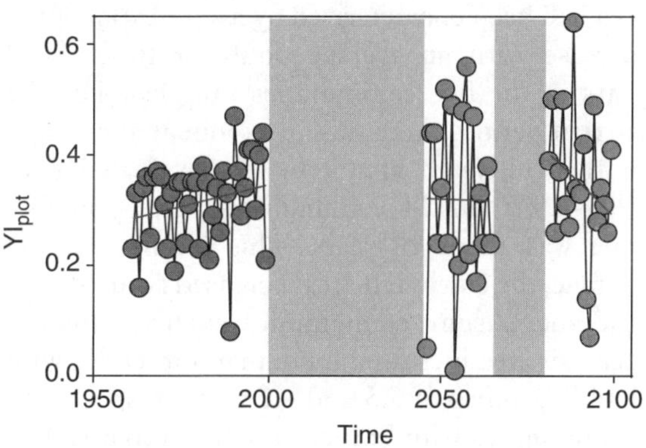

Fig. 4.5 Interannual variability of the quinoa yield index [0–1] at plot scale across the four time-slices 1961–1980, 1981–2000, 2046–2065, and 2081–2100. Red lines show the trends in each time-slice.

2046 to 2065, the mean yield would be slightly lower than the one observed during 1981–2000. Moreover, this value should be associated with a larger between-year variation reaching 50% of the mean. Thus, in some years, there would be very low yields as a consequence of severe water limitations. It should be noted that both values of mean and between-year variation were likely not well estimated statistically, due to the low number of years in the time-slices.

Keeping this caveat in mind, in the projected future, the positive effect of the continued warming should boost forward the potential of carbon assimilation through photosynthesis, and frost occurrence should continue to decline drastically until its complete disappearance by the end of the twenty-first century (Tables 4.2 and 4.3). Despite these positive effects of climate warming, the simulated quinoa yield index remained stable in the second time-slice, before decreasing in the third time-slice (Fig. 4.5). In the third and last time-slice, the between-year variation of the drought limitation had a greater effect than both warming factors f_2 and f_3. At the field plot scale, this resulted in a negative trend for crop productivity, with an average decline in quinoa yield of 17% per decade.

Model results at landscape level

From 1960 onward, the increase of cultivated area had been characterized by a slow exponential increase with an average doubling time of 20 years in the five communities sampled. For the current period, there was an exponential increase in the cultivated area remotely evaluated over 1998–2007 on 14 communities. The doubling time was about 10 years. This means that, on average, the cultivated area increased twofold over a decade. In some communities, such as Otuyo or Lia, this area increased more than threefold, with doubling times of 5.5 and 6.2 years, respectively. Communities with limited suitable crop area had doubling time that may reach 40 years, suggesting a saturation level. For 1980, we estimated the percent suitable crop area covered by quinoa to be 9.5, 5.5, and 6% in *cerros*, *faldas*, and *pampas*, respectively. In 2000, these percent areas decreased to 7.5% in *cerros* but increased to 7.9% and 25% in *faldas* and *pampas*. Finally, for the year 2040, the projected areas covered by the quinoa crop are 5%, 27%, and 50% for *cerros*, *faldas*, and *pampas*, respectively. The projected time change for this last land unit is presented in Fig. 4.6a.

We extrapolated toward 2040 a drastic decline of the pasture lands so that by the end of the period, a suitable area in the *pampas* is composed of about 50% in crop and 50% fallow. Figure 4.6b and c shows the resulting extrapolations for the whole area with a decline of quinoa crop in *cerros*, a slight increase in *faldas*, and an exponential increase in *pampas*.

The consequences of these LULC changes and those associated with the change in the land biological potential for the yield index are plotted in Fig. 4.7. Pooling *faldas* and *pampas* together, we simulated first a sharp increase in the yield index. This is related to the rapid encroachment of quinoa in the *pampas* and also to the positive warming effect and to a nonsignificant change

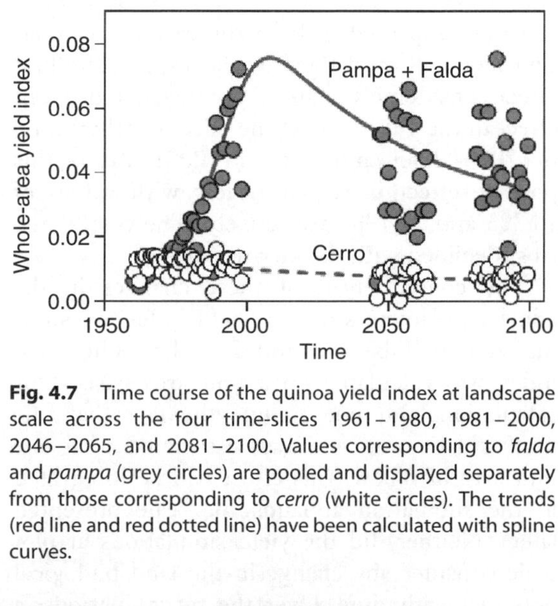

Fig. 4.7 Time course of the quinoa yield index at landscape scale across the four time-slices 1961–1980, 1981–2000, 2046–2065, and 2081–2100. Values corresponding to *falda* and *pampa* (grey circles) are pooled and displayed separately from those corresponding to *cerro* (white circles). The trends (red line and red dotted line) have been calculated with spline curves.

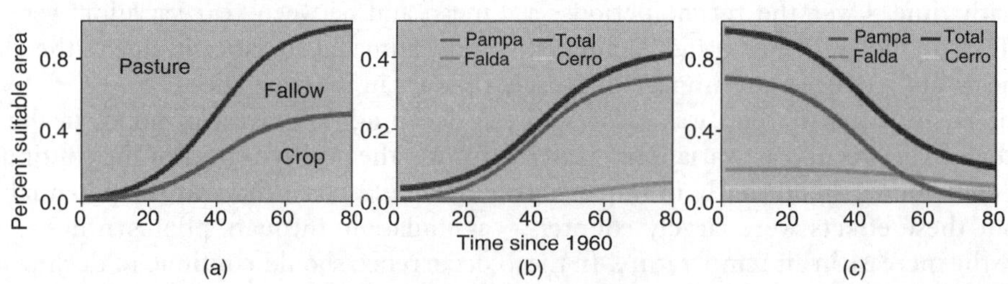

Fig. 4.6 Scenarios of land use and land cover changes used in our simulations. (a) The drastic land use change in the *pampa* land unit with projected crop cover reaching 50% of the suitable area at the 2040 horizon. (b) The change in the crop area aggregated for the whole study area, showing a sharp increase in the *pampas*, a moderate increase in the *faldas*, and a slight decline in the *cerros*. (c) Decline in the percent area devoted to pasture in the different land units.

in water limitation. In the second and third time-slices, we simulated primarily a decline in the average quinoa yield but with especially large between-year variations. In the *cerros*, the yield follows the decline of the cropped area with an enlargement of the between-year variations in both remote time-slices.

DISCUSSION

It is commonly accepted that drought is a multi-scalar phenomenon as the periods when soil water limitations impact the various components of crop productivity distinctly differ. Drought is a phenomenon that may occur simultaneously across various timescales, for example, a short period of particular dryness embedded within longer-term droughts. The SPEI results highlight the limit of a drought metric based on only potential evaporation and rainfall amounts at monthly scale compared to results based on the simulation of a soil water balance model driven by daily values of climate parameters and particularly of rains. However, for all the time-windows we retained (3, 6, and 24 months), we found a decrease of the return period of drought events, a change in the onset of these events toward a shortening of the wet season, and an increase of their duration. All these changes appeared particularly accelerated in the recent period (1973–2009). Air warming was observed especially since the year 1980 and onward, and the associated air dryness increased the potential evaporation. Thus, both the increase in PET and the slight decline in the amount of the rain (Vuille et al. 2003; Bradley et al. 2006) drastically modified the climate water balance toward more frequent and severe droughts. The occurrence of large drought periods in 1983 and 1992, with durations of 28 and 33 months, respectively, may in some ways presage the future climate. Morales et al. (2012) stated that "a high concentration of extreme dry events has occurred during the last 70 years with four of the 12 driest years since AD 1300."

There is a large sphere of uncertainty associated with climate projections, and unknown forces in the future may affect the composition of the atmosphere and feedbacks from the land surface. Over the next three decades, projections of global mean temperature rise are largely insensitive to contrasted emission scenarios. Nevertheless, it is clear that present and future predictability of climate variations and change is not the same everywhere and that gaps in knowledge are revealed by a lack of agreement between climate models in some regions, including projections of rainfall patterns over large areas of South America (Rada et al. 1997; Boulanger et al. 2007). The Altiplano region is poorly resolved in current GCMs, given their coarse resolution relative to the narrowness of the Andean cordilleras. Not surprisingly, direct analysis of GCM outputs results in slight precipitation decline or increase over the Altiplano, in contrast with a robust warming trend expected for the rest of the twenty-first century and already fully documented since 1980 onward. Glaciological studies have shown that the temperature in the tropical Andean cordillera has risen between 0.10 and 0.11°C per decade since 1939 and the ongoing warming trends ranged between 0.32 and 0.34°C per decade from 1974 to 1998 (Vuille and Bradley 2000). In our climate scenario, the observed warming trends were poorly represented if we compare 1981–2000 to the baseline data of 1960–1980. However, for the next time-slices, we projected an increase of 2.1 and 4.0°C in 2046–2065 and 3.0 and 5.1°C in 2081–2100 for both minimal and maximal air temperatures. According to other GCM runs, temperatures are likely to increase in the range 1.3–1.6°C by the year 2030 and between 4.8 and 6°C by the year 2100 (Garreaud et al. 2009).

A compendium of results from GCM runs is ambiguous about rain projections (Vuille et al. 2003; Urrutia and Vuille 2009; Seth et al. 2010; Thibeault et al. 2012). To circumvent the problem with rain projections, Minvielle and Garreaud (2011) proposed an interesting scheme and exploited a rather strong relationship between mid-tropospheric zonal winds and precipitation over the central Andes to project changes in regional rainfall by the end of the twenty-first century. Owing to a projected weaker easterly flow, the central Andes will likely exhibit a decrease in precipitation toward the end of the century that, when averaged over all stations, could reach 15.3 mm/month for the A1B scenario, using 1948–2007 as baseline. In our simulations, the increase of *PET* with the concomitant slight decrease in the amount of rain and change in their distribution induces large between-year variation of the quinoa crop yield and a significant decline of this yield in the remote time-slice 2081–2100.

Reynolds et al. (2004) conducted a comprehensive simulation analysis to explore how plants respond to variations in amounts of rain and availability of soil water in areas with limited water. They concluded that the so-called pulse-reserve model, which relates plant productivity linearly to annual precipitation as in rain-use efficiency approaches (Le Houérou and Hoste 1977; Huxman et al. 2004), is inadequate. Our simulation results showed that rainfall characteristics (e.g., seasonality, frequency, and intensity of rain events) and their consequences on availability of soil water are important for plant growth, and notably crucial for quinoa growth (Geerts et al. 2006). Rainfall seasonality is a singularly important driving variable, as it accounts for much of the difference among water-limited areas and is likely to be quite sensitive to the on-going global climate change already observed over the southern Altiplano (Thibeault et al. 2010). Small changes in seasonality could drastically impact quinoa yield.

In the context of crop productivity, the so-called land degradation results from a mismatch between land quality and land use. Rather than retaining disciplinary-oriented meanings, which could be prone to misinterpretation among disciplines, we prefer to use the idea of decline in the land biological potential. While there is a clear distinction between soil and land, the term "land" refers to an ecosystem perspective comprising soil, landscape, terrain, vegetation, and climate. Mechanisms that initiate decline in biological potential include physical and biological processes. Important among physical processes are a change in soil properties leading to erosion, desertification, and unsustainable use of natural resources (see Reynolds et al. 2007 for a survey including our study area). Biological processes include reduction in total and biomass carbon, fertility depletion, and decline in biodiversity. At least two distinct schools of thought have emerged regarding the prediction, severity, and impact of hypothetical land degradation on productivity. One school of thought believes that it is a serious threat posing a major challenge to humans in terms of its adverse impact on crop productivity and environment quality. Most ecologists, soil scientists, and agronomists primarily support this argument. The second school of thought, propounded primarily by economists, asks why, if decline in the biological potential of land is such a severe issue, market forces have not taken care of it? Supporters argue that producers have vested interest in their land and will not let it degrade to a point that would be detrimental to their profits. There are a number of factors that perpetuate this debate as illustrated by Jacobsen (2011) and Winkel et al. (2012).

In the what-if exercise presented here, we adopt a neutral conservative position by which a lumped metric is included for describing a decline in the biological potential of land. In this metric, both physical and biological factors are aggregated in one continuously changing growth efficiency parameter. Thus, we do not retain the discrete terminology used by specialists, who propose terms such as "slight," "moderate," "severe," and "very severe" to quantify the severity of soil degradation (Lal 1997). Others have used the terms "light," "moderate," "strong," and "extreme." All of these terms are difficult to compare. Our continuously changing parameter parallels the rate of encroachment of the recent and rapid expansion of the quinoa crop and the subsequent decline of the *pampas*. This expansion is done through agricultural intensification, and based mainly on the mechanization of tillage and seed sowing operations. The mechanization of these processes causes a measurable yield decline at plot scale. The reasons for this include poor seedbed preparation and very irregular seedling emergence, resulting in low plant density compared to fields with traditional manual practices, as those maintained in the *cerros*. Additionally, soil aeolian erosion has also been observed in this area. This constitutes an objective warning indicator often associated with the burial of young seedlings during wind storms and the decline in fertility in sandy soils under arid climatic conditions. The decline in soil fertility after conversion of virgin land to agriculture has been fully documented (Fonte et al. 2012) and typically shows two successive phases: a more or less fast decrease in the first years under cultivation, followed by a new equilibrium. A large socioeconomic survey conducted in a wetter part of the Altiplano revealed the importance of fallow and tillage practices in the soil nutrient loss processes and erosion (Swinton and Quiroz 2003). This study also showed that the probability of a farm experiencing soil nutrient loss over the past 20 years depended on natural factors and also on both the social context and the agricultural management practices. The probability of nutrient depletion increases in sandy soils (Fonte et al. 2012). To sustain soil fertility despite the expansion of the quinoa crop, it is recommended that llama dung be incorporated in the soil, albeit an uncommon practice in the traditional autosubsistence cropping system. This organic resource is the only (if ever) nutrient input applied in this region and it is largely dependent of the *pampas* and the livestock it supports. However, the beneficial effects of animal manure and, more generally, the effective importance of nutrient availability in the soil are still poorly documented in quinoa (Cárdenas and Choque 2008). A major shortcoming is the lack of clear cause–effect relationship between soil fertilization and crop productivity in this particular agrosystem. This is hampered by the observation that inferior soil physical structure or low water availability, both caused by poor seedbed conditions, might interfere with nutrient availability. In turn, low nutrient availability causes low plant density and low grain yield in mechanized field plots. In fact, assessing the relative effects of soil water, soil nutrients, and soil structure on quinoa productivity still remains a challenging task, the same as evaluating the benefit/cost of the various options for crop fertilization in this agricultural system.

In our yield simulations at plot scale, without changes in the biological potential of the land, we observed over the recent past time-slices (1961–1980 and 1981–2000) a positive effect of the climate warming trend that facilitated carbon assimilation and reduced the cost of frost occurrence. This positive linear trend was also associated with a reduced between-year variation. Water limitations contributed significantly in diminishing the yield but these effects were mostly counterbalanced by the warming trend. In the two near future time-slices, we observed a saturation of the crop yield followed by a sharp linear decline. To scale these results up to the landscape level, we took into account the assumption that fallow areas equal the crop areas, a scenario of LULC change extrapolated from changes observed at present, and a decline in the biological potential of land. This decline has been discussed previously. At the regional scale, the main result is a potential decline in yield from 2020 onward, despite the continued saturation of the suitable crop area toward 2040. By the end of our simulation, the average regional yield would be half of the yield obtained at peak production levels.

In our landscape, there are many reasons, likely confounding, why smallholders let the biological potential of their land decline. Some of these reasons relate to the values that local societies place on land resources and their perception of the present vulnerability of these resources. While agronomists and development agents see the soil as a nonrenewable resource, this is not necessarily the case for local farmers (Zimmerer 1993, 1994). In the end, the sustainable use of the soil and land resources will lie in the hands of the farmers themselves who are subject to economic and social pressures from both the community and the market within which they operate (Swinton and Quiroz 2003). Ongoing changes in climatic conditions, crop productivity, and land biological potential are strongly interrelated and each must be addressed in the context of the others for a correct assessment of its impact (Valdivia et al. 2010). This is the challenge for the near future, and for which we must be prepared. A study on natural ecosystems in the northern part of the Altiplano warns again an “Andean tipping point” for climate (Bush et al. 2010). This seems to corroborate the climate projections at regional scale that predict the Andean climate to disappear by the end of the twenty-first century (Williams et al. 2007).

We could question the use of speculative exercises. Testing model simulations against field observations and other model simulations is only part of the model validation process, and a good agreement between simulated and measured values alone does not guarantee the correctness of using model projections (Geerts et al. 2007, 2009; Lebonvallet 2008). To make further use of this approach in diverse ecological situations and hierarchies of scales, three challenges must be overcome. First, additional processes associated with land-use practices and decline in land biological potential need to be incorporated. Second, appropriate spatially explicit simulation approaches and scaling methods should be used to link landscape patterns and processes at multiple spatial scales (e.g., local spot, landscape, and whole region). Correspondingly, multilayer spatial datasets need to be developed and continuously updated (see Pouteau et al. 2011 for a significant account on frost mapping). Third, variables that show large changes associated with climate, for example, warming trends and change in rainfall patterns, should be manipulated in field experiments to evaluate their effects on land productivity. Our results show that such numerical experiments are likely to provide a strong and realistic set of predictions with which to compare actual long-term change in a rapidly changing region. Thus, this study provides a basis for further investigating how abiotic and biotic environmental changes influence quinoa yield. Another arising problem is our limited understanding of what the local-level impacts of climate change on crop yield are likely to be. This relates to the uncertainties involved in GCM outputs at the high spatial resolutions needed for an effective application of such type of work (Challinor et al. 2005; Watson and Challinor 2013). While there are still substantial gaps in our current knowledge and techniques, significant opportunities do exist for improving the production and evaluation of higher-resolution climate change scenarios, as illustrated by the work of Buytaert et al. (2010). Despite its limitations, the type of analysis presented here should be able to provide some insights to help research and development organizations target adaptation options for a sustainable use of local land resources.

ACKNOWLEDGMENTS

This chapter is a tribute to Jean-Pierre Ratte who passed away in 2011. The authors acknowledge the work of Roland Bosseno of the IRD (Institut de Recherche pour le Développement) for land-use change analyses on SPOT images. Thanks also to D. Salas-Mélia from the CNRM (Centre National de la Recherche Météorologique) Meteo France for climate scenarios. This study was carried out with the financial support of the "ANR-Agence Nationale de la Recherche - The French National Research Agency" under the program "Agriculture et Développement Durable," project "ANR-06-PADD-011-EQUECO."

REFERENCES

Boulanger JP, Martinez F, Segura EC. Projection of future climate change conditions using IPCC simulations, neural networks and Bayesian statistics. Part 1: Temperature mean state and seasonal cycle in South America. Climate Dynam 2006;27:233–259.

Boulanger JP, Martinez F, Segura EC. Projection of future climate change conditions using IPCC simulations, neural networks and Bayesian statistics. Part 2: Precipitation mean state and seasonal cycle in South America. Climate Dynam 2007;28:255–271.

Bradley RS, Vuille M, Diaz HF, Vergara W. Climate Change: threats to water supplies in the Tropical Andes. Science 2006;312:1755–1756.

Bush MB, Hanselman JA, Gosling WD. Nonlinear climate change and Andean feedbacks: an imminent turning point? Glob Chang Biol 2010;16:3223–3232.

Buytaert W, Vuille M, Dewulf A, Urrutia R, Karmalkar A, Celleri R. Uncertainties in climate change projections and regional downscaling in the tropical Andes: implications for water resources management. Hydrol Earth Syst Sci 2010;14:1247–1258.

Cárdenas CJ, Choque MW. 2008. Fertilidad, uso y manejo de suelos en la zona del Intersalar, departamentos de Oruro y Potosí. La Paz, Bolivia:Fundación AUTAPO, Programa Quinua Altiplano Sur. 105 p.

Challinor AJ, Wheeler TR, Slingo JM, Hemming D. Quantification of physical and biological uncertainty in the simulation of the yield of a tropical crop using present-day and doubled CO_2 climates. Philos T Roy Soc B 2005;360:2085–2094.

Falvey M, Garreaud RD. 2005. Moisture variability over the South American Altiplano during the South American Low Level Jet Experiment (SALLJEX) observing season. J Geophys Res-Atmos 110:D22105.

Fonte SJ, Vanek SJ, Oyarzun P, Parsa S, Quintero DC, Rao IM, Lavelle P, Donald LS. Pathways to agroecological intensification of soil fertility management by smallholder farmers in the Andean highlands. Adv Agron 2012;116:125–184.

Garcia M, Raes D, Allen R, Herbas C. Dynamics of reference evapotranspiration in the Bolivian highlands (Altiplano). Agr Forest Meteorol 2004;125:67–82.

Garreaud RD. Intraseasonal variability of moisture and rainfall over the South American Altiplano. Mon Weather Rev 2000;128:3337–3346.

Garreaud RD, Aceituno P. Interannual rainfall variability over the South American Altiplano. J Climate 2001;14:2779–2789.

Garreaud RD, Vuille M, Compagnucci R, Marengo J. Present-day South American climate. Palaeogeogr Palaeoclimatol Palaeoecol 2009;281:180–195.

Geerts S, Raes D, Garcia M, Del Castillo C, Buytaert W. Agro-climatic suitability mapping for crop production in the Bolivian Altiplano: A case study for quinoa. Agric Forest Meteor 2006;139(3–4):399–412. DOI: 10.1016/j.agrformet.2006.08.018.

Geerts S, Raes D, Garcia M, Miranda R, Cusicanqui JA, Taboada C, Mendoza J, Huanca R, Mamani A, Condori O, etal. 2007. Simulating yield response of quinoa to water availability with AquaCrop. Agron J 111:499–508.

Geerts S, Raes D, Garcia M, Taboada C, Miranda R, Cusicanqui J, Mhizha T, Vacher J. Modeling the potential for closing quinoa yield gaps under varying water availability in the Bolivian Altiplano. Agr Water Manage 2009;96:1652–1658.

Hertel TW, Burke MB, Lobell DB. The poverty implications of climate-induced crop yield changes by 2030. Global Environ Change 2010;20:577–585.

Huxman TE, Smith MD, Fay PA, Knapp AK, Shaw MR, Loik ME, Smith SD, Tissue DT, Zak JC, Weltzin JF, et al. Convergence across biomes to a common rain-use efficiency. Nature 2004;429:651–654.

Jacobsen SE. The situation for quinoa and its production in southern Bolivia: from economic success to environmental disaster. J Agron Crop Sci 2011;197:390–399.

Kalma JD, Perera H, Wooldridge SA, Stanhill G. Seasonal changes in the fraction of global radiation retained as net all-wave radiation and their hydrological implications. Hydrol Sci J 2000;45:653–674.

Lal R. Degradation and resilience of soils. Philos T Roy Soc B 1997;352:997–1010.

Lebonvallet S. 2008. Implantation du quinoa et simulation de sa culture sur l'altiplano bolivien. Thèse de Doctorat, AgroParisTech, France.

Le Houérou HN, Hoste CH. Rangeland production and annual rainfall relations in Mediterranean basin and in African sahelo-sudanian zone. J Range Manage 1977;30:181–189.

Linacre ET. Simpler empirical expression for actual evapotranspiration rates - discussion. Agr Meteorol 1973;11:451–452.

Little MA. Human populations of the Andes: the human science basis for research planning. Mt Res Dev 1982;1:145–170.

Minvielle M, Garreaud RD. Projecting rainfall changes over the South American Altiplano. J Climate 2011;24:4577–4583.

Morales MS, Christie DA, Villalba R, Argollo J, Pacajes J, Silva JS, Alvarez CA, Llancabure JC, Soliz Gamboa CC. Precipitation changes in the South American Altiplano since 1300 AD reconstructed by tree-rings. Clim Past 2012;8:653–666.

Parry ML, Livermore MTJ. 2002. Climate change, global food supply and risk of hunger. In: Issues in Environmental Science and Technology n°17: Global Environmental Change. Cambridge, UK: The Royal Society of Chemistry. pp. 109-137.

Pouteau R, Rambal S, Ratte JP, Gogé F, Joffre R, Winkel T. Downscaling MODIS-derived maps using GIS and boosted regression trees: the case of frost occurrence over the arid Andean highlands of Bolivia. Remote Sens Environ 2011;115:117–129.

Priestley CHB, Taylor RJ. On the assessment of surface heat-flux and evaporation using large-scale parameters. Mon Weather Rev 1972;100:81–92.

Rada OP, Crespo SR, Miranda FT. Analysis of climate scenarios for Bolivia. Climate Res 1997;9:115–120.

Reynolds JF, Kemp PR, Ogle K, Fernandez RJ. Modifying the 'pulse-reserve' paradigm for deserts of North America: precipitation pulses, soil water, and plant responses. Oecologia 2004;141:194–210.

Reynolds JF, Stafford Smith DM, Lambin EF, Turner II BL, Mortimore M, Batterbury SPJ, Downing TE, Dowlatabadi H, Fernández RJ, Herrick JE, et al. Global desertification: building a science for dryland development. Science 2007;316:847–851.

Ronchail J, Gallaire R. ENSO and rainfall along the Zongo valley (Bolivia) from the Altiplano to the Amazon basin. Int J Climatol 2006;26:1223–1236.

Saxton KE, Rawls WJ, Romberger JS, Papendick RI. Estimating generalized soil-water characteristics from texture. Soil Sci Soc Am J 1986;50:1031.

Seiler C, Hutjes RWA, Kabat P. Climate variability and trends in Bolivia. J Appl Meteorol Climatol 2013;52:130–146.

Seth A, Thibeault J, Garcia M, Valdivia C. Making sense of twenty-first-century climate change in the altiplano: observed trends and CMIP3 projections. Ann Assoc Am Geogr 2010;100:835–847.

Swinton SM, Quiroz R. Poverty and the deterioration of natural soil capital in the Peruvian altiplano. Environ Dev Sustain 2003;5:477–490.

Thibeault J, Seth A, Wang GL. Mechanisms of summertime precipitation variability in the Bolivian Altiplano: present and future. Int J Climatol 2012;32:2033–2041.

Thibeault JM, Seth A, Garcia M. 2010. Changing climate in the Bolivian Altiplano: CMIP3 projections for temperature and precipitation extremes. J Geophys Res-Atmos 115:D08103.

Urrutia R, Vuille M. 2009. Climate change projections for the tropical Andes using a regional climate model: temperature and precipitation simulations for the end of the 21st century. J Geophys Res-Atmos 114:D021108.

Valdivia C, Seth A, Gilles JL, García M, Jiménez E, Cusicanqui J, Navia F, Yucra E. Adapting to climate change in Andean ecosystems: landscapes, capitals, and perceptions shaping rural livelihood strategies and linking knowledge systems. Ann Assoc Am Geogr 2010;100:818–834.

Vergara W, Kondo H, Pérez Pérez E, Méndez Pérez JM, Magaña Rueda V, Martínez Arango MC, Ruíz Murcia JF, Avalos Roldán GJ, Palacios E. 2007. Visualizing future climate in Latin America: results from the application of the Earth Simulator. The World Bank, Latin America and the Caribbean Region. 82 p.

Vicente-Serrano SM, Gouveia C, Camarero JJ, Beguería S, Trigo R, López-Moreno JI, Azorín-Molina C, Pasho E, Lorenzo-Lacruz J, Revuelto J. Response of vegetation to drought time-scales across global land biomes. Proc Nat Acad Sci 2013;110:52–57. http://dx.doi.org/10.1073/pnas.1207068110.

Vuille M, Ammann C. Regional snowfall patterns in the high, arid Andes. Climatic change 1997;36(3–4):413–423. DOI: 10.1023/A:1005330802974.

Vuille M, Bradley RS. Mean annual temperature trends and their vertical structure in the tropical Andes. Geophys Res Lett 2000;27:3885–3888.

Vuille M, Bradley RS, Werner M, Keimig F. 20th century climate change in the tropical Andes: observations and model results. Clim Change 2003;59:75–99.

Vuille M, Francou B, Wagnon P, Juen I, Kaser G, Mark BG, Bradley RS. Climate change and tropical Andean glaciers: past, present and future. Earth-Sci Rev 2008;89:79–96.

Watson J, Challinor A. The relative importance of rainfall, temperature and yield data for a regional-scale crop model. Agr Forest Meteorol 2013;170:47–57.

Williams JW, Jackson ST, Kutzbach JE. Projected distributions of novel and disappearing climates by 2100 AD. Proc Natl Acad Sci 2007;104:5738–5742.

Winkel T, Bertero HD, Bommel P, Chevarría Lazo M, Cortes G, Gasselin P, Geerts S, Joffre R, Léger F, Martinez Avisa B, et al. The sustainability of quinoa production in southern Bolivia: from misrepresentations to questionable solutions. Comments on Jacobsen (2011, J. Agron. Crop Sci. 197: 390–399). J Agron Crop Sci 2012;198:314–319.

Zimmerer KS. Soil erosion and social (dis)courses in Cochabamba, Bolivia: perceiving the nature of environmental degradation. Econ Geogr 1993;69:312–327.

Zimmerer KS. Local soil knowledge: answering basic questions in highland Bolivia. J Soil Water Conserv 1994;49: 29–34.

Chapter 5

The Potential of Using Natural Enemies and Chemical Compounds in Quinoa for Biological Control of Insect Pests

Mariana Valoy[1], Carmen Reguilón[2], and Griselda Podazza[1]

[1]*Instituto de Ecología, Fundación Miguel Lillo, Tucumán, Argentina*
[2]*Instituto de Entomología, Fundación Miguel Lillo, Tucumán, Argentina*

INTRODUCTION

Over the centuries, agriculture has transformed natural landscapes into agricultural landscapes, and modified not only plant diversity but also animal diversity, including insect diversity. Insect life depends on plant populations, as they feed, reproduce, oviposit, and shelter in plants. Changes in the structure and diversity of vegetation, land use, cultivated area, type of crop management, physical environment, and other factors influence the behavior of insects, turning many of them into agricultural pests.

Agricultural intensification is one of the factors that modify an ecosystem's ability to manage pest populations by means of natural enemies (Altieri and Letourneau 1982; Oerke et al. 1994; Matson et al. 1997; Wilby and Thomas 2002; Bianchi et al. 2006; Parsa 2010). There are three hypotheses regarding the influence of agricultural intensification on pest populations and their natural enemies. First, with agricultural intensification, natural enemy populations are decimated by pesticides and/or decreased by the simplification of local ecosystem structures and landscapes driven by reduction of plant diversity (DeBach and Rosen 1991; Andow 1991; Bianchi et al. 2006). Second, crop intensification also contributes to the increase and persistence of pests by providing concentrated resources and an optimal environment for insects (Root 1973). Third, crop intensification entails the use of insecticides and techniques to enhance productivity, ultimately weakening plant defenses against herbivores (Harris 1980; Throop and Lerdau 2004). In this context, it has been proposed that ecological knowledge, conservation, and handling of biotic interactions (Médiène et al. 2011) are essential elements for reducing chemical use and promoting ecosystem services, thereby slowing down the loss of diversity in agroecosystems (Robinson and Sutherland 2002; Benton et al. 2003; Bianchi et al. 2006; Farwig et al. 2009).

Since ancient times, Andean agroecosystems have had a wide variety of cultivated species (Kraljevic 2006). Among these species, the Andean grains stand out, of which quinoa (*Chenopodium quinoa* Willd.) is one of the most representative species. Traditionally, quinoa cultivation has been intended for local consumption, with long periods of fallow land. In cultivated areas, direct sowing, manure application, threshing, and cleaning were all performed manually (Jacobsen 2011). This type of subsistence

Quinoa: Improvement and Sustainable Production, First Edition. Edited by Kevin Murphy and Janet Matanguihan.

agriculture was characterized by covering a small area of cultivated land and sowing seed in mountain slopes. However, due to the high international demand for quinoa (Jacobsen 2011), its cultivation has been extended to the plains. This marks the beginning of a semi-intensive agriculture, going from a crop rotation system, which is a key component for sustainable agriculture, to a marked tendency toward intensified monoculture (CIRNMA 2009).

The intensification of quinoa production has also increased pests and diseases and is hastening soil deterioration to the detriment of the environment and sustainability of the quinoa crop itself (Mujica 1993; Campos et al. 2012).

A study assessed the effect of crop intensification on insect pest populations and showed an increase in the population density of the quinoa moth (*Eurysacca melanocampta*) beyond the economic damage threshold, causing a yield reduction and economic loss (Campos et al. 2012). Use of farm machinery, such as a plough and sowing machine, produces a loose soil substratum that helps ticona larvae (*Copitarsia* sp.) and quinoa moth (*Eurysacca quinoae*) to penetrate the soil easily and complete their life cycles (Rasmussen et al. 2003; Sigsgaard et al. 2008).

At present, there has been little attention given to the study of entomofauna associated with quinoa compared to other Andean crops such as corn and potato (Zanabria and Mujica 1977; Sánchez and Vergara 1991; Yábar et al. 2002; Valoy et al. 2011). Because of this, there is a critical lack of knowledge about insect diversity and the role they play in quinoa cropping systems. However, there is a recent movement toward implementation of integrated pest management strategies in quinoa (Palacios et al. 1999; Dangles et al. 2009, 2010).

In this chapter, existing information on harmful and beneficial insects of quinoa is discussed. The short-term and long-term effects of changes in quinoa production are also examined. Possible solutions to the challenges posed by current quinoa farming are explored within the context of sustainable agroecosystems.

INSECTS IN QUINOA

Insects associated with quinoa consist of diverse phytophagous and entomophagous organisms. The abundance and occurrence of these insect populations, and the intensity of herbivory damage, may fluctuate according to quinoa phenological stages and the prevailing environmental conditions throughout the growing period (Mamani Quispe 2009; Valoy et al. 2011). Other factors involved are location and extent of crops, cultivars used, presence or absence of other crops nearby, and presence of native vegetation patches, among others (Costa et al. 2009a, 2009b).

In order to implement integrated management programs with emphasis on biological control, it is fundamental to identify and study the insects comprising the agroecosystem. Native and exotic insect species have been reported associated with quinoa (Ortíz Romero 1993). These are members of different functional herbivorous groups such as leaf miners, leaf cutters, leaf chewers, sap suckers, stem borers, defoliators, and grain and panicle consumers (Alata 1973; Ortiz and Sanabria 1979; Bravo and Delgado 1992; Ortiz 1997; Zanabria and Banegas 1997; Rasmussen et al. 2003). Moreover, many of these herbivores are vectors of pathogenic microorganisms (FAO 1993).

The Neotropical region, an ecozone that includes South and Central America, the Mexican lowlands, the Caribbean islands, and southern Florida, has 74 insect species associated with quinoa. These insect species are distributed in 25 families belonging to five orders: Lepidoptera, Hemiptera, Coleoptera, Diptera, and Thysanoptera (Barrientos Zamora 1985; FAO 1993; Lamborot et al. 1999; Hidalgo and Jacobsen 2000; Rasmussen et al. 2001; Yábar et al. 2002; Saravia and Quispe 2005; Costa et al. 2007, 2009a, 2009b; Valoy et al. 2011; Campos et al. 2012). Most of these insect records pertain to quinoa crops located in Bolivia, Peru, and Ecuador, whereas in the more southern portion of the Andes, information is scarce, with a few descriptive papers (Lamborot et al. 1999; Reguilón et al. 2009; Valoy et al. 2011). A literature review revealed that in 21 years (1987–2012) there are only 20 papers published on insects linked to Andean crops such as quinoa, amaranth, and potato. Twelve of these papers mentioned insects of quinoa, including harmful and beneficial ones (Valoy, unpublished data) (Table 5.1 and Figs 5.1–5.3).

Insect pests of quinoa

Among the insect pests of quinoa, several Lepidopteran species collectively known as the quinoa moth cause the greatest damage. Among these moths, *E. melanocampta* and *E. quinoae* (Lepidoptera: Gelechiidae) are regarded as key pests, due to the direct damage their larvae make to the plant. Even though they affect all developmental stages of the plant, most damage is inflicted during the grain ripening state (Rasmussen 2003). *E. melanocampta*, commonly known as “quinoa kuro,” causes between 35% and 60% of plant damage (Ochoa Vizarreta and Franco 2013). Its distribution covers xeric environments of the Andes Mountains (1,900–4,350 masl) from Argentina and Chile to northern Colombia (Povolny and Valencia 1986; Povolny 1986, 1997). Adults have crepuscular or nocturnal habit and typically lay their eggs in the glomeruli or inflorescence axils. When first-stage larvae emerge, they chew leaves and inflorescences. The more advanced larval stages subsequently roll leaves for use as shelter as they wait to feed during the grain-filling and ripening stages (Rasmussen 2003; Ochoa Vizarreta and Franco 2013). The biology and morphology of *E. quinoae* (commonly named “kona kona” or “grain grinder”) is very similar to *E. melanocampta*, but distinguished from each other by differences in their wing spot patterns (Ochoa 1990; Povolny 1997). *E. quinoae* has often been described as a pest of more restricted distribution than *E. melanocampta*, although recent papers mentioned *E. quinoae* as a key pest in Peru (Rasmussen et al. 2000; Campos et al. 2012) (Fig. 5.1).

Another Lepidopteran group that may occasionally cause important losses in quinoa is the “ticona complex,” a group composed of cutworms belonging to family Noctuidae: *Copitarsia turbata*, *Feltia* sp., *Heliothis* sp., and *Spodoptera* sp. (Blanco 1982; Aroni 2000; Chambilla et al. 2009). The most representative species of this complex is *C. turbata* (Larrain 1996). Its larvae are stemborers, usually causing damage in the beginning of the season by cutting the stem of young plants. In cases of severe infestation, they may even eat leaves, inflorescences, and grains, aside from shoots and stems (Vela and Quispe 1988; Zanabria and Banegas 1997).

Rasmussen et al. (2003) listed several phytophagous insects that affect quinoa, indicating which are pests and potential pests. Included in the list are the orders Orthoptera (Gryllidae) and Diptera (Agromizidae), together with Coleoptera (Bruchidae, Curculionidae, and Tenebrionidae), for a total of 54 species (Table 5.1 and Fig. 5.1).

South of the Neotropical region, in Brazil, the *Atta* sp. (Formicidae: Hymenoptera) and *Diabrotica speciosa* (Genn.) (Coccinellidae: Coleoptera) species have been reported in quinoa (Cabrera Almendros and Oliveira 2011). In the Amaicha del Valle, Tucumán, Argentina, Valoy et al. (2011) carried out studies in an experimental quinoa plot and reported the presence of three families, four genera, and seven species of phytophagous insects that have not been mentioned previously in quinoa grown in the Neotropical region (Table 5.1 and Fig. 5.1).

Beneficial insects in quinoa

Natural enemies of phytophagous insects are also called beneficial or entomophagous insects. Most phytophagous insects have more than one and often many natural enemies that regulate their populations. A large number of beneficial insects attack several species of herbivorous insects in quinoa, while there are entomophagous groups that are more specific (Table 5.2). Generally, natural enemies of phytophagous insects are classified as parasitoids, predators, or pathogens. Specifically, parasitoids and predators are categorized as entomophagous and these will be discussed in more detail in this chapter. Another group of natural enemies of insects are entomopathogens, which are microorganisms that cause diseases on insects. These are closely associated with insect pests and have been used to control insect pests in quinoa but are discussed only briefly in this chapter.

Table 5.1 Insect species associated with quinoa crops.

Order	Family	Genus/species	Source
Coleoptera	Bruchidae	*Acanthoscelides diasanus*	Rasmussen et al. (2003)
	Chrysomelidae	*Acalymma demissa*	Rasmussen et al. (2003)
		Calligrapha curvilinear	Rasmussen et al. (2003)
		Diabrotica decempunctata	Rasmussen et al. (2003)
		Diabrotica sicuanica	Rasmussen et al. (2003)
		Diabrotica sp.	Rasmussen et al. (2003)
		Diabrotica speciosa	Rasmussen et al. (2003), Cabrera Almendros and Oliveira (2011)
		Diabrotica undecimpunctata	Rasmussen et al. (2003)
		Diabrotica viridula	Rasmussen et al. (2003)
		Epitrix subcrinita	Rasmussen et al. (2003)
		Epitrix yanazara	Rasmussen et al. (2003)
		Epitrix sp.	Saravia and Quispe (2005)
	Coccinellidae	*Eriopis connexa*	Yábar et al. (2002), Valoy et al. (2011)
		Eriopis peruviana	Costa et al. (2007)
		Eriopis sp.	Valoy et al. (2011)
		Hippodamia convergens	Hidalgo and Jacobsen (2000), Yábar et al. (2002), Valoy et al. (2011)
	Curculionidae	*Adioristus* sp.	Rasmussen et al. (2003)
	Formicidae	*Atta* sp.	Rasmussen et al. (2003)
	Meloidae	*Epicauta adspersa*	Valoy et al. (2011)
		Epicauta langei	Valoy et al. (2011)
		Epicauta latitarsis	Rasmussen et al. (2003)
		Epicauta marginata	Rasmussen et al. (2003)
		Epicauta sp.	FAO (1993), Hidalgo and Jacobsen (2000), Saravia and Quispe (2005)
		Epicauta willei	Rasmussen et al. (2003)
		Meloe sp.	Rasmussen et al. (2003)
		Tetraonix sp.	Valoy et al. (2011)
	Melolonthidae	*Ancistrosoma vittigerum*	Valoy et al. (2011)
	Melyridae	*Astylus atromaculatus*	Valoy et al. (2011)
		Astylus luteicauda	Hidalgo and Jacobsen (2000), Rasmussen et al. (2003)
		Astylus laetus	Rasmussen et al. (2003)
	Tenebrionidae	*Pilobalia decorata*	Rasmussen et al. (2003)
Diptera	Agromyzidae	*Liriomyza huidobrensis*	Rasmussen et al. (2003), Saravia and Quispe (2005)
	Tachinidae	*Phytomyptera* sp.	Rasmussen et al. (2001)
		Unknown	Valoy et al. (2011)
Hemiptera	Aphidae	*Acrytosiphum kondoi*	Barrientos Zamora (1985)
		Aphis craccivora	Rasmussen et al. (2003)
		Aphis gossypii	Rasmussen et al. (2003)
		Macrosiphum euphorbiae	Barrientos Zamora (1985), Yábar et al. (2002), Rasmussen et al. (2003), Saravia and Quispe (2005), Campos et al. (2012)
		Myzus persicae	Yábar et al. (2002), Rasmussen et al. (2003), Saravia and Quispe (2005), Costa et al. (2007)
		Myzus sp.	FAO (1993)

(continued)

Table 5.1 (*Continued*)

Order	Family	Genus/species	Source
	Cicadellidae	*Anacuerna centrolinea*	Saravia and Quispe (2005)
		Bergallia sp.	Rasmussen et al. (2003)
		Borogonalia impressifrons	Rasmussen et al. (2003)
		Empoasca cisnova	Rasmussen et al. (2003)
		Empoasca hardini	Rasmussen et al. (2003)
		Empoasca sp.	Saravia and Quispe (2005)
		Paratanus exitiousus	Rasmussen et al. (2003)
		Paratanus yusti	Rasmussen et al. (2003)
		Paratanus sp.	Rasmussen et al. (2003)
	Coreidae	*Leptoglossus* sp.	Valoy et al. (2011)
	Nabidae	*Nabis* sp.	Valoy et al. (2011)
	Pentatomidae	*Nezara viridula*	Valoy et al. (2011)
Hymenoptera	Encyrtidae	*Copidosoma gelechiae*	Hidalgo and Jacobsen (2000)
		Copidosoma koehleri	Hidalgo and Jacobsen (2000)
		Copidosoma sp.	Valoy et al. (2011)
	Ichneumonidae	Unknown	Rasmussen et al. (2001), Valoy et al. (2011)
	Vespidae	*Spodoptera eridania*	Valoy et al. (2011)
Neuroptera	Chrysopidae	*Chrysoperla argentina*	Valoy et al. (2011)
		Chrysoperla externa	Valoy et al. (2011)
Lepidoptera	Gelechiidae	*Eurysacca media*	Lamborot et al. (1999)
		Eurysacca melanocampta	FAO (1993), Hidalgo and Jacobsen (2000), Rasmussen et al. (2001, 2003), Saravia and Quispe (2005), Costa et al. (2007, 2009a, 2009b)
		Eurysacca quinoae	Campos et al. (2012), Rasmussen et al. (2001, 2003)
		Scrobipalpula absoluta	Barrientos Zamora (1985)
	Geometridae	*Perizoma sordescens*	Rasmussen et al. (2003), Saravia and Quispe (2005)
	Noctuidae	*Agrotis ipsilon*	Rasmussen et al. (2003)
		Agrotis malefica	Rasmussen et al. (2003)
		Agrotis sp.	Rasmussen et al. (2003), Saravia and Quispe (2005)
		Copitarsia consueta	Barrientos Zamora (1985), Rasmussen et al. (2003)
		Copitarsia turbata	Saravia and Quispe (2005), Rasmussen et al. (2003)
		Dargida graminivora	Rasmussen et al. (2003)
		Feltia experta	Rasmussen et al. (2003)
		Feltia sp.	Rasmussen et al. (2003), Saravia and Quispe (2005)
		Heliothis titicaquensis	Rasmussen et al. (2003), Saravia and Quispe (2005)
		Heliothis zea	Rasmussen et al. (2003)
		Peridroma interrupta	Rasmussen et al. (2003)
		Peridroma saucia	Rasmussen et al. (2003)
		Pseudaletia unipunctata	Rasmussen et al. (2003)
		Spodoptera eridania	Rasmussen et al. (2003)
		Spodoptera frugiperda	Barrientos Zamora (1985), Rasmussen et al. (2003)
		Spodoptera sp.	Saravia and Quispe (2005)
	Psychidae	*Oiketicus kirbyi*	Valoy (personal communication)
		Oiketicus geyeri	Valoy (personal communication)
	Pyralidae	*Achyra similalis*	Lamborot et al. (1999)
		Herpetogramma bipunctalis	Rasmussen et al. (2003)
		Hymenia recurvalis	Saravia and Quispe (2005)
		Pochyzancla bipunctalis	Saravia and Quispe (2005)
		Spoladea recurvalis	Rasmussen et al. (2003)
Thysanoptera	Thripidae	*Frankliniella tabaci*	Rasmussen et al. (2003)
		Frankliniella tuberosi	Saravia and Quispe (2005)
		Frankliniella sp.	Rasmussen et al. (2003), Campos et al. (2012)

Fig. 5.1 Phytophagous insects associated with quinoa in the Neotropical region. (a) Lepidoptera, larvae feeding on quinoa leaf; (b) *Leptoglossus* sp. in quinoa leaf; (c,d) *Oiketicus kirbyi*, neotenic larvae coming out of bag or basket; (e) *O. kirbyi* bag or basket in quinoa branch; (f) adult *Eurysacca* sp.; (g) *Eurysacca* sp. larvae feeding on quinoa leaf; (h) Pyralidae, larvae; (i) *Nezara viridula*, adult in quinoa panicle; and (j) adult *Epicauta adspersa* in quinoa leaf. (*See color insert for representation of this figure.*)

Fig. 5.2 Entomophagous insect on quinoa in Amaicha del Valle. (a) Parasitoid *Copidosoma* sp. adult; (b) *Copidosoma* sp. parasitism in *Eurysacca* larvae; (c) *Eurysacca* damaged pupa; (d) parasitoid Ichneumonidae; (e) unidentified parasitoid; (f–l) predator insects; (f) *Eriopis connexa*, immature stage; (g) *Chrysoperla argentina*, adult; (h) *C. externa*, immature stage; (i) *C. externa*, adult; (j) *C. argentina*, preying on *Spodoptera frugiperda* (Lepidoptera) eggs; (k) *C. argentina*, immature stage preying on aphids; and (l) *C. externa*, immature stage preying on aphids. (*See color insert for representation of this figure.*)

Table 5.2 Beneficial insect species (parasitoids and predators) associated with quinoa crops in the Andes.

Order	Family	Genus	Species	Feeding guild	Host/prey
Hymenoptera	Braconidae	*Aphidius*	*Aphidius* sp.	Parasitoid	*Eurysacca* sp.
		Cotesia	*Cotesia* sp.	Parasitoid	*Eurysacca* sp.
		Meteorus	*Meteorus chilensis*	Parasitoid	*C. turbata, Erusacca* sp.
		Apanteles	*Apanteles* sp.	–	*E. melanocampta*
		Earinus	*Earinus* sp.	–	*E. melanocampta*
		Microplitis	*Microplitis* sp.	Parasitoid	*Eurysacca melanocampta*
	Ichneumonidae	*Deleboea*	*Deleboea* sp.	Parasitoid	*Eurysacca melanocampta*
		Diadegma	*Diadegma* sp.	Parasitoid	*E. melanocampta*
		Venturia	*Venturia* sp.	Parasitoid	*Eurysacca melanocampta*
	Encyrtidae	*Copidosoma*	*Copidosoma gelechiae*	Parasitoid	*E. melanocampta*
			Copidosoma koehleri	Parasitoid	*Eurysacca* sp.
	Vespidae	–	Unknown	Predator	*Eurysacca* sp. and *Copitarsia*
	Sphecidae	*Ammophila*	*Ammophila sabulosa*	Parasitoide	*C. turbata*
		Sphex	*Sphex* sp.	Parasitoide	*C. turbata*
	Trichogrammatidae	–	Unknown	–	*C. turbata, S. frugiperda*
Diptera	Tachinidae	*Dolichostoma*	*Dolichostoma* sp.	Parasitoid	*Eurysacca melanocampta*
		Phytomiptera	*Phytomiptera* sp.	Parasitoid	*E. quinoae, E. melanocampta*
Coleoptera	Coccinellidae	*Eriopis*	*Eriopis connexa*	Predator	*Eurysacca* sp., *aphids aphids*
			Eriopis peruviana	Predator	*Eurysacca* sp., *aphids*
			Eriopis sp.	Predator	*Eurysacca* sp., *aphids*
		Hippodamia	*Hippodamia convergens*	Predator	*Eurysacca* sp., *aphids*
	Carabidae	*Notobia*	*Notobia (Anisotarsus) schnusei*	Predator	*Eurysacca* sp.
			Notobia (Anisotarsus) laevis bolivianus	Predator	*Eurysacca* sp.
		Meotachys	*Meotachys* sp.	Predator	*Eurysacca* sp.
Neuroptera	Chrysopidae	*Chrysoperla*	*Chrysoperla argentina*	Predator	*Eurysacca* sp., *C. turbata, aphids*
			Chrysoperla externa	Predator	*Eurysacca* sp., *C. turbata, aphids*
			Chrysoperla sp.	Predator	*Eurysacca* sp.
Hemiptera	Nabidae	*Nabis*	*Nabis* sp.	Predator	*C. turbata*
	Miridae	*Rhinacloa*	*Rhinacloa* sp.	Predator	*C. turbata*
	Lygaeidae	*Geocoris*	*Geocoris* sp.	Predator	*C. turbata*
	Coreidae	*Ledptoglossus*	*Ledptoglossus* sp.	*Predator*	*C. turbata*

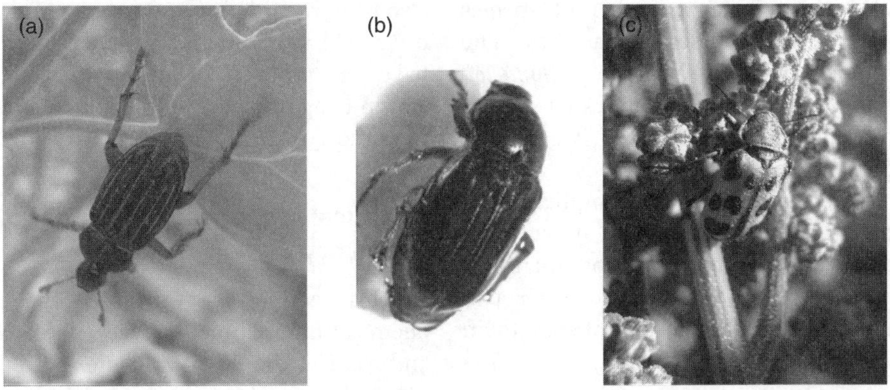

Fig. 5.3 Coleoptera species in quinoa crops. (a) Carabidae; (b) *Ancistrosoma vittigerum*; and (C) *Astylus atromaculatus*.

Parasitoids

Parasitoid groups are mainly distributed in two insect orders, Hymenoptera (wasps) and Diptera (flies). In terms of number of species, the order Hymenoptera is dominant among entomophagous insects and has been widely used for biological control of crop pests. Within this group, there are interesting, specific biological adaptations related to their reproductive success in using other insect bodies that they parasitize. Mainly Hymenopteran parasitoids, and in a few cases, Dipteran parasitoids, have been reported in quinoa (Valoy et al. 2011) (Table 5.2) and parasitize egg, larval, and pupal stages (Fig. 5.2).

Predators

Predators belong to the most varied taxonomic groups of insects and possess diverse adaptations and behaviors for prey capture. From the ecological point of view, the most important predator species belong to the orders Neuroptera, Coleoptera, Diptera, Hemiptera, Hymenoptera, Odonata, and Mantodea (Fig. 5.2).

Coleopterans from the Coccinellidae and Carabidae families are among the predators of quinoa pests (Table 5.2). These families are composed of species that are considered of great importance in agriculture for pest control. Both larvae and adults are active predators of small herbivores; they are generally quite voracious and have high fertility. In Argentinian quinoa crops, green lacewings (Neuroptera: Chrysopidae), *Eriopis* sp., and *Eriopis connexa* (Germar) (Coleoptera: Coccinellidae) have been observed, all of them in the beginning of the flowering stage (Valoy et al. 2011). These data suggest that such predators might be preying on eggs and larvae of Lepidopteran species, such as *Eurysacca* sp., whose larvae are also abundant in this phenological stage of quinoa. Moreover, the presence of the predator *Hippodamia convergens* (Guérin-Méneville) (Coleoptera: Coccinellidae) coincides with aphid infestation during the milk and dough stages of quinoa (Valoy et al. 2011) (Fig. 5.2). In the Peruvian Andes, the abundance of predators fluctuates as plant growth progresses. This was observed in Coccinellidae (*H. convergens* and *E. connexa*) species, whose population peaked in the middle of crop development, whereas other predators such as Syrphidae were abundant during the latest stages of the crop (Yábar et al. 2002). In the Peruvian Altiplano, the following species of carabid predators were found preying on *E. melanocampta* larvae: *Notobia (Anisotarsus) schnusei* Van Emden, *Notobia (Anisotarsus) laevis bolivianus* Van Emden, and *Meotachys* sp. near harvest, during which time these carabid predator populations increase (Loza and Bravo 2001).

While phytophagous species prevail within the order Hemiptera, other entomophagous groups within this order have also been noted.

There is also an important number of species that comprise different beneficial families. The Lygaeidae (*Geocoris* sp.), Miridae (*Rhinacloa* sp.), and Nabidae (*Nabis* sp.) families have been reported in quinoa (Rasmussen 2003; Valoy et al. 2011) (Fig. 5.2). Popularly known as "assassin bugs," the adults as much as the nymphal stages are predators; they possess a pick or rostrum usually bent or folded toward the abdomen that they use to catch their prey. They are very active and voracious; their coloring camouflages them from their prey.

Entomopathogens

In natural pest control, entomopathogens can play an important role, especially when epizootics occur. However, little is known about the pathogens causing diseases on insects linked to quinoa. Even though it is generally known that most common insect diseases are caused by bacteria, fungi, viruses, protozoa, and nematodes, relatively few pathogen species have been identified compared with the number of entomophagous insects. However, there are some entomopathogens on which we have more information. For instance, *Bacillus thuringiensis* Morrisoni and the fungi *Beauveria bassiana* (Bals.-Criv.) Vuill. and *Metarhizium anisopliae* Sorokin attack diverse insect species. Entomopathogens have already been mass-produced for field application as microbiological insecticides or bioinsecticides. In quinoa, bioinsecticides with granulosis virus have been tested against *E. melanocampta*, reaching levels of 50% control (Calderón et al. 1996; Zanabria and Banegas 1997). Other viruses have been isolated from *E. melanocampta* larvae and tested as bioinsecticides, where preliminary screening has shown the presence of nucleopolyhedrosis virus (NPV) (Rasmussen unpublished data). One drawback of using entomopathogens is their inability to search for host, unlike entomophagous insects which can search for prey.

In summary, even though studies on biodiversity in quinoa crops in the Neotropical regions have been oriented toward identifying insect pest species, the presence of natural enemies have also been recorded. Groups belonging to 5 insect orders, 11 families, and 19 genera with 24 species have been reported in various studies and present a promising prospect for biocontrol of insect pests in quinoa.

Chemical responses of quinoa to insect herbivory

In this section, we introduce the concept of chemical interactions between plants and insects and discuss the chemical potential of quinoa to repel the attack of herbivorous insects. Many species of the *Chenopodium* genus possess secondary metabolites (SMs) with potential effects on herbivorous insects, such as flavonoids, glycosides, flavonols, phenolic amides, coumarins, alkaloids, lignin, phytosteroids (ecdysteroids), phenolics, and saponins (Verma and Agarwal 1985; Jam et al. 1990; Dinan 1992; Gee et al. 1993; Horio et al. 1993; Cuadrado et al. 1995; Berdegue and Trumble 1996; Gallardo et al. 2000; Hernández et al. 2000; Woldemichael and Wink 2001; Zhu et al. 2001a, 2001b; Hilal et al. 2004; DellaGreca et al. 2005; Cutillo et al. 2006; Paśko et al. 2008; Kokanova-Nedialkova et al. 2009; Kumpun et al. 2011).

Studies have been conducted to characterize the primary and SMs of quinoa, in order to explain their nutritional and medical properties including the chemical response of quinoa to unfavorable conditions during cultivation (González et al. 1989; Ruales and Nair 1993; Cuadrado et al. 1995; Dinan et al. 1998; Woldemichael and Wink 2001; Hilal et al. 2004; Costa et al. 2007; Paśko et al. 2008; Kokanova-Nedialkova et al. 2009; Jancurová et al. 2009; Kuljanabhagavad and Wink 2009; Rosa et al. 2009; Dowd et al. 2011; Kumpun et al. 2011; Campos et al. 2012). Information about entomofauna associated with quinoa cultivation is also available, especially concerning plant–insect interactions (Yábar et al. 2002; Rasmussen et al. 2003; Valoy et al. 2011; Campos et al. 2012) and chemical interactions (Costa et al. 2009a).

SMs present in quinoa cultivars have the potential to counteract damage caused by herbivores, such as phenolic acids (gallic acid, hydroxybenzoic acid, vanillic ester glucosidic acid, ferulic acid, cinnamic acid, and phytic acid), tannins, flavonoids (isoflavone, glycosylated kaempferol, rutin, and quercetin), as well as 20-hydroxyecdysone and saponins (De Simone et al. 1990; Bi et al. 1997; Lattanzio et al. 2000; Zhu et al. 2001a, 2001b; Sánchez-Hernández et al. 2004; Dini et al. 2004; Cutillo et al. 2006; Paśko et al. 2008; Kokanova-Nedialkova et al. 2009; Kumpun et al. 2011).

In general, plants possess biomechanical and biochemical strategies to resist and/or endure stress. Biomechanical strategies consist of tissue architecture (presence of glandular trichomes and high density of nonglandular trichomes), whereas biochemical strategies constitute metabolites synthesized during plant growth or in response to external stimuli. These strategies provide tissues with resistance, elasticity, hardness, protection, attractive colors, and smells for dispersion of fruit and flowers. The signal compounds also serve as stress warning responses when exposed to excess heavy metals, salinity, UV radiation, and changes in temperature and humidity (Bi and Felton 1995; Ramakrishna and Ravishankar 2011; Podazza et al. 2012; Gómez-Caravaca et al. 2012; Whitney and Feder 2013). These strategies allow quinoa plants to endure herbivore pressure, either because the tissue surface possesses structures that prevent adequate adhesion or tissues at the biochemical level may be unpleasant for herbivory. Volatile compounds, known as herbivore-induced plant volatiles (HIPV), may also be released as part of the biochemical strategies (Howe and Jander 2008; Ponzio et al. 2013).

Volatile compounds work against herbivores, attract parasitoids, and function as chemical signals that can "warn" neighboring plants of the presence of herbivores (Paré and Tumlinson 1999; Li et al. 2012; Ponzio et al. 2013). In this context, the plant's capacity to avoid and/or recover from herbivore damage will depend on herbivory intensity and frequency, phenology and plant health status, chemical signals received from neighboring plants, and abiotic factors such as photoperiod seasonal changes, bioavailability of nutrients, and water (Banerji 1980; Bi and Felton 1995; Lichtenthaler 1996; Rastrelli et al. 1998; Sánchez-Hernández et al. 2004; Costa et al. 2009a, 2009b; Wink 2006; Wise and Abrahamson 2007; Anttila et al. 2010).

The ability of plants to respond to biotic stress requires the redistribution of metabolic resources and nutrients at the expense of energy meant for development and performance (Nabity et al. 2006; Paśko et al. 2008; Wink 2009; Wink and Schimmer 2009; Gómez-Caravaca et al. 2012). In an agriculture-based economy, the deviation of resources in cultivated species from energy to defense against herbivory can lead to economic losses. In this context, it is extremely important to design pest control management systems that consider the ability of cultivated species to defend themselves against insect feeding without affecting yield and, at the same time, allow farmers to reduce the use of harmful pesticides (Griffiths 1999; Wink 2006).

Insects have developed different physiological and feeding behavior strategies adapted to plant defense mechanisms. They can select tissues to feed on based on the balance among chemical defenses, nutritional compounds, and the antioxidant content of plant tissues. This balance can enhance or decrease the toxicity of plant tissues in response to herbivory (Berdegue and Trumble 1996; Bi et al. 1997; Varanda and Pais 2006).

When classified according to behavioral strategies, insects can be generalists (polyphagous) that feed on plants from different families or specialists (monophagous and oligophagous) that feed on one or few plants belonging to the same family (Fürstenberg-Hägg et al. 2013). Specialist insects develop physiological strategies that allow them to perceive repellent or attractant compounds, select plants and plant tissues for feeding or oviposition, and metabolize plant toxins. In order to detect volatile compounds, they possess specific adaptations located in mouthparts and antennae, which function as chemoreceptors. For example, to avoid toxic compounds, they have specific elicitors in salivary secretions. Other insects use SMs ingested in food for survival and reproduction, and they even have digestive enzymes such as polyphenol oxidase, peroxidases, and oxidoreductases that can reduce the toxic effect of some metabolites (Simmonds 2003; Lattanzio et al. 2006) (Fig. 5.4).

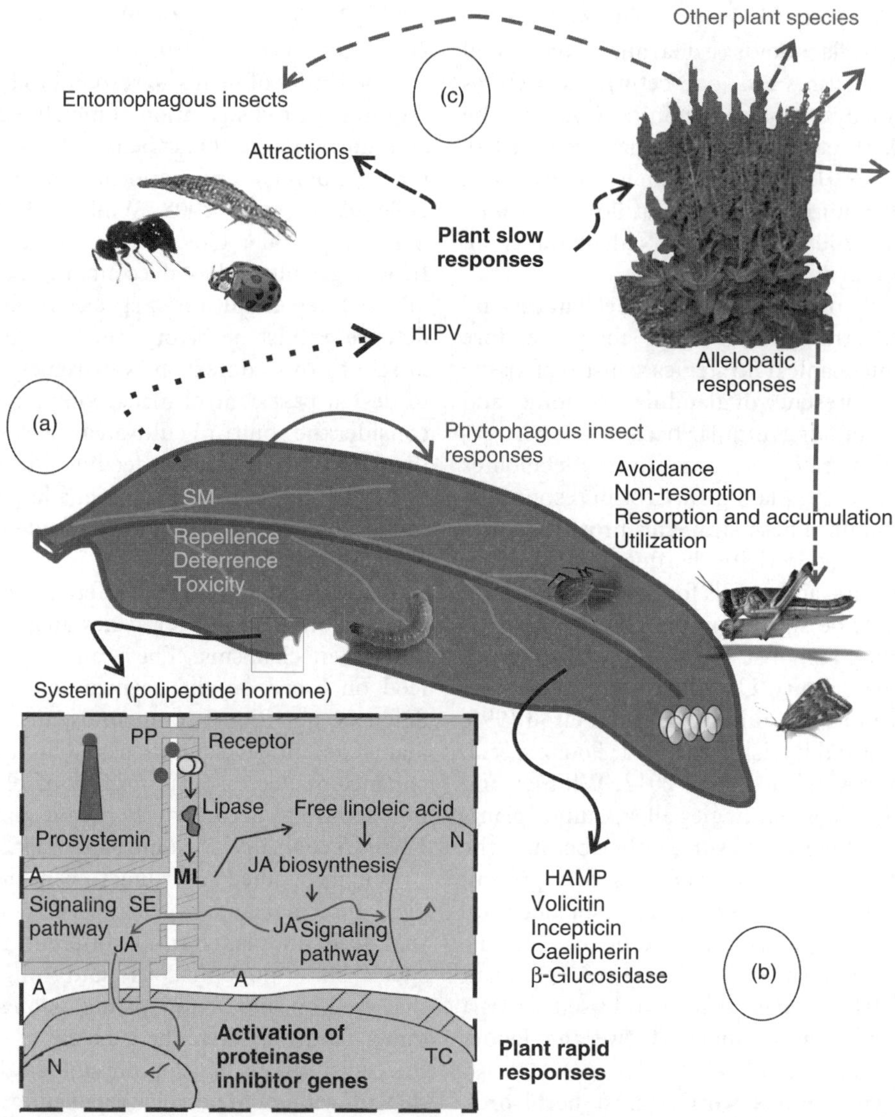

Fig. 5.4 Plant–insect biochemical interactions, with plant chemical responses after attack of phytophagous insects. (a) Function of inducible or constitutive secondary metabolites (SMs) such as phenolics, terpenes, and alkaloids; (b) plant rapid response related to damage, jasmonic acid signal, and herbivory-associated molecular patterns (HAMP); (c) plant slow response, herbivore-induced plant volatiles (HIPVs), monoterpenes act as signals to other plants and entomophagous insects.

In this respect, herbivorous species belonging to the genus *Eurisacca* would be considered specialists. However, their physiological characteristics should be studied more to expand our knowledge and to help develop management strategies at the molecular level to discourage attacks from these pests.

Quinoa secondary metabolites

Terpenoid compounds

Saponins are terpenoid compounds and could be considered as constitutive chemical defenses that deter herbivory by insects. These compounds have a structure composed of a lipophilic nucleus (steroid or triterpene) and a hydrophilic moiety (different glycosylation patterns), which defines their membranolytic properties affecting insect intestinal mucus and damage membrane functionality (Gee et al. 1993; Wink 2006; Kuljanabhagavad and Wink 2009; Dowd et al. 2011). Saponins are found in branches, leaves, flowers, and fruits of different quinoa varieties. However, it has been observed that saponin content fluctuates among varieties. For instance, varieties with yellow seeds such as "Marangani" have higher saponin content than those with white seeds, because of high hederagenin and oleanolic acid content. Differences in saponin concentration due to phytolaccagenic, deoxyphytolaccagenic, and serjanic acid content might also be influenced by water availability (Cuadrado et al. 1995; Woldemichael and Wink 2001; Yábar et al. 2002; Kokanova-Nedialkova et al. 2009; Gómez-Caravaca et al. 2012). Nevertheless, in spite of varying saponin content in quinoa varieties, insect pests would most likely be controlled by their natural enemies than by the presence of saponin (Yábar et al. 2002).

Other metabolites of the steroid type found in quinoa seeds are ecdysteroids, also called phytoecdysteroids. They are known as 20-hydroxyecdysone, makisterone, and kancollosterone (Dinan 1992) and are particularly predominant in quinoa, with amounts ranging between 450 and 1300 μg.g-1DW ecdysteroid equivalents reported (Dinan 1992; Dinan et al. 1998). The importance of ecdysteroids lies in their structural similarity to insect steroids, hormones that regulate biochemical and physiological processes associated with reproduction, embryo maturation, development, and metamorphosis (Dinan 2001; Thummel and Chory 2002). Phytoecdysteroids are present in numerous families of the plant kingdom and play a protective role against nonspecialist herbivores and also against soil nematodes. These compounds affect molting of nematodes and can even be lethal. The presence of phytoecdysteroids in quinoa has led to the development of breeding programs with the goal of increasing these compounds in quinoa lines and varieties. Aside from helping control pest outbreaks, these compounds also have nutritional benefits to humans. They can supplement the caloric intake of the human diet and have remarkable chemical stability during by-products manufacturing (Kumpun et al. 2011).

The terpenes found in quinoa are limonene, α-terpinolen, β-phellandrene, α-terpinene, aromatic monoterpenoids such as *p*-cymene, trans-carveol, carvone and pinocarvone, piperitone, and aromatic sesquiterpenes such as β-elemene and β-caryophyllene. Limonene, α-terpinolen, and β-phellandrene are accumulated in plant tissues against herbivory damage (Lerdau et al. 1994; Loughrin et al. 1994). Moreover, aromatic sesquiterpenes such as β-elemene and β-caryophyllene might be associated with HIPV response, as it is known that many volatile compounds can originate from membrane lipids (during the "quick response") or they can be terpenoid-type (during the "slow response"). In the "quick response" situation, volatile compounds derived from cell membrane fatty acids act as precursors of signals for the synthesis of protease inhibitors, which interfere with the digestion of plant tissues. For example, response to herbivory injury is triggered when tissues release a polypeptide hormone called systemin, which is then coupled to a plasma membrane receptor. A lipase is then induced to release linoleic acid molecules inside the cell, which are precursors of jasmonic acid synthesis. Jasmonic acid is a signal molecule that simultaneously activates genes for protease inhibitors. Thus, this suite of events makes the tissue unpalatable to insects (Sánchez-Hernández et al. 2004; Fig. 5.4). In the "slow response," volatile terpenes, usually monoterpenes, are released. The volatiles can attract predators and parasitoids or act as warning signals to neighboring plants so that they activate their defenses (Paré and Tumlinson 1999; Howe and Jander 2008; Li et al. 2012).

On the other hand, the salivary secretions of many orders, including Lepidoptera, Diptera, and Orthoptera, have a mixture of compounds related both to feeding and oviposition and also function as elicitors or effectors of plant reactions. Some of these compounds have been identified as β-glucosidase, violicitin, incepticin, and caelipherin and are useful to characterize herbivory-associated molecular patterns (HAMPs). Furthermore, it should be noted that microorganisms participate in this interaction, being present both in plant and insect surfaces and in insect salivary secretions, and also cause a signal sequence that is a component of the microorganism-associated molecular patterns (MAMPs). The joint action of compounds coming from plants, insects, and microorganisms can suppress the toxic effects of some compounds synthesized by plants and enable herbivory, as well as induce the release of volatile compounds that can serve as attractants and oviposition stimulants for parasitoids and as warning signals to neighboring plants. These reactions would also be linked to P450s, a versatile enzyme system present in plants and specialist insects, involved both in plant SM biosynthesis and in insect detoxification catabolic systems (Schuler 1996) (Fig. 5.4).

Phenolic compounds

Among the phenolic compounds in quinoa, some metabolites act as signaling molecules and are responsible for raising the alarm against stressors such as UV-B radiation, herbivores, and pathogens. Other phenolic compounds defend plant tissues against resulting oxidative stress. Within this group, there are some pigments, such as flavonoids and anthocyanins, that not only prevent UV-B damage but also act as optical signals to attract pollinators and dispersal vectors because of their bright colors (Bi and Felton 1995; Simmonds 2003; Lattanzio et al. 2006; Wink and Schimmer 2009; Prado et al. 2012).

Phenolic compounds accumulate in vacuoles and are linked to epidermal and sub-epidermal cell walls of leaves, stems, cutin, and wax deposits from outer surfaces of plant organs. Phenolic compounds can even form complexes with DNA molecules in order to provide oxidative protection (Simmonds 2003; Lattanzio et al. 2006). The role of phenolic content in plant–insect interactions is directly related to PAL activity (phenylalanine ammonia lyase), a key enzyme in the synthesis of phenolics. Therefore, analysis of PAL activity may serve as a response indicator to herbivory. Moreover, the phenolic profile in tissues may serve as an estimate of herbivory occurrence.

Although the anti-herbivory role of phenols and polyphenols has not been studied in quinoa, it has been reported that these compounds have anti-herbivory properties in Amaranthus, another genus belonging to the Chenopodiaceae family (Niveyro et al. 2013). On the other hand, quinoa seed has compounds with anti-herbivory properties, such as rutin, orientin, vitexin, morin, hesperidin, neohesperidin, cinamic acid, caffeic acid vanillic acid, and gallic acid (tannin synthesis) and kaempferol and quercetin glycosides (Dini et al. 2004; Paśko et al. 2008). For example, rutin acts as a phagostimulant, whereas kaempferol has genotoxic properties, acts as feeding deterrent, and can affect the emergence of aphids. Vanillic acid glucose ester acts with greater intensity on broods of aphids, whereas tannins decrease the palatability of plant tissues because of their astringent properties or hinder digestion by forming complexes with proteins and/or inactivate digestive enzymes (Bi et al. 1997; Lattanzio et al. 2000, 2006; Simmonds 2003; Paśko et al. 2008; Wink and Schimmer 2009; Steffensen et al. 2011). Phytic acid in quinoa chelates many ions and interferes with their incorporation into the insect diet (De Simone et al. 1990; Bi and Felton 1995; Bi et al. 1997; Zhu et al. 2001b; Dini et al. 2004; Paśko, et al. 2008; Wink and Schimmer, 2009; Steffensen et al. 2011; Niveyro et al. 2013).

In summary, there are various chemical compounds produced in quinoa as a response to herbivory. These present an exciting new area for study, especially assessing their potential use for biological control.

POTENTIAL OF BIOLOGICAL CONTROL IN QUINOA

At present, most of the information on biological control in the Andean quinoa crop comes from studies on entomophagous insects (parasitoids and predators). Cultural practices and host plant resistance are also important factors to consider in establishing a phytosanitary management protocol for quinoa. It is encouraging to know that in the field, up to 45% of parasitoid and predator species exert natural biological control in key quinoa pests such as the quinoa moth (*Eurysacca* sp.) and ticona complex (*Heliothis*, *Copitarsia*, and *Spodoptera* sp.) (Rasmussen et al. 2003). Fluctuations in key and potential pest populations in Andean crops follow a similar temporal pattern in different regions and are associated with crop phenology. However, relative densities of key pests, such as quinoa moths and ticona complex, exhibit variations between different Andean zones. In the Northern Altiplano, pest densities are low, with 1–6 larvae/plant; in the Central Altiplano, 7–15 larvae/plant have been recorded; and in the Southern Altiplano, pest densities have been recorded from 9 to over 45 larvae/plant (Saravia and Quispe 2005), indicating that these pests could be a limiting factor in the southern areas for quinoa production.

The percentage of parasitism on *E. melanocampta* in the field during a complete cycle of quinoa cultivation has been measured by Mamani (1998). The percentage of parasitism ranged from about 25% during the soft dough stage to 45% during the milk grain phenological phase, reaching 80% during physiological maturity (Mamani 1998). The main parasitoids of this Lepidopteran pest are *Copidosoma gelechiae* Howard (Encyrtidae) and *Diadegma* sp. (Ichneumonidae) (Rasmussen et al 2003).

In addition to parasitoids, natural biological control relies on a complex of insect predators that regulate the populations of phytophagous pests of quinoa. Among the predator insect groups, the Coccinellidae (Coleoptera) and Chrysopidae (Neuroptera) species have a more significant role in pest control in agroecosystems. The Chrysopidae is the most important predator genus used in biological control and can be bred on a large scale and commercialized for release in the field (Tauber et al. 2000). The larvae of these predator families are extremely active, fast moving, and have considerable ability to search for prey. They are polyphagous predator species and feed on a large number of economically important insect pests. They can also feed during different life stages of the insect prey, such as the egg, larvae (nymphs), and adult stages.

It is important to have the correct taxonomic identification of predator species in quinoa, particularly of Chrysopidae (Reguilón et al. 2006, 2009), as they are only mentioned at the genus level in studies conducted in the Neotropical region. Correct species identification is the first step to study the biology of natural enemies and their predation capacity to develop methods for mass rearing and design biological control systems to be applied in quinoa.

We followed the course of a quinoa crop in Tucuman, Argentina for two periods (from November 2008 to March 2010), which allowed us to establish the dynamics of two *Chrysoperla* species associated with quinoa (Reguilón, personal communication). *Chrysoperla externa* Hagen and *Chrysoperla argentina* González Olazo-Reguilón (González Olazo and Reguilón 2002) were collected throughout the crop phenology (Fig. 5.5). This is the first report of *C. argentina* in quinoa. As shown in Figure 5.5, the presence of *C. argentina* was evident on all crop survey dates. The highest predator population densities occurred in the months of November and March and coincided with the presence of herbivorous lepidopteran genus *Eurysacca* and Pyralidae family. Lacewings such as *Eriopis* sp. and *E. connexa* (Coleoptera: Coccinellidae) were also observed at the beginning of flowering and during the flowering stage, though in lower densities (Valoy et al. 2011). These data suggest that predators in these phenological stages were actively preying on eggs and larvae of the Lepidoptera species mentioned. Aphids were recorded in milky and dough stage, coinciding with the presence of the predator *H. convergens* (Valoy et al. 2011).

POTENTIAL FOR ECOLOGICAL MANAGEMENT OF QUINOA

The use of biological control agents (parasitoids, predators, or pathogens) is a necessary component of integrated pest management to regulate populations of insects that cause economic damage

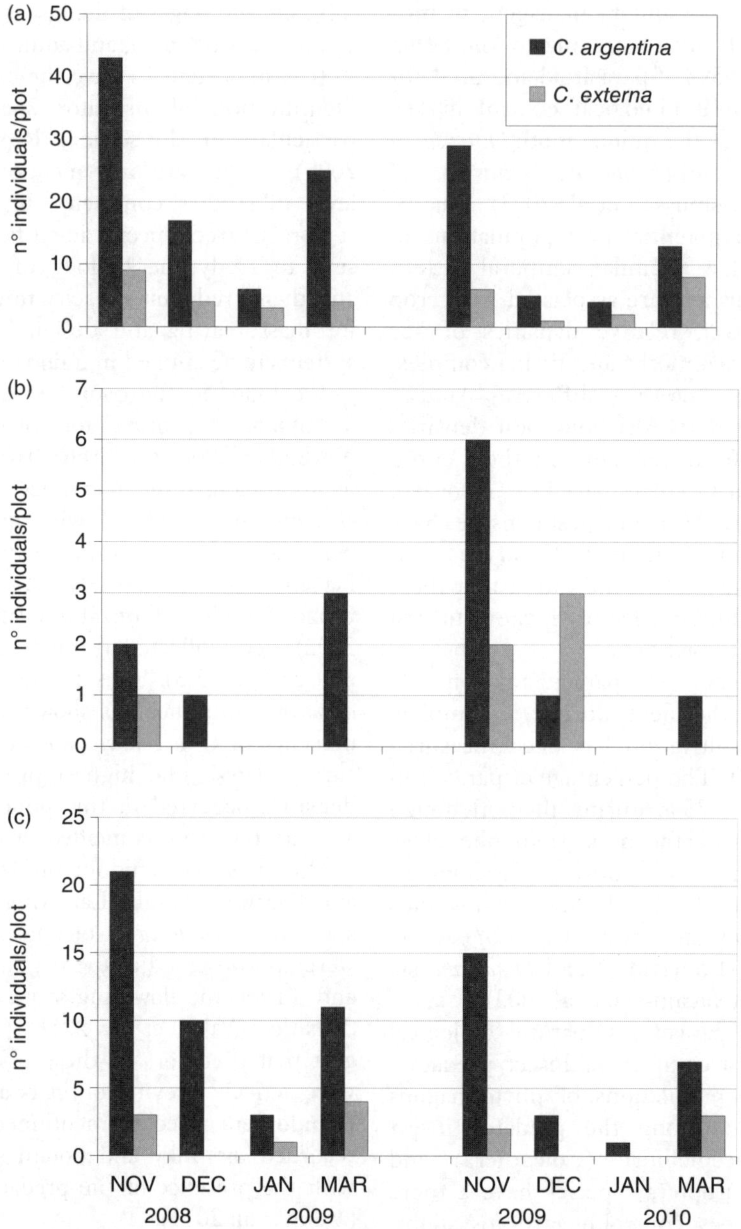

Fig. 5.5 *Chrysoperla* species associated with quinoa in Amaicha del Valle, Tucuman, Argentina. The population dynamics of *Chrysoperla externa* Hagen (light grey bar) and *Chrysoperla argentina* González Olazo-Reguilón (dark bar) species for two periods (from November 2008 to March 2010). A: eggs; B: larvae; C: adult.

to crops (Kogan and Shenk 2002; Romero 2004; Hruska 2008) and to reduce the use of agrochemicals. However, this practice frequently does not consist of a comprehensive agroecosystem management, which should, in all cases, include the application of ecological principles taking into account all key aspects of insect–plant interactions, besides the use of biocontrol agents. It is especially important to understand how the dynamics of species interactions may affect herbivore outbreaks and the expression of pest resistance within an agricultural system. In this context, one of the advantages of plants in natural populations is phenotypic and genetic heterogeneity, which translates into variability in expression of individual plant defenses (Whitham et al. 1984; Nyman 2010). Such heterogeneity can reduce the likelihood of herbivores developing counterdefenses, thus allowing plants to preserve their defense mechanisms over time (Letourneau 1997).

Ecological principles, from their origin in natural systems to their application in agroecosystems, take into consideration several concepts. One of these concepts, and perhaps the most comprehensive one, postulates preservation of biodiversity as the cornerstone for operation and sustainable use of agricultural systems (Altieri et al. 1983; Altieri 2009; Butler et al. 2007; Attwood et al. 2008) as it considers the preservation of beneficial biotic interactions and ecosystem services (Shennan 2008; Letourneau et al. 2009; Kremen and Miles 2012). In addition, agroecosystem management is compatible with the system community approach, which decentralizes attention focused solely on pests while understanding the temporal dynamics of nonpest species (Ordano et al. 2013) that are also part of the agroecosystem (Griffiths 1999; Wink 2006).

The use of intercropping or mixed cropping systems is one of the practices that has been proposed as a means of preserving biodiversity (Vandermeer 1989; Altieri and Nicholls 2004; Perfecto et al. 2009; Lithourgidis et al. 2011). Different cultivars or types of crops could be intercropped. The companion plants may or may not have commercial value (Parker et al. 2013) but act as repellents or as shelter for beneficial insects. This has been tested on numerous occasions and types of crops but with different results. In a review, Poveda et al. (2008) found that diversification of vegetation schemes tested in agriculture in the past decade served to reduce herbivore densities in approximately half of the cases included in the review.

In Latin America, a significant portion of the tropical crops are produced in intercropping or mixed cropping systems. However, this type of management is uncommon in the Andean cropping systems (Altieri 1999; Altieri and Toledo 2001). There are only a few cases where the effect of corn-and-potato mixed systems has been tested (Raymundo and Alcazar 1983; Thiery and Visser 1986; Lal 1991; Rhoades and Bebbington 1990; Silwana and Lucas 2002; Gianoli et al. 2006; Seran and Brintha 2010) and where several types of biological control have also been carried out (Weber 2012; Kroschel et al. 2012). The effect of agricultural intensification on changing the status of insect pests has also been explored in these mixed systems (Risch 1980; Trenbath 1993; Smith and McSorley 2000; Mojena et al. 2012).

Traditionally, quinoa crops are grown in monoculture or polyculture systems side by side with corn, bean, or potato crops that extend to areas where the vegetation is more diverse, consisting of eucalyptus, turnip, and Poaceae, Asteraceae, and Lamiaceae species. The placement of the quinoa crop in relation to other crops is important as it has been shown that in Cusco, Peru, the number of species of parasitoids of quinoa moth (*E. melanocampta*) present in the agroecosystem increased as the diversity of the surrounding crops also increased (Costa et al. 2009a, 2009b). The results of this study should spur further research on this topic so that it can be incorporated in pest management strategies.

Currently, several quinoa-producing countries are changing their crop management practices, gravitating toward monoculture, partially or completely replacing crop rotation and resting or fallow periods (Nieto-Cabrera et al. 1997; Fonte et al. 2012; Soto et al. 2012). The practice of sowing different species that are annually rotated in the same plot is also slowly being phased out

(Clades 1992). Increasing crop intensification of quinoa in Bolivia and Peru has replaced the existing quinoa cultural management, leading to problems of desertification, decline in soil fertility, and loss of natural vegetation. In Bolivia, the plowing of large tracts of land has led to loss of native vegetation such as "thola" (*Pharastrepia* sp.), which is used as feed for llamas. Thola may also be a significant element of the agroecosystem as it is a shelter for beneficial species. Moreover, this species possesses anti-herbivory properties. Together with another native species (*Minthostachys* sp.), *Pharastrepia* sp. is used to manufacture natural insecticides for application in organic quinoa production (Gallegos et al. 1982; Jaldin 2010).

Complementary to these management strategies, the intrinsic characteristics of quinoa should be known, such as its properties and mechanisms against insect pests and how these characteristics vary among selected cultivars. One of the mechanisms acting on herbivore populations is the emission of volatile compounds (terpenes and green leaf lipid-derived volatiles) that can repel herbivores, attract entomophagous insects (predators and parasitoids), or act as warning signals to neighboring plants. Quinoa has monoterpenes and sesquiterpenes (Whitman et al. 1990) that could be studied further to evaluate whether they participate in the exchange of chemical signals or as defense mechanism to avoid damage by herbivore feeding or oviposition. It has also been reported that these chemicals decrease the ability of the males and females of the quinoa moth to connect with each other (Costa et al. 2009a, 2009b). In addition, the study of phenolic composition of different quinoa varieties would help determine which varieties have the potential to be more repellent to insects and thus incur less damage.

Given the current demand for quinoa worldwide, it is particularly urgent to start implementing ecological crop management, with biological control as an integral component. Ecological management of the quinoa crop would also consider restoring traditional agricultural practices, such as the use of organic fertilizers, to mitigate soil degradation while taking into account that soil nutritional quality influences the ability of plants to withstand herbivore attacks (Altieri et al. 2012; Ghorbani et al. 2008).

In conclusion, it is vital to conduct research and promote the sustainable management of quinoa to preserve plant and animal biodiversity where quinoa is grown and to protect the biotic interactions between quinoa and other Andean crops. We are convinced that conscious efforts to manage the quinoa crop in an ecologically sustainable framework would preserve this Andean legacy for generations to come.

REFERENCES

Alata J. Lista de insectos y otros animales dañinos a la agricultura en Perú. Lima, Perú: Ministerio De Agricultura; 1973. p 177p.

Altieri MA. Agroecology, small farms, and food sovereignty. Mon Weather Rev 2009;61(3):102–113.

Altieri M, Ponti L, Nicholls CI. Soil fertility, biodiversity and pest management. In: Gur GM, Wratten SD, Snyder WE, editors. Biodiversity and insect pests: key issues for sustainable management. Willey-Blackwell: West Sussex, UK; 2012. p 72–84.

Altieri MA, Nicholls CI. Biodiversity and pest management in agroecosystems. 2nd ed. Boca Raton: CRC Press; 2004. p 236p.

Altieri MA, Letourneau DK. Vegetation management and biological control in agroecosystems. Crop Prot 1982;1:405–430.

Altieri MA, Toledo VM. The agroecological revolution in Latin America: rescuing nature, ensuring food sovereignty and empowering peasants. J Peasant Stud 2001;38(3):587–612.

Altieri MA. Applying agroecology to enhance the productivity of peasant farming systems in Latin America. Environ Dev Sustain 1999;1(3–4):197–217.

Altieri MA, Letourneau DK, Davis JR. Developing sustainable agroecosystems. Bioscience 1983;33:45–49.

Andow DA. Vegetational diversity and arthropod population response. Ann Rev Entom 1991;36:561–586.

Anttila U, Julkunen-Tiitto R, Rousi M, Yang S, Rantala MJ, Ruuhola T. Effects of elevated ultraviolet-B radiation on a plant–herbivore interaction. Oecologia 2010;164:163–175.

Aroni G. Manejo y producción actual de quinua en Bolivia. In: Jacobsen SE, Portillo Z, editors. Primer taller internacional sobre quinua - recursos genéticos y sistemas de producción. UNALM: Perú; 2000. p 10–14.

Attwood SJ, Maron M, House AP, Zammit C. Do arthropod assemblages display globally consistent responses to intensified agricultural land use and management? Global Ecol Biogeo 2008;17:585–599.

Banerji N. Two new saponins from the root of *Amaranthus spinosus*. J Indian Chem Soc 1980;57(4):417–419.

Barrientos Zamora R. 1985. Dinámica poblacional y ciclos biológicos de insectos en quinua (*Chenopodium quinoa*). Cochabamba (Bolivia): Universidad Mayor de San Simón, Cochabamba (Bolivia), Facultad de Ciencias Agrícolas y Pecuarias Martín Cárdenas. 113p.

Benton TG, Vickery JA, Wilson JD. Farmland biodiversity: is habitat heterogeneity the key? Trend Ecol Evol 2003;18:182–188.

Berdegue M, Trumble JT. Effects of plant chemical extracts and physical characteristics of *Apium graveolens* and *Chenopodium murale* on host choice by *Spodoptera exigua* larvae. Entom Exp Appl 1996;78:253–262.

Bi JL, Felton GW. Foliar oxidative stress and insect herbivory: primary compounds, secondary metabolites, and reactive oxygen species as components of induced resistance. J Chem Ecol 1995;21:1511–1530.

Bi JL, Felton GW, Murphy JB, Howles PA, Dixon RA, Lamb CJ. Do plant phenolics confer resistance to specialist and generalist insect herbivores? J Agric Food Chem 1997;45:4500–4504.

Bianchi F, Booij C, Tscharntke T. Sustainable pest regulation in agricultural landscapes: a review on landscape composition, biodiversity and natural pest control. Proc Royal Soc B 2006;273:1715–1727.

Blanco MC. Evaluación de daños de *Scrobipalpula* sp. y *Perisoma* sp. en el cultivo de quinua en la zona del Cusco. In: Resumen del III Congreso Internacional de Cultivos Andinos. Bolivia: La Paz; 1982. p 133–135.

Bravo R, Delgado P. Colección de insectos en papa, quinoa y pastos cultivados. Puno, Perú: PIWA: Convenio PELT/NADEIC/COTESU; 1992. p 44.

Butler SJ, Vickery JA, Norris K. Farmland biodiversity and the footprint of agriculture. Science 2007;315:381–384.

Cabrera Almendros A, Oliveira NC. 2011. Consumo foliar e sobrevivência de *Diabrotica speciosa* em diferentes linhagens de quinoa. Campo Digital. 6(1):1–6. Available from: http://www.revista.grupointegrado.br/revista/index.php/campodigital/article/view/795/368.

Calderón R, Becerra A, Marquez A. 1996. Expectativas del virus granulosis Phthorimaea operculella para el control de otros lepidopteros. IV Reunión Nacional de la Papa: Compendio de Exposiciones, 8–11 Oct. Cochabamba, Bolivia. pp. 107–108.

Campos E, Bravo R, Valdivia R, Soto J. Plagas insectiles en áreas de intensificación de quinua en Puno. CienciAgro 2012;2(3):379–390.

Chambilla C, Gonzales MA, Jarandilla C, Baltazar B. Estudio de la fluctuación poblacional del complejo ticonas de la quinua (*Chenopodium quinoa* Willd.) bajo condiciones actuales de cambio climático. In: Prácticas y estrategias en respuesta a riesgos climáticos y de mercado en agroecosistemas vulnerables de la Región Andina. La Paz, Bolivia: PROYECTO SANREM′-CRSP; 2009. p 14.

CIRNMA (Centro De Investigación De Recursos Naturales Y Medio Ambiente). 2009. Informe Anual Proyecto ALTAGRO: Programa Orgánico De Quinua. Puno, Perú.

[CLADES]. The Latin American Consortium on Agroecology and Development. Revista Agroecológica y Desarrollo N° 4. Sistemas Agrícolas alternativos. 1992. In Manejo Ecológico del suelo. 1997. CLADES-CIED. Lima-Perú.

Costa F, Cardenas Molina M, Yábar E. 2007. Insectos plaga y enemigos naturales asociados al cultivo de la quinoa (*Chenopodium quinoa* Willd.) en Cusco, Perú. Resumen II, Resumen del Congreso Internacional De La Quinoa, 23–26 oct. Iquique, Chile.

Costa JF, Cosio W, Cardenas Molina M, Yábar E, Gianoli E. Preference of quinoa moth *Eurysacca melanocampta* Meryck (Lepidoptera: Gelechiidae) for two varieties of quinoa (*Chenopodium quinoa* Willd) in olfactometry assays. Chilean J Agr Res 2009a;69:71–78.

Costa JF, Yábar E, Gianoli E. Parasitism on *Eurysacca melanocampta* Meyrick (Lepidoptera: Gelechiidae) in two localities at Cusco, Peru. Revista Facultad Nacional De Agron 2009b;62(1):4807–4813.

Cuadrado C, Ayet G, Burbano C, Muzquiz M. Occurrence of saponins and sapogenols in Andean crops. J Sci Food Agr 1995;67:169–172.

Cutillo F, DellaGrecal M, Gionti M, Previtera L, Zarrelli A. Phenols and lignans from *Chenopodium album*. Phytochem Analysis 2006;17(5):344–349.

Dangles O, Carpio FC, Villares M, Yumisaca F, Liger B, Rebaudo F, Silvain JF. Community-based participatory research helps farmers and scientists to manage invasive pests in the Ecuadorian Andes. Ambiom 2010;39:325–335.

Dangles O, Mesias V, Crespo-Perez V, Silvain JF. Crop damage increases with pest species diversity: evidence from potato tuber moths in the tropical Andes. J Appl Ecol 2009;46:1115–1121.

Debach P, Rosen D. Biological control by natural enemies. 2nd ed. New York: Cambridge University Press; 1991. p 386p.

De Simone F, Dim A, Pizza C, Satuminos P, Schettino O. Two flavonol glycosides from *Chenopodium quinoa*. Phytochem 1990;29:3690–3692.

DellaGreca M, D'Abrosca B, Fiorentino A, Previtera L, Zarrelli A. Structure elucidation and phytotoxicity of ecdysteroids from *Chenopodium album*. Chem Biodivers 2005;2(4):457–462.

Dinan L, Whiting P, Scott AJ. Taxonomic distribution of phytoecdysteroids in seeds of members of the Chenopodiaceae. Biochem Syst Ecol 1998;26:553–576.

Dinan L. The association of phytoecdysteroids with flowering in fat hen, *Chenopodium album*, and other members of the Chenopodiaceae. Experientia 1992;48:305–308.

Dinan L. Phytoecdysteroids: biological aspects. Phytochemistry 2001;57:325–339.

Dini I, Tenore GC, Dini A. Phenolic constituents of Kancolla seeds. Food Chem 2004;84:163–168.

Dowd PF, Berhow MA, Johnson ET. Differential activity of multiple saponins against omnivorous insects with varying feeding preferences. J Chem Ecol 2011;37:443–449.

FAO Food And Agricultural Organization. 1993. http://www.Rlc.Fao.Org/Es/Agricultura/Produ/Cdrom/Contenido/Libro14/Cap2.3.Htm.

Farwig N, Bailey D, Bochud E, Herrmann JD, Kindler E, Reusser N, Schüepp C, Schmidt-Entling MH. Isolation from forest reduces pollination, seed predation and insect scavenging in Swiss farmland. Landscape Ecol 2009;24:919–927.

Fonte SJ, Vanek SJ, Oyarzun P, Parsa S, Carolina Quintero D, Rao IM, Lavelle P. Four pathways to agroecological intensification of soil fertility management by smallholder farmers in the Andean highlands. In: Sparks DL, editor. Advances in agronomy. New York: ; 2012. p 116–125.

Fürstenberg-Hägg J, Zagrobelny M, Bak S. Plant defense against insect herbivores. Int J Mol Sci 2013;14:10242–10297.

Gallardo M, González JA, Prado FE. Presence of betalains in *Chenopodium quinoa* Willd. seedlings. Lilloa 2000;40(1):109–113.

Gallegos J, Díaz J, Jayo E. 1982. Evaluación de la planta repelente "Muña" (*Minthostachys* sp.) en el control de *Scrobipulpa* sp. (Lepidoptera: Gelechiidae) en quinoa. In: Resumen de la XXV Convención Nacional De Entomología, 3–7 de Oct, Huaraz, Perú. pp. 35–36.

Gee JM, Price KR, Ridout CL, Wortley GM, Hurrel RF, Johnson IT. Saponins of quinoa (*Chenopodium quinoa*): effects of processing on their abundance in quinoa products and their biological effects on intestinal mucosal tissue. Sci Food Agr 1993;63:201–209.

Ghorbani R, Wilcockson S, Koocheki A, Leifert C. Soil management for sustainable crop disease control: a review. Environ Chem Lett 2008;6(3):149–162.

Gianoli E, Ramos I, Alfaro-Tapia A, Valdéz Y, Echegaray ER, Yábar E. Benefits of a maize–bean–weeds mixed cropping system in Urubamba Valley, Peruvian Andes. Int J Pest Manage 2006;52(4):283–289.

Gómez-Caravaca AM, Ianfelice G, Lavini A, Pulvento C, Fiorenza Carboni M, Marconi E. Phenolic compounds and saponins in quinoa samples (*Chenopodium quinoa* Willd.) grown under different saline and nonsaline irrigation regimens. J Agr Food Chem 2012;60:4620–4627.

Gonzalez JA, Roldan A, Gallardo M, Escudero T, Prado FE. Quantitative determinations of chemical compounds with nutritional value from Inca crops: *Chenopodium quinoa* ('quinoa'). Plant Food Hum Nutr 1989;39(4):3.

González Olazo EV, Reguilón C. Una nueva especie de *Chrysoperla* (Neuroptera: Chrysopidae) para la Argentina, Revista de la Sociedad Entomológica Argentina ISSN 0373–5680. Volumen 2002;61(1–2):47–50.

Griffiths JT. 1999. Predispersal seed predation of Amaranthus spp. and *Chenopodium album* (L). Thesis of faculty of graduate studies of the University of Guelph [Internet]. [cited 2013 mar 10] pp 93. Available from: http://www.collectionscanada.gc.ca/obj/s4/f2/dsk1/tape2/PQDD_0017/MQ47328.pdf.

Harris MK. Arthropod-plant interactions related to agriculture, emphasizing host plant resistance. In: Harris MK, editor. Biology and breeding for resistance to arthropods and pathogens in agricultural plants. Texas: Texas Agricultural Experiment Station, A&M University, College Station; 1980. p 23–51.

Hernández NE, Tereschuk MLK, Abdala LR. Antimicrobial activity of flavonoids in medicinal plants from Tafí del Valle (Tucumán, Argentina). J Ethnopharmacol 2000;73:317–322.

Hidalgo W, Jacobsen SE. 2000. Principales plagas del cultivo de la quinua en la sierra central del Perú y las perspectivas de control integrado. In: Resumen del primer taller internacional sobre quinua-recursos genéticos y sistemas de producción, 10–14 May. Lima, Perú. pp. 49–50.

Hilal M, Parrado MF, Rosa M, Gallardo M, Orce L, Massa EM, Gonzalez JA, Prado FE. Epidermal lignin deposition in quinoa cotyledons in response to UV-B radiation. Photochem Photobiol 2004;79(2):205–210.

Horio T, Yoshida K, Kikuchi H, Kawabata J, Mizutani J. A phenolic amide from roots of *Chenopodium album*. Phytochem 1993;33:807–808.

Howe GA, Jander G. Plant immunity to insect herbivores. Ann Rev Plant Biol 2008;59:41–66.

Hruska AJ. Nuevos temas en la transferencia de tecnologías de manejo integrado de plagas para productores de bajos recursos. Manejo Integrado de Plagas en Mesoamerica: Aportes Conceptuales 2008;32:265.

Jacobsen SE. The situation for quinoa and its production in Southern Bolivia: from economic success to environmental disaster. J Agron Crop Sci 2011;197(5):390–399.

Jaldin R. 2010. Producción de quinua en Oruro y Potosí. La Paz, Bolivia. Programa de investigación estratégica en Bolivia (PIEB).100 p. Available from: http://www.pieb.com.bo/UserFiles/File/enlinea/pd_quinua_jaldin.pdf.

Jam N, Alam MS, Kamil M, Ilyas M, Niwa M, Sakae A. Two flavonol glycosides from *Chenopodium ambrosioides*. Phytochemistry 1990;29:3988–3991.

Jancurová M, Minarovičová L, Dandár A. Quinoa – a review. Czech J Food Sci 2009;27(2):71–79.

Kogan M, Shenk M. Conceptualización del manejo integrado de plagas en escalas espaciales y niveles de integración más amplios. Manejo Integrado de Plagas y Agroecología Costa Rica 2002;65:34–42.

Kokanova-Nedialkova Z, Nedialkov PT, Nikolov SD. The genus Chenopodium: phytochemistry, ethnopharmacology and pharmacology. Pharmacognosy Rev 2009;3(6):280–306.

Kuljanabhagavad T, Wink M. Biological activities and chemistry of saponins from *Chenopodium quinoa* Willd. Phytochem Rev 2009;8:473–490.

Kumpun S, Annick M, Crouzet S, Evrard-Todeschi N, Girault JP, Lafont R. Ecdysteroids from *Chenopodium quinoa* Willd., an ancient Andean crop of high nutritional value. Food Chem 2011;125:1226–1234.

Kraljevic BGJ. Andean grains and tubers: ancient crops for better livelihoods today. In: Bala Ravi S, Hoeschle-Zeledon I, Swaminathan MS, Frison E, editors. Hunger and poverty: the role of biodiversity – report of an international consultation on the role of biodiversity in achieving the UN Millenium Development goal of freedom from hunger and poverty. M.S. Swaminathan Research Foundation: Chennai, India; 2006. p 113.

Kremen C, Miles A. Ecosystem services in biologically diversified versus conventional farming systems: benefits, externalities, and trade-Offs. Ecol Soc 2012;17(4):40.

Kroschel J, Mujica N, Alcazar J, Canedo V, Zegarra O. Developing integrated pest management for potato: experiences and lessons from two distinct potato production systems of Peru. In: Sustainable potato production: global case studies. The Netherlands: Springer; 2012. p 419–450.

Lal L. Effect of intercropping on the incidence of potato tuber moth, *Phthorimaea operculella* (Zeller). Agr Ecosyst Environ 1991;36:185–190.

Lamborot L, Guerrero MA, Araya JE. Lepidópteros asociados al cultivo de quinoa (*Chenopodium quinoa* Willd.) en la zona central de Chile. Boletín De Sanidad Vegetal: Plagas 1999;25:203–207.

Larrain P. Biologia de *Copitarsia turbata* (Lepidoptera: Noctuidae) bajo ambiente controlado. Agricultura Tecnica 1996;56:220–223.

Lattanzio V, Arpaia S, Cardinali A, Di Venere D, Linsalata V. Role of endogenous flavonoids in resistance mechanism of vigna to aphids. J Agr Food Chem 2000;48:5316–5320.

Lattanzio V, Lattanzio VMT, Cardinali A. Role of phenolics in the resistance mechanisms of plants against fungal pathogens and insects. In: Imperato F, editor. Phytochemistry: advances in research. Research Signpost: Kerala; 2006. p 23–67.

Lerdau M, Litvak M, Monson R. Plant chemical defense: monoterpenes and the growth-differentiation balance hypothesis. Tree 1994;9(2):58–61.

Li T, Holopainen JK, Kokko H, Tervahauta AI, Blande JD. Herbivore-induced aspen volatiles temporally regulate two different indirect defences in neighbouring plants. Funct Ecol 2012;26:1176–1185.

Lichtenthaler HK. Vegetation stress: an introduction to the stress concept in plants. J Plant Physiol 1996;148:4–14.

Loughrin JH, Manukian A, Heath RR, Turlings TCJ, Tumlinson JH. Diurnal cycle of emission of induced volatile terpenoids by herbivore-injured cotton plants. Proc Natl Acad Sci U S A 1994;91:11836–11840.

Letourneau DK, Jedlicka JA, Bothwell SG, Moreno CR. Effects of natural enemy biodiversity on the suppression of arthropod herbivores in terrestrial ecosystems. Ann Rev Ecol Evol Syst 2009;40:573–592.

Letourneau DK. Plant-arthropod interactions in agroecosystems. In: Jackson LE, editor. Ecology in agriculture. San Diego, CA: Academic Press; 1997. p 239–290.

Lithourgidis AS, Dordas CA, Damalas CA, Vlachostergios DN. Annual intercrops: an alternative pathway for sustainable agriculture. Aust J Crop Sci 2011;5:396–410.

Loza LA, Bravo R. Poblaciones de carábidos (Coleoptera) en agroecosistemas del Altiplano Peruano. Revista Peruana De Entomología 2001;42:79–87.

Mamani D. Control Biológico en forma natural de la Polilla de la Quinoa (*Eurysacca melanocampta Meyrick*) por Parasitoides y Perspectivas de Cría para su manipulación en el Altiplano Central. Tesis Ing. Agr. Facultad de Agronomía, UMSA. 1998. La Paz. Bolivia. 91 pp.

Mamani Quispe JR. 2009. Evaluar la dinámica poblacional de la polilla de la quinua (*Eurysacca melanocampta* Meyrick) y complejo ticona en cuatro variedades de quinua en la comunidad Chinchaya del Departamento de La Paz, Bolivia. 84 p. Tesis n° 22. UMSA, Facultad de Agronomía.

Matson PA, Parton WJ, Power A, Swift MJ. Agricultural intensification and ecosystem properties. Science 1997;277:504.

Médiène S, Valantin-Morison M, Sarthou JP, De Tourdonnet S, Gosme M, Bertrand M, Estarde J, Aubertot A, Rusch N, Motisi C, et al. Agroecosystem management and biotic interactions: a review. Agron Sustain Dev 2011;31(3): 491–514.

Mojena M, Bertolí M, Zaffaroni E. Evaluaciones de plagas insectiles en agroecosistemas de intercalamiento de maiz (*Zea mays*, L) y frijol (*Phaseolus vulgaris*, L) con yuca (*Manihot esculenta*, Crantz). Curr Agr Sci Tech 2012;6(1).

Mujica A. Manual del cultivo de quinua. Proyecto TTA-AID-INIA, PIWA. 1993. 32–46 pp. Lima, Peru.

Nabity PD, Heng-Moss TM, Higley LG. Effects of insect herbivory on physiological and biochemical (oxidative enzyme) responses of the halophyte *Atriplex subspicata* (Chenopodiaceae). Environ Entom 2006;35:1677–1689.

Nieto-Cabrera C, Francis C, Caicedo C, Gutiérrez PF, Rivera M. Response of four Andean crops to rotation and fertilization. Mt Res Dev 1997;17:273–282.

Niveyro SL, Mortensen AG, Fomsgaard IS, Salvo A. Differences among five amaranth varieties (Amaranthus spp.) regarding secondary metabolites and foliar herbivory by chewing insects in the field. Arth-Plant Int 2013;7:235–245.

Nyman T. To speciate, or not to speciate? Resource heterogeneity, the subjectivity of similarity, and the macroevolutionary consequences of niche width shifts in plant feeding insects. Biol Rev 2010;85(2):393–411.

Ochoa Vizarreta R, Franco JF. Morfología y biología de la polilla de la quinua *Eurysacca melanocampta* Meyrick 1917 (Lepidoptera: Gelechiidae), De Cusco (Perú). Bioma 2013;1(4):35–38.

Ochoa R. 1990. Ciclo biológico de la polilla de la quinoa (*Eurysacca melanocampta* Meyrick). Thesis, Cusco (Perú) UNSAAC, Thesis For Biologie, UNSAAC, Cusco, Perú.

Oerke EC, Dehne HW, Schonbeck F, Weber A. Crop production and crop protection: estimated losses in major food and cash crops. Amsterdam: Elsevier Science; 1994. p 808p.

Ordano M, Guillén L, Rull J, Lasa R, Aluja M. Temporal dynamics of diversity in a tropical fruit fly (Tephritidae) ensemble and their implications on pest management and biodiversity conservation. Biodiver Conserv 2013;22(6–7):1557–1575.

Ortiz R, Sanabria E. Plagas. En: Quinua y kañiwa cultivos Andinos. Editorial IICA, Bogotá, Colombia. Serie Libros Y Materiales Educativos; 1979. p 121–136.

Ortíz Romero R. [Internet]. 1993. Insectos plaga en quinua. Available from: http://www.rlc.fao.org/es/agricultura/produ/cdrom/contenido/libro14/cap2.3.htm

Ortiz R. 1997. Plagas de cultivos Andinos. En: 2do. Seminario Internacional De Especies Andinas, Una Riqueza No Explotada Por Chile, Calama, Chile.

Palacios M, Tenorio J, Vera M, Zevallos F, Lagnaoui A. 1999. Population dynamics of the Andean potato tuber moth, *Symmetrischema tangolias* (Gyen), in three different agroecosystems in Peru. Lima, Peru: International Potato Center Program Report 1997–1998. pp. 153–160.

Paré PW, Tumlinson JH. Plant volatiles as a defense against insect herbivores. Plant Physiol 1999;121:325–331.

Parker JE, Snyder WE, Hamilton GC, Rodríguez-Saona C. 2013. Companion planting and insect pest control. In: Soloneski S, Larramendy M, editors. Agricultural and biological sciences weed and pest control - conventional and new challenges. Intech 10.5772/55044

Parsa S. Native herbivore becomes key pest alter dismantlement of a traditional farming system. Am Entom 2010;56(4):242–251.

Paśko P, Sajewicz M, Gorinstein S, Zachwiejal Z. Analysis of selected phenolic acids and flavonoids in *Amaranthus cruentus* and *Chenopodium quinoa* seeds and sprouts by HPLC. Acta Chromat 2008;20(4):661–672.

Perfecto I, Vandermeer JH, Wright AL. Nature's matrix: linking agriculture, conservation and food sovereignty. London, UK: Earthscan; 2009. p 233.

Podazza G, Arias M, Prado FE. Cadmium accumulation and strategies to avoid its toxicity in roots of the citrus rootstock Citrumelo. J Hazard Mater 2012;215–216:83–89.

Ponzio C, Gols R, Pieterse CMJ, Dicke M. Ecological and phytohormonal aspects of plant volatile emission in response to single and dual infestations with herbivores and phytopathogens. Funct Ecol 2013;27:587–598.

Poveda K, Gomez MI, Martinez E. Diversification practices: their effect on pest regulation and production. Revista Colombiana De Entomologia 2008;34:131–144.

Povolny D, Valencia L. Una palomilla de papa nueva para Colombia. Memorias Del Curso Sobre Control Integrado De Plagas De Papa, Bogota, Colombia 1986;113:33–35.

Povolny D. Gnorimoschemini of southern South America. II: The genus *Eurysacca* (Lepidoptera: Gelechiidae). Steenstrupia 1986;12(1):1–47.

Povolny D. *Eurysacca quinoae* sp. N. - a new quinoa-feeding species of the tribe Gnorimoschemini (Lepidoptera: Gelechiidae) from Bolivia. Steenstrupia 1997;22:41–43.

Prado FE, Rosa M, Prado C, Podazza G, Interdonato R, González JA, Hila M. UV-B radiation, its effects and defense mechanisms in terrestrial plants. In: Ahmadand P, Prasad MNV, editors. Environmental adaptations and stress tolerance of plants in the era of climate change. New York: Springer; 2012. p 57–84.

Ramakrishna A, Ravishankar GA. Influence of abiotic stress signals on secondary metabolites in plants. Plant Signal Behav 2011;6(11):1720–1731.

Rasmussen C, Jacobsen SE, Lagnaoui A, Esbjerg P. 2000. Plagas de quinua (*Chenopodium quinoa* Willd.) en la zona Andina. In: II Congreso Internacional De Agricultura En Zonas Aridas, 16–21 oct, Iquique, Chile. 42 p.

Rasmussen C, Jacobsen SE, Lagnaoui A. Las polillas de la quinoa (*Chenopodium quinoa* Willd) en el Perú: *Eurysacca* (Lepidoptera: Gelechiidae). Revista Peruana De Entomología 2001;42:57–59.

Rasmussen C, Lagnaoui A, Esbjerg P. Advances in the knowledge of quinoa pests. Food Rev Int 2003;19:61–75.

Rastrelli L, Aquino R, Abdo S, Proto M, De Simone F, De Tommasi N. Studies on the constituents of *Amaranthus caudatus* leaves: isolation and structure elucidation of new triterpenoid saponins and ionol-derived glycosides. J Agr Food Chem 1998;46:1797–1804.

Raymundo SA, Alcazar J. 1983. Effects of polyculture (mixed cropping) on the incidence and severity of potato pests and diseases (*Solanum tuberosum*, Diabrotica spp., *Phytophthora infestans*). In: International Congress Research for the Potato in the Year 2000, 22–27 Feb, Lima, Peru.

Reguilón C, González Olazo EV, Nuñez Campero SR. Morfología de los estados inmaduros de *Chrysoperla argentina* (Neuroptera: Chrysopidae). Acta Zoológica Lilloana 2006;50(1–2):31–39.

Reguilón C, Valoy M, González JA. 2009. Neurópteros asociados a un cultivo experimental de quinoa (*Chenopodium quinoa* Willd) en una zona de valles intermontanos (Tucumán, Argentina). En: XXXII Congreso Argentino de Horticultura, 23 al 26 de Sept, Salta, Argentina.

Rhoades RE, Bebbington AJ. Mixing it up: variations in Andean farmers' rationales for intercropping of potatoes. Field Crops Res 1990;25(1):145–156.

Risch S. The population dynamics of several herbivorous beetles in a tropical agroecosystem: the effect of intercropping corn, beans and squash in Costa Rica. J Appl Ecol 1980;17:593–611.

Robinson RA, Sutherland WJ. Postwar changes in arable farming and biodiversity in Great Britain. J Appl Ecol 2002;39(1):157–176.

Romero F. 2004. Manejo integrado de plagas - las bases, los conceptos, su mercantilización. Universidad Autónoma Chapingo. Colegio de Postgraduados-Instituto de Fitosanidad. Montecillo.

Root R. Organization of a plant-arthropod association in simple and diverse habitats: the fauna of collards (*Brassica oleracea*). Ecol Monogr 1973;43:96–124.

Rosa M, Hilal M, González JA, Prado FE. Low-temperature effect on enzyme activities involved in sucrose–starch partitioning in salt-stressed and salt-acclimated cotyledons of quinoa (*Chenopodium quinoa* Willd.) seedlings. Plant Physiol Biochem 2009;47:300–307.

Ruales J, Nair BM. Saponins, phytic acid, tannins and protease inhibitors in quinoa (*Chenopodium quinoa*, Willd) seeds. Food Chem 1993;48:137–143.

Sánchez G, Vergara C. 1991. Plagas de los cultivos Andinos. Universidad Agraria La Molina, Departamento de Entomología, Lima, Perú. 46 p.

Sánchez-Hernández C, Martínez-Gallardo N, Guerrero-Rangel A, Valdés-Rodríguez S, Délano-Frier J. Trypsin and a-amylase inhibitors are differentially induced in leaves of amaranth (*Amaranthus hypochondriacus*) in response to biotic and abiotic stress. Physiol Plantarum 2004;122:254–264.

Saravia R, Quispe R. 2005. Manejo integrado de las plagas insectiles del cultivo de la quinua. Fascicle 4 In: PROINPA y FAUTAPO Ed. Serie de Módulos Publicados en Sistemas de Producción Sostenible en el Cultivo de la Quinua: Módulo 2. Manejo agronómico de la Quinua Orgánica. Fundación PROINPA, Fundación AUTAPO, Embajada Real de los Países Bajos. La Paz, Bolivia. pp. 53-86.

Schuler MA. The role of cytochrome P450 monooxygenases in plant-insect interactions. Plant Physiol 1996;112:1411–1419.

Simmonds MSJ. Flavonoid–insect interactions: recent advances in our knowledge. Phytochemistry 2003;64:21–30.

Seran TH, Brintha I. Review on maize based intercropping. J Agron 2010;9(3):135–145.

Shennan C. Biotic interactions, ecological knowledge and agriculture. Phil Trans Royal Soc B-Biol Sci 2008;363(1492): 717–739.

Sigsgaard L, Jacobsen SE, Christiansen JL. Quinoa, (*Chenopodium quinoa*) provides a new host for native herbivores in northern Europe: case studies of the moth, *Scrobipalpa atriplicella*, and the tortoise beetle, *Cassida nebulosa*. J Insect Sci 2008;8:50–54.

Silwana TT, Lucas EO. The effect of planting combinations and Weeding on the growth and yield of component crops of maize/bean and maize/pumpkin intercrops. J Agr Sci 2002;138:193–200.

Smith HA, McSorley R. Intercropping and pest management: a review of major concepts. Am Entom 2000;46(3):154–161.

Soto J, Valdivia E, Valdivia R, Cuadros A, Bravo R. Descripción de sistemas de rotación de cultivos en parcelas de producción de quinua en cuatro zonas (siete distritos) del Altiplano Peruano. CienciAgro 2012;2(3):391–402.

Steffensen SK, Pedersen HA, Labouriau R, Mortensen AG, Laursen B, de Troiani RM, Noellemeyer EJ, Janovska D, Stavelikova H, Taberner A, Christophersen C, Fomsgaard IS. Variation of polyphenols and betaines in aerial parts of young, field-grown Amaranthus genotypes. J Agric Food Chem 2011;59:12073–12082.

Tauber MJ, Tauber CA, Daane KM, Hagen KS. Commercialization of predators: Recent lessons from green lacewings (Neuroptera: Chrysopidae). Am. Entomol 2000;46:26–37.

Thiery D, Visser JH. Masking of host plant odour in the olfactory orientation of the Colorado potato beetle. Entom Exp Appl 1986;41(2):165–172.

Throop HL, Lerdau MT. Effects of nitrogen deposition on insect herbivory: implications for community and ecosystem processes. Ecosystem 2004;7:109–133.

Thummel CS, Chory J. Steroid signaling in plants and insects: common themes, different pathways. Genes Dev 2002;16:3113–3129.

Trenbath BR. Intercropping for the management of pests and diseases. Field Crop Res 1993;34(3):381–405.

Valoy ME, Bruno MA, Prado FE, González JA. Insects associated to a quinoa crop in Amaicha Del Valle, Tucumán, Argentina. Acta Zoologica Lilloana 2011;55(1):16–22.

Vandermeer JH. The ecology of intercropping. New York, USA: Cambridge University Press; 1989. p 248p.

Varanda EM, Pais MP. Insect folivory in *Didymopanax vinosum* (Apiaceae) in a vegetation mosaic of Brazilian cerrado. Braz J Biol 2006;66(2B):671–680.

Vela A, Quispe A. Plagas de los cultivos de papa y maiz. Impresiones y Publicaciones Martinez Camañón. Perú: Cajamarca; 1988. p 155 p.

Verma S, Agarwal P. Phytochemical investigation of *Chenopodium album* Linn. and *C. murale* Linn. Nat Acad Sci Lett 1985;8:137–138.

Weber DC. Biological control of potato insect pests. In: Giordanengo P, Vincent C, Alyokhin A, editors. Insect pests of potato: global perspectives on biology and management. USA: Elsevier; 2012. p 399–424.

Whitham TG, Williams AG, Robinson AM. The variation principle: individual plants as temporal and spatial mosaics of resistance to rapidly evolving pests. In: Price P, Slobodchikoff C, Gaud W, editors. A new ecology: novel approaches to interactive systems. New York, USA: Wiley; 1984. p 15–52.

Whitman DW, Blum MS, Alsop DW. Allomones: chemicals for defense. In: Evans DL, Schmidt JO, editors. Insect defenses: adaptive mechanisms and strategies of prey and predators. Albany, New York: State University of New York Press; 1990. p 289–351.

Whitney HM, Feder W. Biomechanics of plant–insect interactions. Curr Opin Plant Biol 2013;16:105–111.

Wilby A, Thomas M. Natural enemy diversity and pest control: patterns of pest emergence with agricultural intensification. Ecol Lett 2002;5:353–360.

Wink M. Importance of plants secondary metabolites for protection against insects and microbial infections. In: Rai M, Carpinella MC, editors. Naturally occurring bioactive compounds. Amsterdam: Elsevier; 2006. p 251–268.

Wink M, Schimmer O. Molecular modes of action of defensive secondary metabolites. Ann Plant Rev 2009;39:21–161.

Wink M. Introduction: biochemistry, physiology and ecological function of secondary metabolites. Ann Plant Rev 2009;40:1–19.

Wise MJ, Abrahamson WG. Effects of resource availability on tolerance of herbivory: a review and assessment of three opposing models. Am Nat 2007;169(4):443–454.

Woldemichael GM, Wink M. Identification and biological activities of triterpenoid saponins from *Chenopodium quinoa*. J Agr Food Chem 2001;49:2327–2332.

Yábar E, Gianoli E, Echegaray ER. Insect pests and natural enemies in two varieties of quinoa (*Chenopodium quinoa*) at Cusco, Perú. J Appl Ent 2002;126:275–280.

Zanabria E, Mujica S. Plagas de la quinoa. Fondo Simón Bolivar, Puno, Perú: Universidad Nacional Técnica Del Altiplano; 1977. p 129–142.

Zanabria E, Banegas M. Entomología económica sostenible. Puno, Perú: Aquarium Impresiones; 1997. p 201.

Zhu N, Kikuzaki H, Vastano BC, Nakatani N, Karwe MV, Rosen RT, Ho C. Ecdysteroids of quinoa seeds (*Chenopodium quinoa* Willd.). J Agr Food Chem 2001a;49:2576–2578.

Zhu N, Sheng S, Li D, Lavoie EJ, Karwe MV, Rosen RT, Ho CT. Antioxidative flavonoid glycosides from quinoa seeds (*Chenopodium quinoa* Willd). J Food Lipids 2001b;8(1): 37–44.

Chapter 6
Quinoa Breeding

Luz Gomez-Pando
Universidad Nacional Agraria La Molina-Agronomy Faculty, Lima-Peru

HISTORY – DOMESTICATION PROCESS

Quinoa was probably among the earliest cultivated plants in the Andes region. The earliest archeological evidence indicates that the process of quinoa domestication began around 5000 BC in Ayacucho, located in the central highlands of Peru (Lumbreras et al. 2008) and in several other locations of South America. Remnants of *Chenopodium* sp. seeds were found in the Chinchorro Complex of Chile dating to 3000 BC, in Indian graves at Tarapaca, Calma, Calchaqui-Diaguita, Tiltil, and Quilagua in Chile (Bollaert 1860; Tapia 1979). The progress of quinoa cultivation has been studied intensively in the Lake Titicaca basin in Peru and Bolivia and it was found as a part of an agricultural complex that developed during the regional Formative Period, from 1800 BC to AD 500 (Bruno and Whitehead 2003). Seeds and fruit of cultivated species of the genus *Chenopodium* were identified from botanical remnants recovered at the archeological site of Punta de la Peña 4, layer 3, dated to ca. 760–560 BC at Sierra, Catamarca, Southern Argentinean Puna (Rodríguez et al. 2006).

In view of the archeological findings, quinoa improvement began many thousands of years ago when men and women started selecting seeds and plants. After centuries of natural and human selection, the different quinoa phenotypes and genotypes show adaptation to different parts of the Andean region, similar to the way that selection has affected hundreds of annual and perennial plant species worldwide (Duvick 1996). Seeds are the only available evidence to show the improvement of plant characteristics by natural and human selection through time. The major changes reported from the wild types and the domesticated varieties of chenopods are in the morphology of seeds, specifically those related to seed coat thickness, seed size, margin configuration, and surface patterns. The occurrence of a thin seed husk and the truncated margin among the domesticated chenopod seeds reflects an adaptive response to the selective pressures of human domestication, in terms of seedling viability and reduced seed dormancy (Murray 2005). Another character affected by human manipulation is the color of seeds, as there was a decrease in the number of black seeds among samples of varying ages as determined by C^{14} dating (Tapia 1979).

Farmers from the Highland Andean region required a complex combination of traits from the quinoa that they grew, traits needed to meet their food needs, and agronomic characteristics to enable them to grow in countless microclimates, from driest to wettest, coldest to hottest, and lowest to highest elevation. The products of natural and human selection are five quinoa ecotypes, each adapted to a particular growing environment. The Salares ecotype is recommended for the salt plateau (Salares) of southern Bolivia; the Altiplano ecotype for the high plateau

Quinoa: Improvement and Sustainable Production, First Edition. Edited by Kevin Murphy and Janet Matanguihan.

(Altiplano) of Peru and Bolivia surrounding Lake Titicaca; the Valley ecotype for the low valleys of the highland below 3,800 masl; the Sea-level ecotype for sea level or low altitudes in central Chile and high altitudes in southern Chile; and the Subtropics ecotype for the eastern, subtropical slopes of the Andes.

As a result of natural and human selection, complex populations were developed as cultivars, some of which are still used at present. The variability of these cultivars is seen in traits such as duration of the growth cycle, dormancy (time elapsed between physiological maturity and sprouting), disease resistance, and yield stability because of tolerance to frost, salt, and drought. On the other hand, quinoa cultivars share common characteristics such as plant height, seed color, and quality characteristics related to quinoa end use and commercialization. Besides important agronomic criteria, consumption criteria were also considered by farmers in their selection, such as suitability for traditional dishes or drinks and secondary use of the leaves as vegetables or animal feed.

Together with potato and maize, quinoa became an important food source for the Inca Empire with almost 15 million inhabitants in the Andean region. After the Spanish conquest of the region around 1532 AD, quinoa was replaced by crops such as barley, wheat, broad beans, peas, and oats. From that time on until the 1960s, almost 500 years since Pizarro's conquest, quinoa remained as a neglected crop in the highlands. During this period, quinoa had received little attention from researchers or scientists, and almost no commercial advancement. Although it was largely lost to the outside world, quinoa did not become extinct because farmers maintained this crop mainly in the Peruvian and Bolivian Altiplano (Cusack 1984; National Research Council 1989; Mujica 1992; Jacobsen and Stolen 1993).

Beginning in the 1960s, interest in quinoa increased, eventually leading to the establishment of several new breeding programs in different countries. Several factors have contributed to this increased interest in quinoa, namely awareness of the role of quinoa in food security in the Andean South America, its unique potential as a crop for marginal soils worldwide, and a growing health food export market. Genetic improvement of quinoa started in Bolivia in 1965 at the Experimental Station Patacamaya supported by FAO-OXFAM (Oxford Famine Relief) and the Bolivian Government (Gandarillas 1979a). In Peru, it started almost at the same time in the Technical University of Altiplano in Puno.

COLLECTION OF GENETIC RESOURCES

Crop genetic improvement in the Andean region started in the 1960s with the collection and characterization of quinoa germplasm preserved by the farmers. Gandarillas (1968) made one of the first descriptions of quinoa germplasm of Bolivia, Peru, and Ecuador. On the basis of the descriptors such as plant habit, type of inflorescence, leaf shape, seed, and leaf dentation number, Gandarillas (1968) described 17 different races of quinoa: four races in the North of Cusco (Pichincha, Ancash, Cajamarca, Junin), three races in Cusco (Cusco, Sicuani, Puca), four races around the basin of Lake Titicaca (Copacabana, Dulce, Achacachi, Puno), four races in the Andean valley located at the southeast of Lake Titicaca (Potosi, Sucre, La Glorieta, Cochabamba), and two races around the Poopo Lake basin (Real, Challapata). On the basis of these characteristics and research conducted on quinoa genetic diversity, it is accepted that the region with the highest quinoa diversity is located between Cuzco (Peru) and Poopo Lake (Bolivia) (Wilson 1988; Christensen et al. 2007).

Quinoa germplasm and that of other *Chenopodium* species are mainly preserved *ex situ*. In Bolivia, around 5,000 accessions are preserved in several institutions, such as the Torolapa Centre of the National Institute of Agricultural and Forestry Innovation, the Choquenaira Experimental Station of the Universidad Mayor de San Andrés-UMSA, the Biotechnology and Phytogenetic Resources Research Centre (CIBREF) of the Technical University of Oruro (UTO), the Tiahuanacu Academic Unit of the Bolivian Catholic University-UCB, the Kallutaca

Experimental Centre of Public University El Alto-UPEA, and the Centre of Communal Research and Promotion-CIPROCOM. These germplasm collections are preserved mainly in the INIAF Toralapa Centre that currently maintains 3,121 accessions, both cultivated and wild, which were gathered from communities in the Altiplano and the valleys in the Bolivian departments of La Paz, Oruro, Potosi, Cochabamba, Chuquisaca, and Tarija. It also has germplasm from Peru, Ecuador, Colombia, Argentina, Chile, Mexico, and the United States. Eight wild species of quinoa were identified among 270 accessions collected (Rojas et al. 2010).

In Peru, around 5,351 accessions are also maintained in several institutions. The Institute of Agricultural Research (INIA) at the Experimental Station of Illpa preserves 536 accessions. The National University of the Altiplano (UNA) in Puno has a collection of 1,873 accessions that were collected in Bolivia (457), Altiplano (990), Valley (357), Colombia (1), Ecuador (18), Chile (8), and the United States (2) (Bravo and Catacora 2010). At La Molina Agrarian University, 2,942 accessions are preserved, collected from the departments of Cajamarca, La Libertad, Ancash, Junín, Ayacucho, Arequipa, Apurimac, Cusco, and Puno, as well as from Bolivia (Gómez and Eguiluz 2011). Other universities housing quinoa germplasm are the National University of the Centre in Junín, the National University of San Cristobal de Huamanga in Ayacucho, and the National University of San Antonio Abad in Cusco (Mujica 1992; Bonifacio et al. 2004).

Aside from Bolivia and Peru, there are other countries that also preserve quinoa germplasm. In Ecuador, around 642 accessions are preserved at the Santa Catalina Experimental Station of the National Institute of Agricultural Research-INIAP. In Chile, 25 accessions are maintained at the Austral University and 59 accessions in the North of Chile (Fuentes et al. 2006). In addition, 164 quinoa accessions are maintained in the USDA National Plant Germplasm System (NPGS) (Christensen et al. 2007). The reduction in the genetic diversity of quinoa in countries of the Andean region could be a result of founder effects associated with the dispersal of the crop from its center of origin or response to selective adaptation to other ecological regions. In the past centuries, the loss of biodiversity probably was due to the systematic decline and, in some areas, the disappearance of the quinoa crop and its use outside the Peruvian and Bolivian Altiplano and the salt flats of Bolivia.

The *in situ* conservation systems for quinoa are found mainly in Bolivia and Peru. Germplasm is preserved by cultivation and utilization of quinoa and wild types through traditional systems in farmers' communities. In Peru, it is kept in the *aynokas*, ancient systems maintained for several different purposes such as food security, adequate management of soils and pests, conservation of genetic diversity *in situ*, and rational use of plant diversity (Ichuta and Artiaga 1986). In Bolivia, five microcenters have been identified where farmers still keep customs and traditions such as biological indicators or signs during the crop year that are used in crop management, rituals, and festivals. The quinoa crop has a spatial and temporal distribution in traditional farming systems called *aynoqas*, *sayañas*, *huyus*, and *jochiirana*. The social system of community organizations annually elects families called *yapu campus* or farm caretakers, who are responsible for the care of crops against adverse weather such as hail, frost, and floods (Rojas et al. 2010b).

In these *in situ* conservation systems, cultivated quinoa and wild relatives are preserved. These systems are widely distributed in the Andean region and are known by different names such as *mandas* and *laymes* (Mujica and Jacobsen 2000). The related wild species growing alongside cultivated quinoa in the Altiplano and the valleys are *Chenopodium carnosolum*, *Chenopodium petiolare*, *Chenopodium ambrosioides*, *Chenopodium hircinum*, and *Swaeda foliosa* (Bonifacio 2003; Mujica and Jacobsen 2005). The wild relatives are sometimes found in isolation, either at the edges of the farmers' fields or in places considered sacred (Gentilwasi o Phiru). Still, these wild relatives of quinoa are cared for by farmers and used as food, medicine, or for ritual purposes, especially in times of drought or extreme weather disasters characteristic of the Andean highlands of Peru and Bolivia. The young leaves of wild quinoa are used as a salad and the precooked, toasted seed is made into flour.

Domesticated chenopods are distinguished from wild *Chenopodium* species on the basis of several characteristics. Domesticated chenopods have (i) more compacted inflorescences; (ii) lost their natural seed shattering mechanisms; (iii) uniform maturation of fruit; and (iv) reduced germination dormancy (Smith 1984; Gremillion 1993; Bruno and Whitehead 2003).

Quinoa germplasm has been generally characterized and evaluated using the morphological characteristics of vegetative and reproductive organs and agronomic characters such as plant height, life cycle, yield potential, susceptibility to stress, and resistance or tolerance to pathogens and diseases. In addition, quality characters of the seeds such as size and color as well as protein, oil, and saponin content have been used. In some cases, variation in these characters was measured in different years and locations to determine the performance of the same genotype in different environments (genotype × environment interaction). This work has permitted, though in a preliminary way, a method of knowing the magnitude of the variation and the sources of genes for many characters, thus facilitating breeding programs. A number of this type of research has permitted the formation of core collections to simplify the management and enhance the utilization of quinoa genetic resources (Ortiz et al. 1998).

At present, quinoa germplasm is divided into five ecotypes (Tapia et al. 1980). These ecotypes differ in adaptation to altitude, tolerance to drought and salinity, and photoperiod response. These populations have most likely received different selection pressures from their environments as well as during the domestication process.

1. *Valley Type*: These quinoas have evolved at valleys situated at altitudes from 2,000 to 3,800 masl. Plants of this ecotype are generally 2–3 m in height and have branched stems, a life cycle of more than 210 days, low saponin content, and some resistance or tolerance to downy mildew (*Peronospora variabilis*).
2. *Altiplano Type*: These quinoas originate around the Lake Titicaca basin that has an altitude of 3,800 to 4,000 masl and is characterized by adverse conditions (drought, frost, and hail). The plant height ranges from 1 to 1.8 m and the life cycle from 120 to 210 days. In general, they do not have branched stems and the seeds are rich in saponin. These quinoas vary in their reaction to downy mildew and can be tolerant, resistant, or highly susceptible.
3. *Salar Type*: These quinoas have evolved in the salt flats of the southern altiplano in Bolivia with altitudes around 4,000 masl, low rainfall (300 mm), and soils having a pH above 8.0. The plant characteristics are similar to the Altiplano ecotype. In general, they have black seeds with sharp edges and are rich in saponin. Among the Salar ecotypes, there are some sweet, saponin-free genotypes and some "Real"-type quinoa, which have large white seeds.
4. *Sea Level Type*: These quinoas originate from southern Chile, at latitudes around 40° S. They are mostly unbranched and will flower in long days. They have small, yellow translucent seeds with high saponin content. They are reported to be resistant to fungal diseases such as downy mildew (Fuentes et al. 2009).
5. *Subtropical Type*: These quinoas are found at the subtropical Yungas of Bolivia. Plants have an intense green coloration that turns orange at maturity and produce very small yellow-orange seeds (Tapia 1982).

After the 1960s, there had been exchanges of germplasm among different organizations such that the quinoa population distribution of the past has been changed in different degrees. Some new sites have been identified in the American and European Test of quinoa, organized by the FAO. Field trials in Italy and Greece have shown promising results with reported seed yield of 2,280 and 3,960 kg/ha, respectively (Mujica et al. 2001).

Using morphological and agronomic characteristics, 95–100% of the quinoa collections of Bolivia had been characterized and evaluated, whereas 12% of the collections had been evaluated using nutritional value and characteristics (Rojas et al. 2010). In a similar way, 100% of the Peruvian collections are being morphologically and agronomically evaluated. In the collection at La Molina National Agrarian University, 43% of the collection was evaluated by quality characteristics. Some of the results of these evaluations are presented in Tables 6.1 and 6.2 and show the wide genetic variability in quinoa collections of Bolivia and Peru.

A limited number of the germplasm studied from Ecuador and Argentina has shown low diversity, indicating that the most probable point of introduction for Ecuadorian accessions was the Altiplano (Peru-Bolivia), whereas for Argentina, the original introduction could have been from the Chilean highlands and coastal zone (south of Chile) (Christensen et al. 2007). The other reasons are probably due to potential genetic drift owing to abandonment of the crop or the isolation of farmer communities.

Table 6.1 Variation in morphological characters of quinoa (*Chenopodium quinoa* Willd.) described in germplasm from diverse origins.

Morphological characters	Bolivia[a]	Peru (UNALM)[c]
Color of the plant before flowering	Green, purple, red, and mixed	–
Color of leaves before flowering	–	Green, purple, mixed, and red
Color of leaf axil	–	Green, purple, red, and pink
Color of stem streaks	–	Yellow, green, purple, pink, and red
Color of the plant at physiological maturity	White, cream, yellow, orange, pink, red, purple, brown, and black	–
Color of the Inflorescence at physiological maturity	–	Yellow-green, yellow, yellow-orange, orange, orange-red, red, red-purple, purple, purple-violet, violet, violet-blue, white, white-gray, white-yellow, white-orange, gray-yellow, gray-orange, gray-red, gray-purple, gray-green, gray-brown, brown, gray, and black
Inflorescence shape	Amaranthiform, glomerulate, or intermediate	Amaranthiform, glomerulate, or intermediate
Inflorescence density	Compact, lax, or intermediate	Compact, lax, or intermediate
Seed color	White, cream, yellow, orange, pink, red, purple, brown, and black[b]	–
Seed color pericarp (fruit coat)	–	Yellow, yellow-orange, orange, orange-red, red, red-purple, white, white-yellow, white-orange, gray-yellow, gray-orange, gray-red, gray-purple, gray-green, gray-brown, brown, gray, and black
Seed color episperm (seed coat)	–	Yellow, yellow-orange, orange, red-purple, purple, white, white-yellow, white-orange, white-gray, gray-yellow, gray-orange, gray-purple, brown, and black

[a]Rojas et al. 2001; Rojas 2003; Rojas et al. 2008; Rojas et al. 2009
[b]Cayoja (1996) reported 66 colors considering different shades
[c]Gómez and Eguiluz (2011)

Table 6.2 Agronomical variation and quality characters of quinoa (*Chenopodium quinoa* Willd.) described in germplasm from diverse origins.

Agronomic and quality characters	Bolivia[a]	Peru (UNALM)[c]	
		Altiplano type	Valley type
Vegetative cycle (days)	110–210	–	–
Date of flowering (days)	–	46–100	50–115
Date of maturity (days)	–	115–195	140–220
Seed yield per plant (g)	48–250	–	–
Seed yield (kg/ha)[b]	–	165–2,975	109–3,531
Seed diameter (mm)	1.36–2.66	1.4–2.2	1.2–2.2
Weight of 100 seeds (g)	0.12–0.60	–	–
Seed protein content (%)	10.21–18.39	7.0–24.4	10.3–18.5
Seed saponin content (%)[b]	–	0–1.42	0–1.57
Starch grain diameter (μm)	1.5–22	–	–

[a]Rojas et al. 2001; Rojas 2003; Rojas et al. 2008; Rojas et al. 2009
[b]Eguiluz et al. (2010)
[c]Gómez and Eguiluz (2011)

Chilean quinoa has been characterized as morphologically diverse and bifurcated into coastal and highland ecotypes. The quinoa is cultivated primarily by indigenous Aymara in the northern Chilean Altiplano. Chilean quinoa is characterized by a broad range of morphological diversity that likely resulted from artificial and natural selection and genetic drift as landraces were introduced to south-central Chile via trade and migration of indigenous peoples. Little focus on germplasm collection, conservation, and evaluation has been placed on the quinoa populations found in coastal lowlands of south-central Chile (Fuentes et al. 2009). Bhargava et al. (2007a) evaluated quinoa lines from different origins under subtropical North Indian conditions and reported that Chilean lines were more adapted to countries having a monsoon climate such as India, with markedly cold winters and hot summers. In another study, Gęsiński (2008) reported that conditions in Europe are favorable to *Chenopodium quinoa* for green matter and seed yield, and the most favorable seed yield was recorded in Greece.

There are several studies that have used molecular markers in quinoa to characterize germplasm diversity. According to Rojas et al. (2010), 86% of Bolivian germplasm has been characterized using molecular markers. Fuentes et al. (2009) indicated that highland and coastal quinoas from Chile shared 21.3% of alleles. On the other hand, highland quinoas contained 28.6% of unique alleles, whereas coastal quinoa had 50% of unique alleles. Wilson (1988) and Christensen et al. (2007) demonstrated a high genetic similarity between highland and coastal quinoas.

GOALS AND METHODS OF QUINOA BREEDING

The main aim of quinoa breeders is the development of cultivars adapted to diverse agroclimatic regions with high seed yield and good quality components for food and industry use. These desirable cultivars should combine valuable characteristics for the farmers, industry, and consumers.

Requirement of the farmers

Higher yield

The primary objective of most breeding programs is to improve quinoa yield, as the yield of quinoa in the Andean region is generally low. In the period between 2005 and 2012, the yield in Bolivia ranged from 570 to 642 kg/ha; in Peru, it ranged from 958 to 1,163 kg/ha, whereas in Ecuador, it ranged from 63 to 848 kg/ha (FAOSTAT 2014).

Yield stability is one of the most important objectives of any quinoa breeding program.

In some cases, quinoa genotypes that have been selected for good yield at a given location will not perform well at another location. Thus, it is important to identify genotypes that give good yields in a broad range of environments. In the Andean region, where subsistence agriculture is prevalent, a cultivar with the ability to give a consistent yield under variable climatic conditions is preferred to a cultivar that has the potential for outstanding yields but only under favorable growing conditions.

To assess the possibility of genetic improvement of seed yield, it is necessary to determine the extent and nature of genetic variation, genotype (G) × environment (E) interactions, and heritability. According to Bertero et al. (2004), the extremely high levels of environmental variation encountered within the quinoa growing region require breeding and testing strategies structured to accommodate the effects of large G × E interactions. Mujica et al. (2001) reported seed yields of 2,280 and 3,960 kg/ha in Italy and Greece, respectively, from some genotypes selected in the American and European Test of Quinoa organized by the FAO. These findings showed good yield potential for quinoa. Bonifacio (2003) indicated that the improved varieties obtained at the Patacamaya experimental station yielded about 1,200 kg/ha on a commercial scale in the Altiplano, though their potential is much greater (3 t/ha) with increased inputs and the use of modern technology. This is a considerable improvement when compared to 700 kg/ha obtained by the farmers using native varieties. However, acceptance of the improved varieties depends on their adaptation to local conditions and their commercial and culinary characteristics.

It is also necessary to study seed size, plant height, and lodging in relation to seed yield. Correlations among these characters are useful in determining both favorable and unfavorable correlated responses to selection. Bertero et al. (2004) found that simultaneous progress for seed yield and seed size can be expected from selection because no association was found between the average cultivar responses for seed yield and seed size and no associations were found between the G × E interaction effects for both traits. Bhargava et al. (2007a) reported that plant height, leaf area, branches/plant, inflorescence/plant, seed size, 1,000 seed weight, dry weight/plant, and harvest index exhibited significant positive association with seed, and the maximum value was recorded for inflorescence/plant. In another study, Bhargava et al. (2008) used direct and indirect selection for some parameters. They found that stem diameter, chlorophyll a, total chlorophyll, and leaf carotenoid content had high correlated response and relative selection efficiency values for seed yield. Thus, these traits can be used to increase seed yield. Protein and carotenoid contents in seed showed negative correlated response. Relative selection efficiency values for seed yield indicated that direct selection for seed yield would lead to a slight decrease in these quality characters. Bertero and Ruiz (2008) reported a negative association between reproductive efficiencies and panicle biomass. Seed size was not of much significance either in increasing seed yield or any of the quality components. Seed yield can also be increased through indirect selection for stem diameter, whereas leaf pigments are likely to play a major role in enhancement of quality traits such as leaf and seed carotenoid levels.

Plant height

Plant height is generally considered to be a quantitative character, although several reports indicate that reduced height or semidwarf characteristics may be controlled by single recessive genes. There are significant variations in plant height among quinoa ecotypes. Plants with shorter heights are usually found among the Altiplano and Salares ecotypes and the taller plants (more than 2 m high) are found in the Valley ecotypes. Plant height could be reduced by selection among genotypes without a semidwarf allele. Another way to reduce plant height is by the use of mutagens to develop dwarf mutants, as reported by Gomez-Pando and Eguiluz-de la Barra (2013). It is also important to note that there are indications of a positive correlation between plant height and inflorescence length (Ochoa and Peralta 1988; Rojas et al. 2003).

Stalk strength

Major advances in yield have been achieved by the development of cultivars with resistance to lodging, planted in highly productive environments managed with good farming technology. Resistance to lodging ensures good seed filling and minimal harvest loss. In most crops, stem strength must be dealt with as a quantitative trait. Several individual characters influence resistance to lodging, including stem diameter, stem outer wall thickness, plant height, and type of root system. The genetic control of stalk strength is also quantitative.

Life cycle

Another important objective in breeding quinoa is the development of early-maturing cultivars that can be grown in areas where the average number of frost-free days in the growing season is <120–150, such as in the Andean region. Early maturing cultivars could also be less vulnerable to inclement weather and have more potential to fit into multiple cropping systems in the coastal areas. According to Mujica (1988), the number of days to flowering has a heritability of 0.82. Late-maturing genotypes frequently produce higher seed yield but run a higher risk of encountering frost, drought, or hail damage to an unacceptable degree.

Resistance to biotic stresses

Quinoa improvement programs in the Andean region have made it a priority to breed for pest and disease resistance, especially because farmers in the highlands can increase their income by selling organic quinoa. Many programs concentrate on resistance to the fungus *P. variabilis*, the causal agent of quinoa downy mildew. Downy mildew causes serious yield losses throughout the Andean region and other locations in the world. At a number of locations, the focus has been on screening germplasm for resistance, and genetic materials with different levels of resistance have been identified. Ochoa et al. (1999) determined the factors of resistance and virulence groups in the quinoa-downy mildew pathosystem. Using 60 accessions of quinoa in Ecuador and 20 isolates of *P. variabilis*, they identified three factors of resistance and four virulence groups or races of the pathogen. The specific interaction between fungal isolates and host genes indicated the presence of major genes. Accessions ECU-291, ECU-470, ECU-379, and ECU-288, were proposed as a set of preliminary differentials to identify pathogenic races of downy mildew in quinoa. However, virulence tests showed that the variability in *P. variabilis* is much more complex than could be revealed using the Ecuadorian *C. quinoa* differentials used in the study (Danielsen et al. 2000). In India, 34 quinoa accessions were evaluated for response to downy mildew. Seven accessions were found to be resistant, suggesting physiological specialization of current pathotypes for *C. quinoa* (Kumar et al. 2006).

Insects can be very damaging in quinoa, especially when the insects feed on floral structures or seeds, thus reducing seed yield and quality. Breeding for resistance to insects must consider the genetics of the pest and the host because complex interactions occur between insects and plants that may be morphological, biochemical, or physiological in nature. Plant resistance may be obtained through nonpreference (suppression of feeding or oviposition), antibiosis (adverse effects on normal growth or survival of insects), and tolerance (ability of the plant to survive when insects are present).

Bird predation is another problem in quinoa. Variations in seed saponins, which cause differences in bird feeding preferences, have been reported to deter bird feeding and thus decrease inflorescence damage. Achenes with high amounts of saponin are often less damaged. Developing lines with tolerance to bird predation is a complex process, apparently requiring a combination of many characters that are independently inherited but contribute to overall tolerance.

Resistance to abiotic stresses

Quinoa is cultivated under a wide range of environments in the Andean region. Quinoa has been noted to have tolerance to drought and salt stresses (Jacobsen et al. 2001; Jacobsen and Mujica 2002; Jacobsen et al. 2003; Jacobsen et al. 2005; Jacobsen et al. 2007; Gomez-Pando et al. 2010; Ruiz-Carrasco et al. 2011; Verena et al. 2013). It is important to understand the genetic principles and the mechanisms governing tolerance to these factors and apply this knowledge to develop improved varieties of quinoa. There are two approaches to achieve drought- and salt-tolerant varieties: breed cultivars with a plant growth type that escapes the water and temperature stress conditions and/or breed for characteristics that contribute to physiological drought and salt tolerance.

Plant morphology

In some cultivars, it will be necessary to modify the architecture of plants to allow for mechanized harvesting. The areas devoted to quinoa cultivation have been expanding rapidly and traditional ways of harvesting quinoa must be replaced by mechanical methods. Nonbranched or simple stem plants with an adequate plant height and unique terminal inflorescence and uniform seed maturity will be suitable for mechanized agriculture.

Harvest index

The harvest index measures the photosynthetic capacity and the effective translocation of assimilates to the seeds. The harvest index can be modified by agronomic practices and environmental growing conditions (Bertero and Ruiz 2010). Rojas et al. (2003) reported a variation in the harvest index of quinoa from 0.06 to 0.87. Traditional cultivars, especially valley ecotypes, have low harvest index due to the high amount of branching relative to plant height and more partitioning of assimilated products to the vegetative parts rather than the seeds.

Requirements of the industry and consumers

Within the quinoa germplasm, significant differences have been found in some qualitative traits that can be used in breeding to improve the nutritional quality of quinoa.

Protein content and composition

The protein concentration in quinoa seeds, as seen in other crops, is linked with soil nitrogen (N) availability and uptake, N transport and assimilation in the vegetative structures, direct N transport and remobilization from plant structures into developing seeds, carbohydrate deposition in the developing endosperm, and the number and size of seeds per unit area. These complex relationships may cause a wide range in protein percentages for the same genotype grown in different parts of the same field, in different fields, in different regions, or in different years. The range of seed protein content reported was from 7% to 22% (Koziol 1992; Prakash et al. 1993; Wright et al. 2002; Repo-Carrasco et al. 2003; Bhargava et al. 2007a; Gómez and Eguiluz 2011). High protein percentage generally results from low carbohydrate deposition in the kernel so that increased yield is often associated with lower protein content. However, Bhargava et al. (2007a) reported insignificant correlation and low values of direct path among seed yield and protein content in quinoa lines of different origins and remarked that this result could help the development of cultivars with good yield potential and high seed protein content.

Quinoa proteins have a remarkably high biological value. However, raising the level of some amino acids is not yet possible at the moment, due to the lack of knowledge regarding the variability of amino acid composition among genotypes and the effect of environment. Even with this lack of knowledge, this is a promising area of future research because the potential to increase the level of some amino acids exists.

Seed characteristics

Color

Until the 2000s, white- or cream-colored quinoa seeds had been preferred by consumers and the industry. In the past decade, however, quinoa with darker-colored seeds had been incorporated to the market probably due to the high association of seed color with carotenoid content. The results of Bhargava et al.'s study (2007a) showed that accessions with dark seed coat color have high seed carotenoid content.

Size of seeds

Another potential goal in breeding quinoa is to develop cultivars with homogeneous seed size in the inflorescence. The size of quinoa seed varies from <1.4 mm diameter to more than 2.0 mm, and it is very common to find such variability in a single inflorescence (Gómez and Eguiluz 2011). Highly significant correlations between seed diameter and seed weight have been determined in quinoa (Ochoa and Peralta 1988; Cayoja 1996; Rojas et al. 2003). Temperature and photoperiod after anthesis also affect seed diameter to a considerable extent. Short photoperiod and cool temperatures after anthesis promote larger seed diameters, whereas long photoperiods and high temperature after anthesis negatively affect seed size (Bertero et al. 1999; Bhargava et al. 2007a). Large seed size is preferred when the seed is used directly as pearled seeds or flakes, whereas seed size is immaterial when seeds are to be used for flour.

Seed saponin content

The saponin content has different implications for quinoa growers, consumers, and the industry. For the consumers, saponin gives seeds a bitter taste and must be removed from the seed, thus increasing the time for food preparation. On the other hand, under some circumstances, growers will prefer bitter quinoa as the saponin could reduce damage due to birds and certain insects. For the industry, saponin removal requires specific equipment and a good source of quality water, thus increasing processing costs. Recently, several uses have been found for saponin, namely as an organic detergent, as foam fire extinguishers, and as deodorant and other uses in the cosmetic industry. To fulfill these differing requirements for saponin content, there are several accessions in the gene banks without saponin (sweet), with very low content (medium sweet), and with high levels (bitter) (Ward 2000a; Malt et al. 2006).

Further investigation into the genetic inheritance of several quinoa characteristics will be useful in developing methodologies for cultivar development and for determining short- or long-term breeding goals. Genetic studies in quinoa have provided knowledge regarding the inheritance of some qualitative and quantitative characters such as plant color, axillary pigmentation, inflorescence type, saponin content, seed color and type, genetic and cytoplasmic male sterility, earliness, and plant height (Gandarillas 1968, 1974, 1979a, 1979b, 1986; Rea 1969; Espindola 1980; Bonifacio 1990, 1995; Saravia 1991; Ward 2000b).

Methods in genetic improvement

Reproductive biology

Quinoa is predominantly autogamous, though some out-crossing is also possible. The sexual reproduction system of quinoa provides opportunities to produce new combinations of genes or mutations, passing on to the succeeding generations traits that might provide superior adaptability to changing environments. Even though quinoa is mainly grown from seed, *in vitro* vegetative propagation of quinoa has also been reported (Ruiz 2002).

Gynomonoecy is the predominant breeding system in quinoa. The inflorescence bears three basic flower types, namely, hermaphrodite, chlamydeous female, and achlamydeous female. There can be five flower types based on flower size. On the other hand, there can be 10 flower types based on the proportion of hermaphrodite and female flowers and their arrangement and also on the number of divisions of the dichasium on the glomerule (Leon 1964; Rea 1969; Bhargava et al. 2007b).

Quinoa is predominantly autogamous, though some out-crossing is also possible. In studies conducted in Peru and Bolivia, the amount of cross pollination of different genotypes varied from 1.5% at a distance of 20 m to 9.9% at a distance of 1 m (Gandarillas and Tapia 1976). Some studies in the United States showed similar percentages of cross pollination, with out-crossing exceeding 10% (House 1982; Jennings et al. 1981). In Argentine-Mendoza, out-crossing rates of 17.36% were observed (Silvestri and Gil 2000). The amount of cross pollination depends on (i) the proportion of the different types of flowers in the inflorescence; (ii) the proportion of hermaphrodite flowers on a plant, which can vary from 2% to 99%; (iii) the number of androsterile hermaphrodite flowers; (iv) the presence of self-incompatibility and protogyny; and (v) the environment and presence of insects. Temperatures lower or higher than 30°C retard anthesis and affect pollen viability. Wind pollination studies of fertile quinoa (CO 407) with an orange panicle crossed with a near-isoline (CO 407R) with a red panicle showed that pollen can move as far as 36 cm. In South America, pollen can be transported by some insects such as thrips and green aphids (*Aphis* sp.). However, in the San Luis Valley, Colorado, no insect activity in the flowers was observed, although various *Diptera* spp. visit quinoa flowers and may add to wind distribution of pollen (Rea 1969; Gandarillas 1979a; Aguilar 1980; Johnson and Ward 1993; Lescano 1994; Silvestri and Gil 2000). However, there are also reports of some extreme cases of complete self-pollination through cleistogamy (Nelson 1968) and obligatory out-crossing due to self-incompatibility and male sterility (Nelson 1968; Gandarillas 1969; Simmonds 1971), indications that quinoa has a fairly versatile breeding system.

Male sterility was reported in many quinoa accessions collected from the Central highlands and from those growing around the Lake Titicaca basin of Peru-Bolivia and northern Bolivia. The inheritance of male sterility has been found to be controlled by nuclear genes and cytoplasmic factors. Cytoplasm was classified as normal (N) and sterile (S) and the nuclear genes as fertile (Ms) and sterile (ms). Sterile cytoplasm from the cultivar "Apelawa" has been denoted as A-cytoplasm, whereas the older source and that in PI 510536 are referred to as C-cytoplasm (Gandarillas 1969; Saravia 1991; Ward and Johnson 1993; Ward 1998). The anthers of normal fertile quinoa are generally bright lemon yellow (Rea 1969), but the anthers of quinoa possessing a gene for male sterility are whitish-yellow. Plants with male sterile cytoplasm produce flowers without anthers and show a prominent exertion of stigmas, whereas male sterile plants from PI 510536 have shrunken anthers that do not produce pollen (Ward and Johnson 1993; Ward 1998).

The cultivars "Amachuma" and Apelawa are two potential sources of male sterile quinoa plants. Male sterility in the Amachuma type appears to be a simply inherited genetic trait. On the other hand, the Apelawa type is a cytoplasmic male sterile and has been transferred into four additional background genotypes. Progeny from the crosses of Apelawa with the weedy species *Chenopodium berlandieri* had partially restored male fertility (Ward and Johnson 1993).

Polyploidy level

Polyploidy in quinoa is also an important factor in plant breeding because it can influence reproductive compatibility, fertility, and phenotypic traits. The base chromosome number of the genus *Chenopodium* is $n = x = 9$ chromosomes. Quinoa has $2n = 36$ chromosomes and is considered an allotetraploid. It is the result of the cross between two diploid species with a subsequent doubling of chromosomes leading to the development of a fertile allotetraploid (Gandarillas 1986; Simmonds 1971; Maughan et al. 2004, 2006).

Allelic segregation analysis showed disomic inheritance with independent assortment at homologous loci. Several sources have reported classic Mendelian ratios for qualitative traits segregating in the F_2 generation for one or two gene pairs (Cardenas and Hawkes 1948; Gandarillas and Luizaga 1967; Gandarillas 1968, 1979b, 1986; Simmonds 1971; Saravia 1991; Bonifacio 1990, 1991). Ward (2000b) reported that allelic segregation analysis in the F_1 and F_2 ratios indicate both disomic-digenic and tetrasomic inheritance in two of the three traits studied, as well as distorted F_2 ratios suggesting erratic multivalent formation at meiosis.

QUINOA BREEDING METHODS

Several breeding methods have been used with quinoa, such as selection (individual selection and/or mass selection), introduction of foreign germplasm, hybridization, backcrossing, and induction of mutation.

Selection

The early stages of quinoa improvement began in the different areas of the Andean region, driven by selection of individuals from the natural populations or land races developed throughout the centuries. Both individual selection and mass selection had been applied.

Individual selection

Individual selection consists of selecting individual plants with one or more outstanding characteristics from a landrace population and sowing a single inflorescence per row. The offspring are evaluated and selected in and between rows for characters under consideration. A modification of this method, named inflorescence-row, follows the same procedure described earlier but with one difference – selfing of the inflorescence is controlled in each season. This modified procedure was repeated for two or more cycles, enough to achieve homogeneity (Gandarillas 1979a; Bonifacio 2003). Using this method, the cultivar "Sajama Amarantiforme" was obtained.

The selfing process could take place in all the inflorescence or only in some selected glomerulus or group of flowers. When the whole inflorescence is to be selfed, it is necessary to cover it with a grease-proof paper bag and the leaves removed. When only a small number of glomeruli in the inflorescence are to be selfed, the major part of the inflorescence should be removed, and 8–10 glomeruli in the base of the inflorescence are selected and isolated from the rest by enclosure in a glassine bag. Plants should be checked periodically to avoid damage from diseases and insects because isolated glomeruli are often preferred by larvae that consume growing seeds.

Mass selection

This method consists of the selection of a large number of superior plants with similar phenotype. Their seeds are then harvested and mixed together to constitute the new variety. Mass selection is applied multiple times in the same population to improve the base population performance as mass-selected cultivars may contain considerable genetic variation. Cultivars developed by mass selection have wide adaptability, a wide genetic base, and yield stability over a long period of time. Characters such as plant height, plant and seed color, seed size, disease resistance, saponin content, and life cycle have been considered in the mass selection process. The main characteristic of cultivars developed through mass selection is that they are a mixture of different genotypes, generally uniform in size, seed color, and saponin content, and in other characters preferred by the general market and the industry. In Bolivia, stratified mass selection was used to produce seeds of cultivars developed from the Real landrace and also used to preserve the identity and composition of established cultivars. The most important cultivars obtained using mass selection are "Real" (Bolivia), "Baer" (Chile), "Dulce de Quitopamba" (Colombia), "Pasankalla," "Chewecca," "Blanca de Juli," "Amarilla de Marangani," "Blanca de Junín," "Rosada de Junin," and "Blanca de Hualhuas" (Peru).

Participatory plant breeding (PPB)

Although initially developed for smallholder agriculture in difficult and diverse environments, participatory plant breeding (PPB) appears to be a viable alternative for improving quinoa germplasm in small-scale farmers' fields. PPB involves farmers and other stakeholders in the breeding process. Stakeholders may work with scientists in any of the five basic steps of the plant breeding cycle: setting goals, creating variability, selecting experimental lines/genotypes, testing experimental lines/genotypes, and cultivar release/diffusion. In Ecuador, the PPB processes were applied and the farmers' field selection criteria for quinoa were based mostly on yield, earliness, and plant color. At seed selection, farmers with insecure food sources chose lines based on yield, whereas farmers who had more resources also considered seed size, color, saponin content, and marketability in addition to yield (Ceccarelli et al. 2000; McElhinny et al. 2007).

Introduction of foreign germplasm

For the past several decades, germplasm and cultivars of quinoa have been exchanged among countries of the Andean region. After a series of evaluations and multiplications, some of these introduced materials have been developed as commercial cultivars. On the whole, the introduction of cultivars from the Bolivian southern altiplano to the Bolivian northern altiplano and Peru altiplano had not been successful, due to the high susceptibility of these quinoas, especially the Real race, to downly mildew, although there have been some exceptions. Similarly, accessions from the Northern and Central Highlands did not adapt very well because these accessions are of the Valley ecotype, have a long life cycle, and are susceptible to drought and frost. An exception is the "Sayaña" cultivar that is currently adapted to the surrounding area of Uyuni. Cultivars developed at Patacamaya such as "Sajama," "Kamiri," "Huaranga," and "Chucapaca" were introduced by farmers and research institutions to Peru and had adapted well in Puno. These are used as commercial cultivars mainly because of the high quality of their seed. The cultivar "Pata de Ganso" from Ecuador was developed from introduced Bolivian material.

Quinoa was also introduced to other countries outside of the Andean region. Approximately 300 accessions from Bolivia, Peru, and Chile have been introduced to England (Risi and Galwey 1984; Risi 1986). Outstanding accessions of quinoa from the region selected by different institutions were included in the American and European Quinoa Test, which was conducted in different countries (Mujica et al. 2001). The preliminary results showed that accessions from the Peruvian and Ecuadorian Valley have good tolerance to mildew, whereas material from Bolivia showed high susceptibility to the same disease. Accessions from the Coast matured earlier but were highly susceptible to hail at the highland locations. Adaptation to North Indian subtropical conditions were reported by Bhargava et al. (2007a) who studied 27 germplasm lines *of C. quinoa* and two lines of *C. berlandieri* subsp. *nuttalliae* over a 2-year period. Seed yield ranged from 0.32 to 9.83 t/ha, with higher yields recorded in four Chilean, two US, one Argentinian, and one Bolivian line.

Hybridization

Hybridization has been used to improve quinoa. The main objectives of hybridization are to bring together desired characters from different accessions into one new plant or plant line to improve certain quantitative characteristics.

Selection of parents

Selection of parents is the first and one of the most important steps in a successful quinoa breeding program. The choice of parents will depend on the objectives of the program and the availability of genotypes to meet the specific objectives. Crossing two genotypes with complementary traits is a common practice. Valuable characteristics are scattered among germplasm of different origin such as saponin-free ecotypes that have small to medium seed size and bitter ecotypes that have large seeds; tolerance/resistance to mildew (*P. variabilis*) is present in Valley ecotypes combined with branched stems and a long life cycle; Altiplano ecotypes are susceptible to downy mildew but have a single stem and mature earlier. All these characters can be combined into one genotype using hybridization and subsequent selection. Risi and Galwey (1989) concluded that quinoa characteristics required for temperate conditions are available to a large extent in accessions collected near sea level in southern-central Chile, whereas seed characteristics are scattered throughout the germplasm. Carmen (1984) recommended the utilization of earlier genetic material available from the quinoa collection of the Germplasm Bank of Andean Crops at the Altiplano-Puno National University

(UNAP). Quinoa breeders might focus on introducing resistance to downy mildew from coastal germplasm to Altiplano germplasm. Conversely, coastal cultivars need the introduction of more attractive agronomic characteristics, such as larger seed size, that are present in the Real-type quinoas from the Andean highlands (Fuentes et al. 2009).

Flowering biology

Anthesis in quinoa starts at the apex of each glomerulus or group of flowers. Hermaphrodite and female flowers generally open at the same time. Most flowers open in the morning and the maximum number of flowers open at midday. Rain reduces the number of open flowers. Anther dehiscence occurs from early morning until late afternoon. It is also highest at midday during which large quantities of pollen are produced. Flowers remain open for 5–13 days. In eight landraces and five commercial varieties of quinoa, the average duration of anthesis was 14.5 days and the average duration of dehiscence was 18.2 days. An average of 2.5% floral aberrations had been observed (Ignacio and Vera 1976; Gandarillas 1979a; Lescano 1980).

Emasculation and pollination procedure

Procedures for artificial hybridization of quinoa have been described by Rea (1948), Gandarillas and Luizaga (1967), Gandarillas (1979a), Lescano and Palomino (1976), and Bonifacio (1990, 1995). Emasculation techniques and artificial pollination are often simple procedures in many other crops but are quite laborious in quinoa due to inflorescence characteristics and extremely small flower size. Quinoa crosses can be made in the field and in the greenhouse but it is very important to have favorable environmental conditions for plant growth and development. Crosses between "Pasankalla" × "Salcedo" INIA and "Pasankalla" × "Choclo" made under field conditions resulted in 41% and 63% seed set in the female parent, respectively (Leon 2004, 2005).

Crossing may be simple or reciprocal, but in all cases, it is important to have one or more morphological markers or qualitative characters present in one of the parents, to be used as morphological marker to identify the hybrids in the F_1 generation. The female parental line must have the recessive character and the male parental line the dominant character. It is also important to select parental lines that can withstand handling during the whole process, which can take 10–14 days. The female parent must tolerate repetitive emasculation and pollination and the male parent must provide an adequate amount of viable pollen.

The equipment needed for crossing consist of small scissors with sharp and pointed blades, finely tipped forceps or needles, glassine bags, paper clips or staples, and small tags. The female plant must be prepared by choosing two or three equidistant glomerulus from the base of the inflorescence before flowering time. Leaves of the inflorescence must be removed, retaining only one leaf under each glomerulus and the rest of the inflorescence also removed. The optimum time to emasculate quinoa flowers is when the anthers have obtained full size but are not yet ready to dehisce. Hermaphrodite flowers are chosen for emasculation, which is the process of removing the anthers with a needle before they have shed pollen without damaging the stigmata. After which, the female flowers of the glomerulus are retained. Five anthers in each inflorescence must be removed. The process can take 10–14 days. The inflorescence must be covered with a glassine bag or a paper bag depending of the size of the inflorescence to exclude unwanted pollen once the process has been initiated. The bag should be just wide enough to cover the inflorescence and long enough to be folded and stapled or secured with a paper clip. In the field, crossed inflorescences are sometimes supported by a stake. An identification tag must be used, containing the names of parent lines, the date of emasculation, and the name of the operator.

Emasculation using hot water (42°C), as in the case of sorghum (Alvarez 1993), has not been successful in quinoa. The use of the suction pump recommended for rice (Jennings et al. 1981) has not been tested in quinoa, though presumably it will be ineffective due to the tightly closed perianth lobes enveloping the anthers.

Clear days with temperatures near to 24°C are ideal for gathering viable pollen. Cloudy days with high relative humidity and low temperature, near or below 15°C, are not favorable for anthesis. Quinoa pollen must be collected in a watch glass, paper bag, or Petri dish. The selected inflorescence must be bent over and shaken to collect pollen. Viable pollen can be recognized by the appearance of fine yellowish powder without agglutination, whereas old pollen shows agglutination and the presence of pollen sacs.

The phenomenon of male sterility in quinoa can also be utilized in crossing quinoa. However, using male sterility requires prior development of male sterile lines and maintainer lines, as well as the introduction of male sterility into the selected parents for each cross. Conversely, if there are excellent lines possessing male sterility, they can be used directly as female parents without being emasculated. The process begins with the identification of male sterile plants, in which apical parts of the inflorescence must be removed to eliminate any flower that could have already received foreign pollen. Plants are then isolated for 1 week and flowers forming seed are removed. Male sterile plants (to be used as female parents) are then pollinated with pollen collected from the male parent. Pollination can be repeated for a second time to ensure greater seed production. Ward and Johnson (1993) made crosses using male sterility, where the selected male parent was placed next to the male sterile plant. This method is simple and may be applicable in greenhouses for a single or few crosses where foreign pollen contamination can be easily controlled.

Bulk or mass selection

The F_2 population must be large so that there are enough recombinant plants to develop a new population combining desirable characteristics. Bulk selection, pedigree selection, and the single-seed descent (SSD) method could be used to manage the segregating population.

Bulk selection is a preferred breeding method due to its simplicity and low labor cost. Plants from F_2 to F_6 are harvested in bulk. The following year, a sample of the bulk in each generation is planted in a plot with the expectation that natural selection will reduce the frequency of maladapted types in the population. In quinoa, it is recommended to have more than 30,000 plants in each generation for mass selection. At the F_6 generation, plants are considered to have the desired level of homozygosity and breeders make inflorescence selections. In the next generation, inflorescence rows are planted. Each is treated as a line and yield trials are begun.

In Bolivia, the cultivar Sajama was developed following this method. In the F_2 generation, all the bitter plants were discarded. In the following generations, plant vigor, seed size, compact inflorescence, and sweet seeds were the traits used in the selection procedure (Gandarillas 1979a, 1979b). Other varieties obtained by this method are "Chucapaca," "Huaranga," and "Kamiri Robura" (Bonifacio 2003).

Individual or pedigree selection

Individual or pedigree selection requires the maintenance of records on selected material, from F_2 to F_6 or until the generation when the seed is bulked to start yield trials. Each F_2 plant selected from a cross is identified by the number of the cross and selection number. Inflorescence rows are grown from each F_2 plant and selection is made among and within the F_3 progeny rows. Each plant selected is identified by a number, and seeds from each plant are grown as the F_4 progeny. This procedure can be continued until the desired level of homozygosity is reached, at which point seeds within selected rows are bulked and planted as new lines in yield trials.

In the F_2 or F_3 generation, it is advisable to select for the most heritable characters such as sweet seeds, color of inflorescence, and color of seeds. In the F_3, F_4, or more advanced generations, selection will be made for characters controlled by multiple genes. More recently, this method has been modified by the inflorescence-row systems.

In Bolivia, in a joint effort among the PROINPA 2002-2003, PREDUZA, and McKnight projects, a cross with 25 diverse parental varieties was made to obtain 36 successful crosses that generated true F_1 hybrids. The F_2, F_3, and F_4 generations were managed using two generations per year in a "waliplini" (semi-subterranean greenhouse). The progenies were selected for favorable characteristics, such as resistance to mildew, large seed size, early maturity, white and colored seeds, and plant vigor. Advanced generations were tested in a yield trial in five locations. Line L-26(85) was selected because of its yield, earliness, and seed size. It was released as a new variety with the name "Jach'a Grano," which in the native language means "large seed" (http://www.proinpa.org).

Combination of individual and mass selection

A combination of individual selection and mass selection can also be used in the F_2 population. The combination permits the evaluation of the individual offspring and the broadening of the genetic base of future varieties. The varieties obtained by this combined method are "Sayaña," "Jilata," "Patacamaya," "Ratuqui," "Jumataki," "Intinaira," "Surumi," and "Santamaria" (Bonifacio 2003).

Single-seed descent

In the SSD method, the objective is to maintain a high level of heterogeneity among a larger number of plants in a population as they are brought rapidly toward homozygosity. The basic procedure is to remove one or two seeds from the inflorescence of selected plants from the F_2 to F_6 generations. In quinoa, due to the small size of seeds and the difficulty in the initial establishment of the plants, it is recommended to remove 5–10 seeds in each plant per generation. After planting all seeds, the most vigorous plant is chosen to be advanced. The SSD method advances cycles in the greenhouse and field and reduces the time needed to develop a new cultivar. The F_7 and more advanced generations are grown in the field, and outstanding plants are selected with farmer participation.

Interspecific and intergeneric crosses

Interspecific crosses (simple and reciprocal) are done in quinoa to recombine favorable traits present in different species and concentrate them in the selected offspring. However, hybrid sterility had been observed, with fertility only restored through backcrossing (Bonifacio 1995). It was also possible to obtain hybrids from the intergeneric cross between quinoa and *Atriplex hortensis*, but hybrid sterility was more difficult to overcome. Another intergeneric cross between quinoa and *S. foliosa* showed that high level of frost and salt resistance present in Swaeda (Kauchi) may be incorporated into quinoa (Bonifacio 2003).

Backcross method

The backcross method can be used improve a cultivar that is outstanding in many traits but is deficient in certain qualitative characters. One example is the work done in Bolivia, where the line 1638 (donor parent) was crossed with the cultivar "Patacamaya" (recurrent parent). The donor parent is an accession of the Real race, a Pandela (pink) type with large bitter seeds and the recurrent parent is green in color and has sweet seeds. The resultant F_1 was a Pandela (pink) type with bitter seeds. The F_1 was backcrossed to the Patacamaya cultivar, resulting in progenies that are of a Pandela (pink) type cultivar with large, sweet seeds. Backcrossing has also been used primarily after cultivated and wild progenitors were crossed to clean up the genetic background of the progeny (Bonifacio 2003).

Using heterosis in quinoa

Hybrid vigor in quinoa, as observed by Ward and Johnson (1993), has generated interest in the production of commercial hybrids. Wilson (1990) differentiated groups in quinoa, which are the result of crosses between accessions of very different origin and sources of male sterility. In Colorado crossing trials, quinoas within Wilson's designated groups have shown no heterosis for yield, whereas crosses between his groups have shown heterosis varying between 201% and 491% (Wilson 1990).

Mutagenesis

The advantage of nuclear techniques such as mutagenesis, compared to conventional methods such as hybridization and recurrent selection, is that a single or few traits can be targeted for improvement in local or native cultivars that are already superior in many traits. Hybridization involves the recombination of parental genome but the selection of target traits will require a large amount of time and land. In contrast, the use of mutagenesis to select for target traits can be faster and with less use of land and other resources. The first work on mutagenesis in quinoa was reported in Peru (Gomez-Pando and Eguiluz-de la Barra 2013). Dry seeds of the cultivar Pasankalla were irradiated with gamma rays at doses of 150, 250, and 350 Gy. In the M_1 generation, the germination process was delayed with increasing radiation dose, whereas seedling height, root length, and leaf development were mostly reduced at a dose of 250 Gy. At the 350 Gy dose, no plants survived. In the M_2, the maximum spectrum of chlorophyll mutations corresponded to the 150 Gy dose, whereas the maximum frequency of chlorophyll mutations were at 250 Gy. The chlorine mutation was predominant, followed by the xanthan mutation. In both doses, changes were observed in branch number, pedicel length, plant height, life-cycle duration, stem and foliage color, and leaf morphology, with overall improvements in plant type, especially with increased vigor and yield potential. More than one mutation per plant was found, especially at the 250 Gy. In the M_3, the same spectrum of mutations was observed, along with desirable changes such as reduction in plant height. Some quinoa traditional cultivars can reach more than 2 m, are prone to lodging, and present difficulties in harvesting. In such cultivars, reduced plant height is desirable. Changes in seed color (from red to white) can also be achieved, especially to target markets where the white seed color is preferred (Gomez-Pando and Eguiluz-de la Barra 2013).

Marker-assisted selection (MAS)

Modern techniques that permit the use of markers linked to quantitative trait loci (QTL) in quinoa could be exploited for future cultivar improvement. In a joint effort among the PROINPA, PREDUZA, and McKnight projects, a medium-density genetic map was developed. Hundreds of single-sequence repeat (SSR) and single-nucleotide polymorphism (SNP) markers have been developed by scientists at the Brigham Young University (BYU) and are being mapped to three mapping populations to produce a high-density genetic map for marker-assisted selection (MAS) in Bolivia. Molecular markers for important traits such as resistance to mildew, saponin content, and seed protein content and composition had been mapped using these markers (Maughan et al. 2004; Coles et al. 2005; Mason et al. 2005; Stevens et al. 2006; Jarvis et al. 2008; Rodriguez and Isla 2009), making it possible to use MAS in quinoa breeding programs.

CONCLUSION

Genetic improvement of the quinoa crop can be achieved by various methods such as classical breeding, mutagenesis, and molecular genetics techniques to obtain outstanding genotypes. Quinoa is an underutilized crop that could be a valuable alternative crop to help the world face critical challenges such as climatic change, food security, human nutrition, and overdependence on a few plant species for the world food supply.

REFERENCES

Alvarez A. 1993. Evaluacion de Tecnicas de Hibridacion en el Mejoramiento Genetico de la quinua (*Chenopodium quinoa* Willd). Mg. Sc. Thesis, Universidad Nacional Agraria La Molina, Lima, Peru. p. 51.

Aguilar AP. 1980. Identificación de mecanismos de androesterilidad, componentes de rendimiento y contenido proteico en quinua (*Chenopodium quinoa* Willd). Mg. Sc. Thesis, Universidad Agraria La Molina, Lima-Peru. p. 77.

Bhargava A, Shukla S, Ohri D. Genetic variability and interrelationship among various morphological and quality traits in quinoa (*Chenopodium quinoa* Willd.). Field Crops Res 2007a;101:104–116.

Bhargava A, Shukla S, Ohri D. Gynomonoecy in *Chenopodium quinoa* (Chenopodiaceae): variation in inflorescence and floral types in some accessions. Biologia Bratislava 2007b;62(1):19–23.

Bhargava A, Shuklab S, Ohrib D. Implications of direct and indirect selection parameters for improvement of grain yield and quality components in *Chenopodium quinoa* Willd. Int J Plant Prod 2008;2:184–191.

Bollaert W. 1860. Antiquarian, ethnological and other researchers in New Granada, Ecuador, Peru and Chile. S.N.T: 279.

Bertero HD, King RW, Hall AJ. Photoperiod-sensitive development phases in quinoa (*Chenopodium quinoa* Willd.). Field Crops Res 1999;60:231–243.

Bertero HD, de la Vega AJ, Correa G, Jacobsen SE, Mujica A. Genotype and genotype-by-environment interaction effects for grain yield and grain size of quinoa (*Chenopodium quinoa* Willd.) as revealed by pattern analysis of international multi-environment trials. Field Crops Res 2004;89:299–318.

Bertero HD, Ruiz RA. Determination of seed number in sea level quinoa (*Chenopodium quinoa* Willd.) cultivars. Eur J Agron 2008;28(3):186–194.

Bertero HD, Ruiz RA. Reproductive partitioning in sea level quinoa (*Chenopodium quinoa* Willd.) cultivars. Field Crops Res 2010;118:94–101.

Bonifacio A. 2003. *Chenopodium* spp.: genetic resources, ethnobotany, and geographic distribution. Food Rev Int 19(1):1–7. DOI: 10.1081/FRI-120018863. Available from: 10.1081/FRI-120018863.

Bonifacio A, Mujica A, Alvarez A, Roca W. Mejoramiento genético, germoplasma y producción de semilla. In: Mujica A, Jacobsen S, Izquierdo J, Marathee MP, editors. Quinua: ancestral cultivo andino, alimento del presente y futuro. Chile: FAO. UNA. CIP. Santiago; 2004. p 125–187.

Bonifacio A. 1990. Caracteres hereditarios y ligamiento factorial en la quinua (*Chenopodium quinoa* Willd.) Ing. Agr. Thesis, Cochabamba, Bolivia Universidad Mayor de San Simon. p. 189

Bonifacio A. 1991. Materiales de aislamiento en cruzamientos de la quinua. In: Congreso Internacional sobre Cultivos Andinos, 6to. Quito, Ecuador. pp. 67–68.

Bonifacio A. 1995. Interspecific and intergeneric hybridization in chenopod species. Thesis M.Sc., Provo, Utah Brigham Young University. p. 150

Bravo R, Catacora P. Granos Andinos: avances, logros y experiencias desarrolladas en quinua. In: Bravo R, Valdivia R, Andrade K, Padulosi S, Jager M, editors. Cañihua y Kiwicha en el Perú. Ed FIDA - Bioversity International: Roma, Italia; 2010. p 136.

Bruno MC, Whitehead WT. *Chenopodium* cultivation and the formative period of agriculture at Chiripa, Bolivia. Latin Am Antiq 2003;14(3):339–355.

Cardenas M, Hawkes JG. Número de cromosomas de algunas plantas nativas cultivadas por los indios en los Andes. Rev Agric 1948;4:30–32.

Carmen M. Acclimatization of quinoa (*Chenopodium quinoa* Willd.) and canihua (*Chenopodium pallidicaule* Aellen) to Finland. Ann Agr Fenniae 1984;23:135–144.

Cayoja MR. 1996. Caracterización de variables continuas y discretas del grano de quinua (*Chenopodium quinoa* Willd.) del banco de germoplasma de la Estación Experimental Patacamaya. Tesis de Lic. en Agronomía. Oruro, Bolivia, Universidad Técnica Oruro, Facultad de Agronomía. 129 p.

Ceccarelli S, Grando S, Tutweiler R, Baha J, Martini AM, Salahieh H, Goodchild A, Michael M. A methodological study on participatory barley breeding: I. selection phase. Euphytica 2000;111:91–104.

Christensen SA, Pratt DB, Pratt C, Stevens MR, Jellen EN, Coleman CE, Fairbanks DJ, Bonifacio A, Maughan PJ. Assessment of genetic diversity in the USDA and CIP-FAO international nursery collections of quinoa (*Chenopodium quinoa* Willd.) using microsatellite markers. Plant Genet Res 2007;5:82–95.

Coles ND, Coleman CE, Christensen SA, Jelle EN, Stevens MR, Bonifacio A, Rojas-Beltran JA, Fairbanks DJ, Maughan PJ. Development and use of an expressed sequenced tag library in quinoa (*Chenopodium quinoa* Willd.) for the discovery of single nucleotide polymorphisms. Plant Sci 2005;168(2):439–447.

Cusack DF. Quinoa: grain of the Incas. Ecologist 1984;14:21–31.

Danielsen S, Jacobsen SE, Mujica A. 2000. Susceptibilidad al mildiu (*Peronospora farinosa*) y pérdida de rendimiento en ocho cultivares de quinua (*Chenopodium quinoa* Willd). En: Resumen II Congreso Internacional de Agricultura en Zonas Áridas. Iquique, Chile. p. 59.

Duvick DN. Plant Breeding, an Evolutionary Concept. Crop Sci 1996;36:539–548.

Eguiluz A, Maugham PJ, Jellen E, Gómez L. 2010. Determinación de la diversidad fenotipica de accesiones de quinua (*Chenopodium quinoa* Willd) provenientes de valles interandinos y del altiplano peruano. Available from: http://www.infoquinua.bo/fileponencias/.

Espindola G. 1980. Estudio de componentes directos e indirectos del rendimiento en quinua (*Chenopodium quinoa* Willd.) Ing. Agr. Thesis, Cochabamba, Bolivia Universidad Mayor de San Simon. 91 p.

FAOSTAT. 2014. http://faostat3.fao.org.

Fuentes F, Martinez E, Delatorre J, Hinrichsen P, Jellen E, Maughan PJ. 2006. Diversidad genética de germoplasma chileno de quinua (*Chenopodium quinoa* Willd.) usando marcadores de microsatélites SSR. In: Estrella A, Batallas M, Peralta E and Mazón N, editors. Resúmenes XII Congreso Internacional de Cultivos Andinos. 24 al 27 de Julio de 2006. Quito, Ecuador.

Fuentes FF, Martinez EA, Hinrichsen PV, Jellen EN, Maughan PJ. Assessment of genetic diversity patterns in Chilean quinoa (*Chenopodium quinoa* Willd.) germplasm using multiplex fluorescent microsatellite markers. Conserv Genet 2009;10:369–377.

Gandarillas H, Luizaga J. Numero de cromosomas de *Chenopodium quinoa* Willd. En radículas y raicillas. Turrialba (Costa Rica) 1967;17(3):275–279.

Gandarillas H. 1968. Razas de quinua. Boletin Experimental 34. La Paz, Bolivia: Ministerio de Agricultura y Asuntos Campesinos. Division de InvestigacionesAgrıcolas, Universo. p. 35.

Gandarillas H. Esterilidad genética y citoplasmática de la quinua (*Chenopodium quinoa* Willd.). Turrialba 1969;19:429–430.

Gandarillas H. 1974. Genetica y origen de la quinua. Ministerio de Asuntos Campesinos y Agropecuarios. Instituto Nacional del Trigo. La Paz, Bolivia. Boletin Informativo 9. p. 21.

Gandarillas H, Tapia M. 1976. La variedad de quinua dulce Sajama. In: II Convención Internacional de Quenopodiaceas, Quinua y Cañahua. 26 – 29 abril, Potosi, Bolivia. UBTF, CDOP de Potosi, IICA. Potosi, Bolivia. p. 105.

Gandarillas H. Mejoramiento genético. In: Tapia M, editor. Quinua y Kaniwa. Cultivos Andinos. Bogotá, Colombia: Centro Internacional de Investigaciones para el Desarrollo (CIID), Instituto Interamericano de Ciencias Agricolas (IICA); 1979a. p 65–82.

Gandarillas H. Genetica y origen. In: Tapia M, editor. Quinua y Kaniwa. Cultivos Andinos. Bogotá, Colombia: Centro Internacional de Investigaciones para el Desarrollo (CIID), Instituto Interamericano de Ciencias Agricolas (IICA); 1979b. p 45–64.

Gandarillas H. 1986. Estudio anatomico de los organos de la quinua. Estudio de caracteres correlacionados y sus efectos sobre el rendimiento. Hibridaciones entre especies de la subseccion. Cellulata del Genero *Chenopodium*, La Paz, Bolivia. 48 p.

Gęsiński K. Evaluation of the development and yielding potential of *Chenopodium quinoa* Willd. under the climatic conditions of Europe. In: Part two: Yielding potential of *Chenopodium quinoa* under different conditions. Acta Agrobotanica 2008;61(1):185–189.

Gomez-Pando L, Alvarez-Castro R, Eguiluz-de la Barra A. Effect of salt stress on Peruvian germplasm of *Chenopodium quinoa* Willd.: a promising crop. J Agron Crop Sci 2010;196:391–396.

Gómez L, Eguiluz A. 2011. Catalógo del Banco de Germoplasma de Quinua (*Chenopodium quinoa* Willd), Universidad Nacional Agraria La Molina, p. 183.

Gomez-Pando L, Eguiluz-de la Barra A. 2013. Developing genetic variability of quinoa (*Chenopodium quinoa* Willd.) with gamma radiation for use in breeding programs. Am J Plant Sci 4:349–355. doi:10.4236/ajps.2013.42046. Published Online February 2013 available from: http://www.scirp.org/journal/ajps.

Gremillion KJ. The evolution of seed morphology in domesticated *Chenopodium*: an archeological case study. J Ethnobiol 1993;13(2):149–169.

House LR. El sorgo. México, Gaceta: Guía para su mejoramiento. Universidad Autónoma Chapingo; 1982. p 425.

Ichuta F, Artiaga E. 1986. Relación de géneros en la producción y en la organización social en comunidades de Apharuni, Totoruma, Yauricani-Ilave. Informe para optar el grado de Bachiller en Trabajo Social. Puno, Peru. pp. 15–17.

Ignacio J, Vera R. 1976. Observaciones sobre la intensidad de floraci{on durante las diferentes horas del día efectuados en quinua *Chenopodium quinoa* Willd. Anales de la II Concenci{on Internacional de Quenopodiaceas quinua-cañihua. Potosi Bolibia. p. 102.

Jacobsen S, Stolen O. Quinoa - Morphology, phenology and prospects for its production as a new crop in Europe. Eur J Agron 1993;2(1):19–29.

Jacobsen SE, Quispe H, Mujica A. 2001. Quinoa: an alternative crop for saline in the Andes. In: Scientist and Farmer – Partners in Research for the 21st Century. CIP Program Report 1999–2000. pp. 403–408.

Jacobsen SE, Mujica A. Genetic resources and breeding of the Andean grain crop quinoa (*Chenopodium quinoa* Willd.). Plant Genet Res Newsl 2002;130:54–61.

Jacobsen SE, Mujica A, Jensen CR. The resistance of quinoa (*Chenopodium quinoa* Willd.) to adverse abiotic factors. Food Rev Int 2003;19:99–109.

Jacobsen SE, Monteros C, Christiansen JL, Bravo LA, Corcuera LJ, Mujica A. Plant responses of quinoa (*Chenopodium quinoa* Willd.) to frost at various phenological stages. Eur J Agron 2005;22(2):131–139.

Jacobsen SE, Monteros C, Corcuera LJ, Bravo LA, Christiansen JL, Mujica A. Frost resistance mechanisms in quinoa (*Chenopodium quinoa* Willd.). Eur J Agron 2007;26(4):471–475.

Jarvis DE, Kopp OR, Jellen EN, Mallory MA, Pattee J, Bonifacio A, Coleman CE, Stevens MR, Fairbanks DJ, Maughan PJ. Simple sequence repeat marker development and genetic mapping in quinoa (*Chenopodium quinoa* Willd.). J Genet 2008;87(1):39–51.

Jennings P, Coffman W, Kauffman H. Mejoramiento de Arroz. Centro Internacional de Agricultura Tropical (CIAT). CIAT: Cali; 1981. p 237.

Johnson DL, Ward SM. Quinoa. In: Janick J, Simon JE, editors. New crops. New York: Wiley; 1993. p 219–221.

Koziol MJ. Chemical composition and nutritional value of quinoa (*Chenopodium quinoa* Willd.). J Food Comp Anal 1992;5:35–68.

Kumar A, Bhargava A, Shukla S, Ohri D. Screening of exotic *Chenopodium quinoa* accessions for downy mildew resistance under mid-eastern conditions of India. Crop Prot 2006;25:879–889.

Leon J. 1964. Plantas Alimenticias. Bol. Tecnico. No.6. Instituto Interamericano de Ciencias Agricolas-Zona Andina, Lima, Peru. p. 112.

Leon J. 2004, 2005. Hibridación y comparación de la F1 con sus progenitores en tres cultivares de quinua (*Chenopodium quinoa* Willd.) en Puno, Perú. www.monografias.com/ ... quinua ... /mejoramiento-genetico-quinua-hibrid.

Lescano R, Palomino L. 1976. Metodologia de cruzamiento en quinua. In Segunda convención Internacional de Quenopodiaceas. Universidad Boliviana Tomas Frias, Comité Departamental de Obras Publicas de Postosi, Instituto Interamericano de Ciencias Agricolas, Potosi, Bolivia, pp. 78–80.

Lescano R. 1980. Avances en la genetica de la quinua. In Primera Reunion de Genetica y Fitomejoramiento de la Quinua. Universidad Nacional del Altiplano, Instituto Boliviano de Tecnologia Agropecuaria, Instituto Interamericano de Ciencias Agricolas, Centro Internacional de Investigaciones para el Desarrollo de Puno. pp. 1–9.

Lescano JL. Mejoramiento y fisiologia de cultivos andinos. CONCYTEC, Proyecto FEAS: Cultivos andinos en el Peru. Lima; 1994. p 231.

Lumbreras LG, Kaulicke P, Santillana JI, Espinoza W. 2008. Economía Prehispanica (Tomo 1). In: Contreras C, editor. Compendio de historia economica del Peru. Banco Central de Reserva del Peru. Instituto de Estudios Peruanos. pp. 53–77.

Malt SH, Mittelbach M, Rechberger GN. Tandem mass spectrometric analysis of a complex triterpene saponin mixture of *Chenopodium quinoa*. J Am Soc Mass Spec 2006;17(6):795–806.

Mason SL, Stevens MR, Jellen EN, Bonifacio A, Fairbanks DJ, Coleman CE, McCarty RR, Rasmussen AG, Maughan PJ. Development and use of microsatellite markers for germplasm characterization in quinoa (*Chenopodium quinoa* Willd.). Crop Sci 2005;45(4):1618–1630.

Maughan PJ, Bonifacio A, Jellen EN, Stevens MR, Coleman CE, Ricks M, Mason SL, Jarvis DE, Gardunia BW, Fairbanks DJ. A genetic linkage map of quinoa (*Chenopodium quinoa*) based on AFLP, RAPD, and SSR markers. Theor Applied Genet 2004;109:1188–1195.

Maughan PJ, Kolano BA, Maluszynska J, Coles ND, Bonifacio A, Rojas J, Coleman CE, Stevens MR, Fairbanks DJ, Parkinson SE, Jellen EN. Molecular and cytological characterization of ribosomal RNA genes in *Chenopodium quinoa* and *Chenopodium berlandieri*. Genome 2006;49(7):825–839.

McElhinny E, Peralta E, Mazón N, Danial D, Thiele G, Lindhout P. Aspects of participatory plant breeding for quinoa in marginal areas of Ecuador. Euphytica 2007;153:373–384.

Mujica A. 1988. Parámetros genéticos e índices de selección en quinua (*Chenopodium quinoa* Willd.). Tesis de Doctor en Ciencias. Colegio de Postgrado, Montecillo, México. 113 p.

Mujica A. Granos y leguminosas andinas. In: Hernandez J, Bermejo J, Leon J, editors. Cultivos marginados: otra perspectiva de 1492. Organización de la Naciones Unidas para la Agricultura y la Alimentación FAO: Roma; 1992. p 129–146.

Mujica A, Jacobsen SE. 2000. Agrobiodiversidad de las Aynokas de quinua (*Chenopodium quinoa* Willd.) y la seguridad alimentaria. In: Felipe-Morales C, Manrique A, editors. Proc. Seminario Taller Agrobiodiversidad en la Región Andina y Amazónica. 23–25 Noviembre 1988. Lima: NGO- CGIAR. pp. 151–156.

Mujica A, Jacobsen SE, Izquierdo J, Marathee JP. 2001. Resultados de la prueba americana y Europes de la quinua. FAO, UNA-Puno, CIP. p. 51.

Mujica A, Jacobsen SE. 2005. La quinua (*Chenopodium quinoa* Willd.) y sus parientes silvestres. In: Moraes M, Øllgaard RB, Kvist LP, Borchsenius F, Balslev H, editors. Botánica económica de los Andes Centrales. La Paz: Universidad Mayor de San Andrés. pp. 449–457. Available from http://www.ars-grin.gov/cgi-bin/npgs/html/tax_site_acc.pl?NC7%20Chenopodium%20quinoa.

Murray A. 2005. *Chenopodium* domestication in the South Central Andes: confirming the presence of domesticates at Jiskairumoko (Late Archaic-Formative), Peru. Thesis, Master of Arts in Anthropology, California State University: Fullerton. 98 p.

[NRC] National Research Council. Lost crops of the Incas. Washington, DC: National Academy Press; 1989. p 414.

Nelson DC. 1968. Taxonomy and origins of *Chenopodium quinoa* and *Chenopodium nuttalliae*. Ph.D. Thesis, University of Indiana, Bloomington. 99 p.

Ochoa J, Peralta E. 1988. Evaluación preliminar morfológica y agronómica de 153 entradas de quinua en Santa Catalina, Pichincha. Actas del VI Congreso Internacional sobre Cultivos Andinos. Quito, Ecuador. pp. 137–142.

Ochoa J, Frinking HD, Jacobs T. Postulation of virulence groups and resistance factors in the quinoa/downy mildew pathosystem using material from Ecuador. Pl Path 1999;48(3):425–430.

Ortiz R, Ruiz-Tapia EN, Mujica-Sanchez A. Sampling strategy for a core collection of Peruvian quinoa germplasm. Theor Appl Genet 1998;96:475–483.

Prakash D, Nath P, Pal M. Composition, variation of nutritional contents in leaves, seed protein, fat and fatty acid profile of Chenopodium species. J Sci Food Agric 1993;62:203–205.

PROINPA (http://www.proinpa.org). 2002–2003. Promocion e Investigacion de Productos Andinos Foundation, [BYU] Brigham Young University, The McKnight Foundation: Sustainable production of quinoa (*Chenopodium quinoa* Willd.) a neglected food crop in the Andean region. Annual report. p. 31.

Rea J. 1948. Observaciones sobre biologia floral y studio de saponinas en Chenopodium quinoa Willd. Ministerio de Agricultura. La Paz. Serie Tecnica3. p. 17.

Rea J. Biologıa floral de la quinua (*Chenopodium quinoa* Willd.). Turrialba 1969;19:91–96.

Repo-Carrasco R, Espinoza C, Jacobsen SE. Nutritional value and use of the Andean crops Quinoa (*Chenopodium quinoa*) and Kaniwa (*Chenopodium pallidicaule*). Food Rev Int 2003;19:179–189.

Risi J, Galwey NW. The *Chenopodium* grains of the Andes: Inca crops for modern agriculture. Adv Appl Biol 1984;10:145–216.

Risi J. 1986. Adaptation of the Andean grain crop quinoa (*Chenopodium quinoa* Willd.) for cultivation in Britain. Thesis Ph. D. University of Cambridge, Cambridge, Britain. 338 p.

Risi J, Galwey NW. Chenpodium grains of the Andes: a crop for temperate latitudes. In: Wickens GE, Haq N, Day P, editors. New crops for food and industry. London: Chapman and Hall; 1989. p 222–234.

Rodríguez MF, Rugolo de Agrasar ZE, Aschero CA. El uso de las plantas en unidades domésticas del Sitio arqueológico Punta de la Peña 4, Puna Meridional Argentina. Chungara Revista de Antropología Chilena 2006;38(2):257–271.

Rodriguez LA, Isla MT. Comparative analysis of genetic and morphologic diversity among quinoa accessions (*Chenopodium quinoa* Willd.) of the South of Chile and highland accessions. J Plant Breeding Crop Sci 2009;1(5):210–216.

Rojas W, Cayoja MR, Espindola G. Catálogo de la colección de quinua conservado en el Banco Nacional de Granos Altoandinos. Bolivia: La Paz; 2001. p 128.

Rojas W. Multivariate analysis of genetic diversity of Bolivian quinoa germplasm. Food Rev Int 2003;19(1–2):9–23.

Rojas W, Barriga P, Figueroa H. Multivariate analysis of genetic diversity of Bolivian quinoa germplasm. Food Rev Int 2003;19:9–23.

Rojas W, Mamani E, Pinto M, Alanoca C, Ortuño T. 2008. Identificación taxonómica de parientes silvestres de quinua del Banco de Germoplasma de Granos Altoandinos. Revista de Agricultura Año 60, Nro. 44. Cochabamba, Bolivia. pp. 56–65.

Rojas W, Pinto M, Maman E. 2009. Logro e impactos del Subsistema Granos Altoandinos, periodo 2003 – 2008. En Encuentro Nacional de Innovación Tecnológica, Agropecuaria y Forestal. INIAF. Cochabamba. 29 y 30 de Junio de 2009. pp. 58–65.

Rojas W, Pinto M, Bonifacio A, Gandarillas A. Banco de germoplasma de granos Andinos. In: Rojas W, Pinto M, Soto JL, Jagger M, Padulosi S, editors. Granos Andinos: avances, logros y experiencias desarrolladas en quinua, cañahua y amaranto en Bolivia. Bioversity International: Roma, Italia; 2010. p 24–38.

Ruiz R. 2002. Micropropagación de germoplasma de quinua (*Chenopodium quinoa* Willd.). Tesis Universidad Nacional Agraria La Molina. 100 p.

Ruiz-Carrasco AF, Coulibaly AK, Lizardi S, Covarrubias A, Martinez EA, Molina-Montenegro MA, Biondi S, Zurita- SA. Variation in salinity tolerance of four lowland genotypes of quinoa (*Chenopodium quinoa* Willd.) as assessed by growth, physiological traits, and sodium transporter gene expression. Plant Physiol Biochem 2011;49(11):133–1341.

Saravia R. 1991. La androesterilidad en quinua y forma de herencia. Tesis Ing. Agr., Cochabamba, Bolivia Universidad Mayor de San Simon. 139 p.

Simmonds NW. The breeding system of *Chenopodium quinoa*. I Male sterility. Heredity 1971;27:73–82.

Silvestri V, Gil F. Alogamia en quinua. Tasa en Mendoza (Argentina). Revista de la Facultad de Ciencias Agrarias, Universidad Nacional de Cuyo 2000;32(1):71–76.

Smith BD. *Chenopodium* as a prehistoric domesticate in Eastern North America: evidence from Russelll Cave, Alabama. Science 1984;226(4671):165–167.

Stevens MR, Coleman CE, Parkinson SE, Maugham PJ, Zhang HB, Balzotti MR, Kooyman DL, Arumuganathan K, Bonifacio A, Fairbanks DJ, et al. Construction of a quinoa (*Chenopodium quinoa* Willd.) BAC library and its use in identifying genes encoding seed storage proteins. Theor Applied Genet 2006;112:1593–1600.

Tapia M. Historia y distribución geográfica. In: Tapia M, editor. Quinua y Kaniwa. Cultivos Andinos. Bogotá, Colombia: Centro Internacional de Investigaciones para el Desarrollo (CIID), Instituto Interamericano de Ciencias Agricolas (IICA); 1979. p 11–19.

Tapia ME, Mujica SA, Canahua A. 1980. Origen, distribución geográfica y sistemas de producción en quinua. In: Primera reunión sobre genética y fitomejoramiento de la quinua. Universidad Técnica del Altiplano, Instituto Boliviano de Tecnología Agropecuaria, Instituto Interamericano de Ciencias Agrícolas, Centro de Investigación Internacional para el Desarrollo, Puno, Peru. pp. A1–A8.

Tapia ME. 1982. El medio, los cultivos y los sistemas agrícolas en los Andes del Sur del Peru. Proyecto de Investigación de los Sistemas Agricolas Andinos, Instituto Interamericano de Ciencias Agrícolas, Centro de Investigación Internacional para el Desarrollo, Cusco, Peru. 79 p.

Verena IA, Jacobsen SE, Shabalab S. Salt tolerance mechanisms in quinoa (*Chenopodium quinoa* Willd.). Env Exp Bot 2013;92:43–54.

Ward SM, Johnson D. Cytoplasmic male sterility in quinoa. Euphytica 1993;66:217–223.

Ward SM. A new source of restorable cytoplasmic male sterility in quinoa. Euphytica 1998;101(2):157–163.

Ward SM. Response to selection for reduced grain saponin content in quinoa (*Chenopodium quinoa* Willd). Field Crop Res 2000a;68(2):157–163.

Ward SM. Allotetraploid segregation for single-gene morphological characters in quinoa (*Chenopodium quinoa* Willd.). Euphytica 2000b;116(1):11–16.

Wilson HD. Quinoa biosystematics I: domesticated populations. Econ Bot 1988;42:461–477.

Wilson HD. Crop/weed gene flow: *Chenopodium quinoa* Willd and *C. berlandieri* Moq. Theor Applied Genet 1990;86: 642–648.

Wright KH, Pike OA, Fairbanks DJ, Huber CS. Composition of *Atriplex hortensis*, sweet and bitter *Chenopodium quinoa* seeds. J Food Sci 2002;67:1383–1385.

Chapter 7
Quinoa Cytogenetics, Molecular Genetics, and Diversity

Janet B. Matanguihan[1], Peter J. Maughan[2], Eric N. Jellen[2] and Bozena Kolano[3]

[1]*Department of Crop and Soil Sciences, Washington State University, Pullman, WA, USA*
[2]*Plant and Wildlife Sciences, Brigham Young University, Provo, UT, USA*
[3]*Department of Plant Anatomy and Cytology, University of Silesia, Poland*

INTRODUCTION

Quinoa, *Chenopodium quinoa*, is the domesticated South American member of a complex of allotetraploid taxa ($2n=4x=36$) native to the New World. This complex includes the weedy North American *Chenopodium berlandieri* subsp. *berlandieri*, the weedy and malodorous South American *Chenopodium hircinum*, and the two domesticated Mesoamerican forms of *C. berlandieri* subsp. *nuttalliae*. An additional member of the complex, now extinct but known from archeological remains, was the domesticated eastern North American *C. berlandieri* subsp. *jonesianum* (Smith and Yarnell 2009).

Quinoa's importance as an internationally exported food crop has expanded greatly within the past decade. As cultivation of this crop moves out of the highland Andes and relatively isolated coastal lowlands of central and southern Chile, there is a real danger that the crop will contract new diseases and pick up new pests against which it does not currently harbor genetic resistance. Consequently, existing quinoa breeding programs in the Andean region will need to pay more attention to germplasm diversification. Emerging quinoa breeding programs in other parts of the world, particularly in the Eastern Hemisphere where quinoa's close genetic cousins of the lamb-squarters complex (*Chenopodium album* et al.) have been widespread weeds for thousands of years, will need to be extra vigilant in identifying and exploiting existing genetic variation for pest and disease resistance in *C. berlandieri* and other taxa cross-compatible to *C. quinoa*. In this chapter, we describe the cytogenetic and genomic structure of quinoa – information that points to potentially close genetic relatives and, therefore, exploitable exotic gene sources – along with a series of advanced, DNA-based molecular genetic tools to facilitate and accelerate the transfer of exotic genes into *C. quinoa*.

CYTOGENETICS AND GENOME STRUCTURE OF *CHENOPODIUM QUINOA*

C. quinoa is an allotetraploid species with $2n = 4x = 36$ chromosomes. Its haploid genome size (1C value) has been estimated at between 1.005 and 1.596 pg (Bennett and Smith 1991; Stevens et al. 2006; Bhargava et al. 2007a; Palomino et al. 2008; Kolano et al. 2012a). This discrepancy in genome size among different reports could be due either

Quinoa: Improvement and Sustainable Production, First Edition. Edited by Kevin Murphy and Janet Matanguihan.

to intraspecific polymorphism in genome size or to the use of different measurement methods. Until now only limited intraspecific genome size variation (5.9%) was demonstrated for this species (Kolano et al. 2012a). The cytogenetic characterization of the quinoa karyotype has still been limited, partly due to its somatic chromosome morphology and a scarcity of chromosome markers. Quinoa chromosomes are small, ranging from 0.94 to 1.60 μm, and are mainly metacentric (Bhargava et al. 2006; Palomino et al. 2008). Methods of classical cytogenetics allowed for characterization of the distribution of heterochromatin in quinoa chromosomes. The heterochromatin revealed by C-banding is located preferentially in centromeric and pericentromeric positions in most quinoa chromosomes; however, the size of the heterochromatin blocks differed considerably among chromosomes (Fig. 7.1a, Kolano unpublished). Other types of heterochromatin bands were distinguished using a GC-specific fluorochrome – chromomycin A_3 (CMA). Two CMA^+ bands were observed in one pair of homologous chromosomes in the terminal position. These CMA^+ bands colocalized with secondary constructions and AgNOR bands, indicating that this type of heterochromatin is correlated with 35S rRNA gene loci (Fig. 7.1b, Kolano et al. 2001).

The rRNA genes were the first tandem repetitive sequences mapped to the quinoa karyotype using fluorescence *in situ* hybridization (FISH). The 35S RNA gene loci were localized in the terminal regions of two chromosomes. The 5S rRNA genes were arranged in two pairs of loci – one of them was located in the terminal position and the other interstitially in two different chromosome pairs (Maughan et al. 2006). These results showed that one or more pairs of the 35S rDNA loci were deleted during quinoa evolution because the locus number was lower than the expected additive values of any known diploids. This phenomenon was also observed for other polyploid *Chenopodium* species, such as the tetraploid *C. auricomum* and hexaploid *C. album* (Kolano et al. 2012b). The other tandem repetitive sequence mapped in quinoa chromosomes was a telomeric repeat.

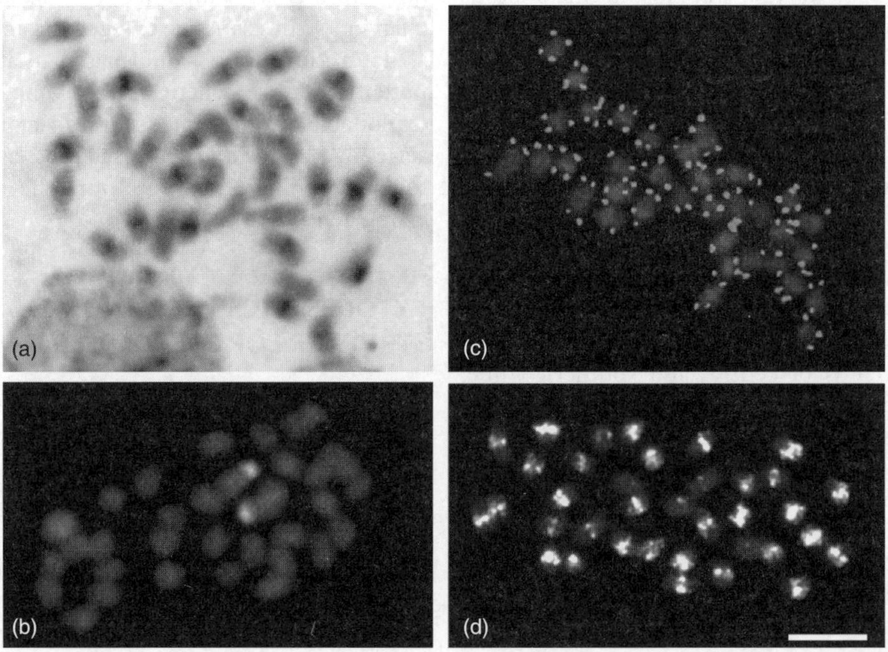

Fig. 7.1 Metaphase plates of *Chenopodium quinoa* (a) after C-banding and (b) stained with CMA_3. Localization of (c) *Arabidopsis*-type telomere repeat and (d) clone 12-13P on metaphase chromosomes of quinoa. Scale bar = 5 μm.

Quinoa has an *Arabidopsis*-type telomere repeat located exclusively in the terminal position in each chromosome arm (Fig. 7.1c, Kolano unpublished). Centromeric and pericentromeric localization in quinoa chromosomes was identified for clone 12-13P (Fig. 7.1d, Kolano et al. 2011). This repetitive sequence showed partial homology to satellite DNA (pBC1447) detected near the centromere of *Beta corolliflora* chromosomes (Gao et al. 2000). Hybridization signals of 12-13P were observed in each chromosome of *C. quinoa*; however, the intensity of the FISH signals differed considerably among chromosomes indicating that varying numbers of 12-13P repeats are present at each locus. The pattern observed on *C. quinoa* chromosomes after FISH with 12-13P was very similar to C-banding results, suggesting that the 12-13P sequence constituted a major part of the heterochromatin of *C. quinoa* (Kolano et al. 2011). The 12-13P sequence also hybridized to centromeric and pericentromeric chromosomal regions of related North American tetraploid *C. berlandieri* as well as European accessions of hexaploid *C. album*; however, the intensity of hybridization signals was somewhat lower (Kolano et al. 2011).

The second group of repetitive DNA, which are dispersed repetitive sequences, comprise mainly mobile elements (retrotransposons and DNA transposons). Kolano et al. (2013) analyzed the chromosomal distribution of reverse transcriptase-coding fragments of LTR retrotransposons and discovered that both Ty1-*copia* and Ty3-*gypsy* retrotransposons were preferentially located in pericentromeric heterochromatin of quinoa chromosomes. Apart from retrotransposons, dispersed distribution in quinoa chromosomes was demonstrated for two repetitive sequences (pTaq10 and 24-18J) without homology to mobile elements. The first sequence, pTaq10, hybridized across the quinoa genome, without a specific chromosome or subgenome distribution pattern (Kolano et al. 2008a). In contrast, the second dispersed repetitive sequence, 24-18J, hybridized primarily to one subgenome (18 chromosomes) of quinoa (Kolano et al. 2011). A similar hybridization pattern was observed for *C. berlandieri*, which supports the hypothesis that *C. berlandieri* and *C. quinoa* are descended from at least one common diploid ancestor (Maughan et al. 2006). Both *C. berlandieri* and *C. quinoa* share the A-genome (*Chenopodium standleyanum*) and the B-genome (*Chenopodium ficifolium*, Old World). After allopolyploid speciation of *C. quinoa* and *C. berlandieri*, at least one of the ancestral subgenomes experienced loss of the 35S rRNA gene locus. Comparison of the chromosomal distributions of 18–24J homologous and rDNA sequences revealed that in quinoa and *C. berlandieri*, rRNA gene loci were retained in the subgenome to which 18–24J abundantly hybridized (Kolano et al. 2011). As probe 18-24J also abundantly hybridized to metaphase chromosomes of a diploid form of Eurasian *C. album*, as well as to one of the subgenomes in allohexaploid *C. album*, we conclude that this subgenome originated in the Eastern Hemisphere (Kolano et al. 2011).

Another item of interest with respect to quinoa cytogenetics is that *C. quinoa* is a polysomatic plant (Kolano et al. 2008b). Polysomaty is the occurrence of cells with different ploidy levels in the same organ or tissue (Joubes and Chevalier 2000). During quinoa seedling development, endopolyploid cells (cells with more than the 4C DNA value) were observed in roots, hypocotyls, and in young cotyledons. Its tissues comprised a mixture of cells with DNA content ranging from 2 to 16 C in varying proportions and the polysomaty patterns corresponded to the developmental stage and the individual organ. Endopolyploidization was not present in nuclei from leaves and the shoot apex (Kolano et al. 2008b).

CROSSABILITY OF QUINOA AND ALLIED TETRAPLOID TAXA

The New World allotetraploid "quinoa complex" consists of three taxa that are cross-compatible – *C. berlandieri*, *C. hircinum*, and *C. quinoa*. This is an important observation because the extensive geographical distribution of weedy *C. berlandieri*, from Alaska and the Canadian Maritimes in the north to Mesoamerica and perhaps the northern Andes in the south, suggests that this taxon possesses a wealth of genetic diversity, including novel pest and disease resistance genes (Jellen, personal observations). Similarly, the native range of *C. hircinum* in South America extends across the temperate and subtropical lowlands of the Argentinian pampas, an area in which it has been in contact with invasive Eurasian pests and pathogens since the period of the Spanish Conquest. Heiser and Nelson (1974) first reported that quinoa × Mexican huauzontle (*C. berlandieri* subsp. *nuttalliae*) domesticated hybrids were fertile; interestingly, the hybrids had dark seed, unlike either parent. Wilson (1980) again reported that quinoa × weedy *C. berlandieri* and quinoa × huauzontle hybrids were fertile. Wilson and Manhart (1993) provided isozyme-based evidence that >30% of the offspring of *C. berlandieri* plants surrounding quinoa fields in the Pacific Northwest carried alleles transferred sexually from quinoa; these hybrids were also partially fertile and displayed heterotic phenotypic effects.

At Brigham Young University (BYU), we are currently analyzing F_3 plants derived from a cross between a large-seeded weedy *C. berlandieri* var. *macrocalycium* (BYU 803, from Maine) and highland quinoa cv. "Ingapirca." The F_2 generation plants were highly diverse and phenotypically vigorous, though between 10% and 30% had varying degrees of pollen sterility, presumably due to summer greenhouse heat stress. In total, 19 of the 115 F_2 plants displayed phenotypic evidence of hybrid breakdown, including leaf whorling, abnormal coloration, or dwarfing. All of the recovered seed had the wild, black-pericarp trait from BYU 803 (Jellen unpublished). More recent DNA sequence evidence, however, suggests that *C. berlandieri* var. *macrocalycium* is a relatively distant member of the allotetraploid complex and may have a different origin or at least introgression from a different diploid species (Jellen and Walsh unpublished).

With respect to crossability between the New World allotetraploid complex and Eurasian-origin *C. album*, Wilson (1981) failed to detect transfer of isozyme marker alleles from *C. hircinum* to sympatric populations of allohexaploid *C. album*. Consequently, as quinoa begins to be cultivated more extensively in areas of the world where lambsquarters is a common weed, transfer of exotic alleles into the quinoa crop from this source is probably not going to be an issue, nor does *C. album* appear to be a promising source of breeding alleles to improve quinoa's biotic stress resistance.

DNA SEQUENCE EVIDENCE FOR QUINOA'S GENOMIC ORIGINS

Maughan et al. (2006) sequenced portions of 35S rRNA gene NOR intergenic spacer (IGS) and 5S rRNA gene nontranscribed spacer (NTS) regions from five cultivars of *C. quinoa* and a weedy accession of *C. berlandieri*. They also performed FISH on metaphase chromosome preparations of both species along with domesticated *C. berlandieri* subsp. *nuttalliae* to verify locus copy numbers. While subsp. *nuttalliae* had either one or two 35S loci – presumably, the ancestral locus number in an allotetraploid – both the weedy accession and *C. quinoa* carried only a single locus. However, for 5S, the subsp. *nuttalliae* genotypes carried three loci, whereas *C. quinoa* and weedy *C. berlandieri* had only two. DNA sequencing experiments identified two NTS sequence variants in *C. quinoa* – apparently, from each of the two loci – one of which was almost an exact match (196/200 bases) with the sequence of the weedy *C. berlandieri* clone. The other NTS sequence was much more divergent (158/200 bases). The IGS region was more divergent among five varieties of quinoa and a single weedy *C. berlandieri*; the latter possessed different numbers of B, C, and H subrepeats than the former, while the five quinoa varieties varied mostly for SNPs. These combined observations suggest that ancestors of quinoa and the weedy accession of *C. berlandieri* lost different 35S loci; this hypothesis could be tested by sequencing the two IGS loci in subsp. *nuttalliae*.

Another gene studied extensively is *Salt Overly Sensitive 1* (*SOS1*), which encodes a plasmalemma Na^+/H^+ antiporter protein (Shi et al. 2000). The two full-length genomic orthologs from quinoa were previously cloned and sequenced (Maughan et al. 2009a) and introns 16 and 17 were selected for further phylogenetic study. Extensive sequencing of introns 16 and 17 of *SOS1* in a number of cultivated and weedy *Chenopodium* taxa and quinoa cultivars at the 2*x*, 4*x*, and 6*x* ploidy levels has confirmed the Eurasian and New World origins of the two constituent subgenomes of *C. quinoa*, *C. berlandieri*, and *C. hircinum* (Walsh et al. manuscript in preparation). The New World subgenome is now referred to as "AA" and the Eurasian subgenome as "BB." These studies have also pointed to a North American origin of the allotetraploid, followed by dispersion – most likely as a wild or weedy ecotype – to South America (*C. hircinum*), with domestication events occurring on both continents.

QUINOA GENETIC MARKERS AND LINKAGE MAPS

The recent awareness of the importance of quinoa in food security of the Andean region as well as a growing export health food market has led to the establishment of several new quinoa breeding programs. In addition to germplasm management, principal objectives of these programs include enhancing grain yield, earliness, disease resistance, drought tolerance, and modulating saponin content (Ochoa et al. 1999). These programs recognize that the development and use of molecular markers is critical to meeting these objectives (Bonifacio, PROINPA personal communication).

Marker-assisted selection methods, including association mapping applied directly to breeding populations (Kraakman et al. 2004; Crossa et al. 2007) and genomic selection (Jannink et al. 2010; Windhausen et al. 2012), are quickly becoming the commercial standard for enhancing breeding efficiency (Eathington et al. 2007; Moose and Mumm 2008). Key to these enhanced breeding methodologies is easy access to numerous, inexpensive, reliable, and easily scored genetic markers. Unfortunately, few researchers have reported the development and use of molecular markers in quinoa. Wilson (1988a) reported the development of the first set of allozyme markers, whereas Fairbanks et al. (1990) identified the first seed proteins variants within the USDA quinoa germplasm collection. Ruas et al. (1999) reported the use of randomly amplified polymorphic DNA (RAPD) markers to detect DNA polymorphisms among several *C. quinoa* cultivars and *Chenopodium* weed species. Maughan et al. (2004) used amplified fragment length polymorphism (AFLP) markers to generate the first low-resolution linkage map of quinoa. Unfortunately, all of these markers have had only limited utility in breeding programs, due at least in part to the inherent problems of cost, reproducibility, and technology transfer of these types of markers, especially to laboratories in the developing world where resources are limiting.

In an attempt to make quinoa marker technology more widely available, Mason et al. (2005) and Jarvis et al. (2008) developed the first large sets (>400) of microsatellite (also termed simple sequence repeats or SSRs) markers for quinoa. Jarvis et al. (2008) also reported the first quinoa linkage genetic map consisting primarily of microsatellite markers. This early map was composed of 38 linkage groups ($n = 18$) and covered just over 900 cM. Microsatellite loci consist of short tandemly repeated nucleotide motifs flanked by conserved sequences (Tautz 1989). Polymorphism is detected as variation in the number of repeat units among individuals using standard polymerase chain reaction (PCR) techniques (Weber and May 1989). Microsatellites are multiallelic and generally more informative, based on polymorphic information content (PIC) values, than RAPD or AFLP markers (Powell et al. 1996). Microsatellite loci, while ubiquitous in eukaryotic genomes (approximately 1 microsatellite per 33 kb in plants; Chawla 2000), are not the most abundant marker type.

Single nucleotide polymorphisms (SNPs) are the most abundant type of DNA polymorphism found in eukaryotic genomes (Garg et al. 1999; Batley et al. 2003) and are the marker of choice in marker-assisted plant breeding programs (Batley and Edwards 2007; Eathington et al. 2007). The high frequency of SNPs in plant genomes is well documented (Russell et al. 2004; Ossowski et al. 2008), with actual SNP densities ranging dramatically depending on the species type (auto- or allogamous), number and genetic diversity of the cultivars being assessed, and whether coding or noncoding regions are being considered. For example, in soybean [*Glycine max* (L.) Merr.], SNPs occur at a frequency of 1 per 2038 bp in coding sequence and 1 per 191 bp in noncoding sequence (Van et al. 2005), whereas in maize (*Zea mays* L.), 1 SNP was observed per 124 bp of coding sequence and 1 per 31 bp in noncoding regions (Ching et al. 2002). In quinoa, Coles et al. (2005) identified 38 single-base changes and 13 insertions–deletions (indels) in 20 EST sequences analyzed across five quinoa accessions, suggesting an average of 1 SNP per 462 bases and 1 indel per 1812 bases. The high frequency of SNPs in most species offers the possibility of constructing extremely dense genetic maps that are particularly valuable for map-based gene cloning efforts and haplotype-based association studies.

Several technical methods have been reported for the initial discovery of SNPs in plant genomes, specifically (i) EST sequencing (Barbazuk et al. 2007); (ii) targeted amplicon resequencing (Bundock et al. 2009); (iii) gene space resequencing with methylation-sensitive digestion (Gore et al. 2009; Deschamps et al. 2010); and (iv) genomic reduction based on restriction site conservation (Maughan et al. 2009b). When combined with next-generation DNA sequencing technologies, these methods can be used to identify large numbers of SNPs with limited technical expertise and at minimal cost. SNPs can then be cost-effectively genotyped using several next-generation technologies, including bead arrays (Shen et al. 2005), nano-fluidic devices (Wang et al. 2009), and genotyping by sequencing (Miller et al. 2007; Maughan et al. 2010; Elshire et al. 2011). In 2012, Maughan et al. reported the use of a reduced representation protocol and genotyping by sequencing to identify >14,000 putative SNPs in five bi-parental quinoa populations. Transition mutations (A/G or C/T) were the most numerous, outnumbering transversions (A/T, C/A, G/C, G/T) by 1.6× margin, which was in accordance with the observation that transition SNPs are the most frequent SNP type reported in both plant and animal genomes and are thought to result from hypermutability effects of CpG dinucleotide sites and deamination of methylated cytosines (Zhang and Zhao 2004; Morton et al. 2006). Maughan et al. (2012) converted 511 of the putative SNP into functional SNP assays. A diversity screen of 113 quinoa accessions using these 511 SNPs clearly revealed the two major quinoa subgroups corresponding to the Andean and coastal quinoa ecotypes (Maughan et al. 2012). Minor allele frequency of the SNPs ranged from 0.02 to 0.50, with an average MAF of 0.28. Linkage mapping of the SNPs in two recombinant inbred line populations (KU-2 × 0654 and NL-6 × 0654) produced an integrated linkage map consisting of 29 linkage groups with 20 large linkage groups, spanning 1,404 cM with a marker density of 3.1 cM per SNP marker. The functional SNP assays were developed using KBioscience KASPar™ genotyping chemistry detected using a Fluidigm integrated fluidic chip (Fig. 7.2). The combination of the KASPar chemistry with the nano-fluidic chip technology (9.7 nL reaction volume) not only significantly reduces the marker data point genotyping costs (~US$0.05) but also significantly increases the speed of genotyping. Indeed, a single Fluidigm 96.96 IFC is capable of producing 9216 PCRs in a single run (~3 h) with little technical expertise.

Given the dramatic decrease in costs and relative ease of genotyping, we anticipate the development of fully saturated genetic maps of quinoa within 12–18 months. These maps should quickly open up the possibility to integrate marker-assisted selection protocols, specifically genomic selection, into accelerated quinoa breeding programs. However, before genomic selection can be fully realized, quinoa breeders must identify a "training population" of quinoa accessions and develop rigorous phenotyping strategies. The training population will need to be both fully genotyped and phenotyped to develop the statistical models necessary to estimate breeding values of genomic regions (Jannink et al. 2010).

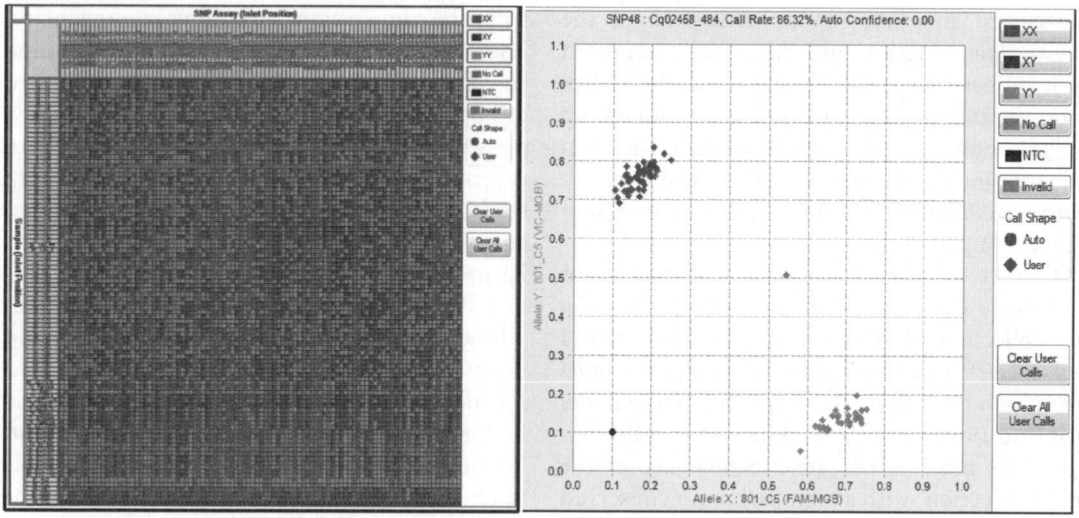

(a) (b)

Fig. 7.2 Example of SNP assays using the KASPar^{TM} genotyping chemistry on the Fluidigm access array in the quinoa RIL mapping population. (a) The genotyping across the 96.96 IFC chip (96 DNA samples on the vertical, 96 SNP assays on the horizontal). (b) Individual SNP loci in a Cartesian graph. A no template control (NTC) and a synthetic heterozygote are identified (*See color insert for representation of this figure.*).

QUINOA DIVERSITY

The phenotypic and genetic diversity of quinoa is borne out by its wide distribution in the Andes, covering Bolivia, Peru, Ecuador, Colombia, and the north of Argentina and Chile, attesting to its adaptation to a range of agroecological conditions. Quinoa can survive in adverse climatic and edaphic conditions and thrive in locations where few crops can (Bonifacio 2003). Ecotypes of quinoa specifically adapted to major ecosystems have been identified: (i) *Valley* (from the inter-Andean valleys); (ii) *Altiplano* (from the highland plateau in Bolivia and Peru); (iii) *Salares* (from the salt flats of Southern Bolivia); (iv) *Sea level* (from Central Chile); and (v) *Subtropical* (from the Bolivian Yungas) (Tapia et al. 1980). Each of these ecotypes is associated with subcenters of diversity that originated around Lake Titicaca (Fuentes et al. 2012). The phenotypic and genetic diversity of quinoa is also shown in the variability in plant color, inflorescence type, growth habit, and chemical composition.

Over the past few years, scientists have characterized the genetic diversity of quinoa to understand its biological diversity as a function of its eco-geographic distribution and to identify genetically distinct groups (del Castillo et al. 2007). This knowledge is a prerequisite in quinoa conservation strategies and effective germplasm management and characterization (Roa 2004). The study of quinoa's genetic diversity is also essential in plant breeding programs, especially in the identification of diverse parental combinations (Fuentes and Bhargava 2011).

Phenotypic diversity

Initially, scientists have studied the genetic diversity in quinoa using morphological markers. Wilson (1988a) combined morphological and isozyme data to elucidate the genetic relationships among the quinoa ecotypes. He used the variation in electrophoretic patterns of 21 isozyme loci together with the morphometric data to compare 98 quinoa populations from South America. His work showed two main groups: a Coastal type from south-central Chile and an Andean type distributed at elevations above 1,800 m from northwest Argentina to southern Colombia. This last group was also divided into northern and southern Andean quinoa. Wilson (1988b) also constructed the first phylogenetic tree of the *Chenopodium* species and his data supported the hypothesis that the Altiplano was the center of origin and diversity of quinoa. A follow-up study, conducted using seed protein variation and morphological markers (Fairbanks et al. 1990), confirmed Wilson's initial conclusion (Jellen et al. 2011).

Ortiz et al. (1998) demonstrated that eight phenotypic descriptors can be used to derive a core collection of 103 accessions (10%) from a pool of 1,029 accessions from the Peruvian quinoa germplasm bank at the Universidad National del Altiplano-Puno (UNAP). The core collection represents most of the genetic diversity of the whole collection, with a high correlation observed between quantitative trait variation and altitude. The authors recommend that a comprehensive quinoa core collection should also include accessions from Bolivian, Ecuador, and Chile, together with cross-compatible, wild *Chenopodium* species sympatric with cultivated quinoa.

Rojas et al. (2000) used morphological traits and agronomic performance of 1,512 accessions of the Bolivian national quinoa collection to analyze genetic diversity. Genotypes were collected in Chile, Argentina, and Peru, as well as native Bolivian materials. Using three multivariate procedures, including principal component, cluster, and discriminate function analyses, quinoa germplasm was classified into seven distinct groups, including five within the Altiplano and two from lower altitude valleys of the eastern Andean mountain range. The traits used in the study did not discriminate for the Chilean lowland germplasm, which grouped with one of the Altiplano clusters.

More recently, Bhargava et al. (2007b) used morphological and quality traits to investigate genetic diversity in *C. quinoa* germplasm. Twenty-nine lines of *C. quinoa* and two lines of *C. berlandieri* subsp. *nuttalliae* were evaluated for 12 morphological and 7 quality traits. Cluster analysis and principal component analysis conducted for the 19 traits revealed a high level of genetic variability existing in the lines tested. Although cluster analysis grouped those lines with greater genetic similarity, it did not group those lines from the same origin, indicating heterogeneity of lines within a geographical region. The authors suggested that heterogeneity, genetic architecture of population, history of selection, and/or developmental traits could account for this population diversity within a geographical region. In another study, Bhargava et al. (2007c) used morphological and qualitative traits to analyze the degree of similarity/dissimilarity among 27 diverse germplasm lines of *C. quinoa* in the Indian subcontinent. These traits were discussed in relation to their utility for plant breeding efforts.

Fuentes and Bhargava (2011) presented the first report on quinoa germplasm grown under lowland desert conditions. A total of 11 morphological descriptors were used on 28 quinoa accessions collected from the northern highlands of Chile, and cluster analysis classified the accessions into 6 discrete groups. Multivariate analysis elucidated the genetic relationships among the accessions, with yield being the most important descriptor for discriminating the accessions used. Significantly, Chilean quinoa under desert conditions did not show particularly extreme values for any of the variables measured.

Curti et al. (2012) characterized 34 quinoa populations from the northwest Argentina region using quantitative and qualitative phenotypic traits. Northwest Argentina represents the southern end of what is known as the *C. quinoa* Andean complex. Quinoa is considered marginal in terms of its cultivation in this region. Curti et al. (2012) analyzed the data using descriptive and multivariate techniques. On the basis of quantitative traits, both the principal component analysis and the cluster analysis differentiated between accessions from the highlands, transition zone, central dry valleys, and eastern valleys. On the other hand, the principal coordinates analysis based on qualitative traits only discriminated accessions from transition zone and eastern valleys. The accessions from the highlands and dry valleys presented the more advanced domesticated traits, whereas accessions from transition zone and eastern valleys showed traits more similar to wild-type-related Chenopods from the Andean region. This is the first phenotypic diversity study of accessions from Northwest Argentina.

Genetic diversity

With the development of molecular markers, the genetic diversity of quinoa germplasm collections has been studied, as well as the genetic diversity of quinoa populations, for both cultivated and wild/weedy populations. Molecular markers provide unique and effective tools for evaluating and characterizing plant genetic diversity in a way that is unaffected by the environment (Gupta and Varshney 2000). Random amplified polymorphic DNA (RAPD) markers were the first markers used to detect DNA polymorphisms among different quinoa accessions (Fairbanks et al. 1993). RAPDs have also been used to identify true hybrids from intergeneric crosses (Bonifacio 1995). The genetic relationships among 19 *Chenopodium* species were investigated by Ruas et al. (1999) using RAPD markers. The *Chenopodium* species used in the study included *C. quinoa* cultivars and weed species. Polymorphisms were detected among these species but accessions clustered according to their species classifications. Wild and crop populations of *C. quinoa* shared a low level of molecular variation, without differentiation between sympatric domesticated and weedy populations, and low levels of intraspecific variation within accessions (Ruas et al. 1999).

In contrast to the study of Ruas et al. (1999) and del Castillo et al. (2007) studied the genetic diversity and relationships among wild and cultivated populations of quinoa collected directly from farmer's fields. Ruas et al. (1999) compared accessions, whereas del Castillo et al. (2007) compared individuals of the cultivated and wild forms of quinoa, growing sympatrically within cultivated fields. del Castillo et al. (2007) sampled from three distinct regions of the altiplano and one inter-Andean valley in Bolivia. Using RAPD markers, the wild and cultivated populations show a significant but very low level of global differentiation. However, there was a strong genetic differentiation among the eight populations in the study, with a strong correlation with the regional ecogeography. The population structure appears related to three major biogeographic zones: (i) the northern and central altiplano, (ii) the interandean valley, and (iii) the southern Salar. A small proportion of the variation was explained by geographical distance.

Aside from RAPD markers, AFLP markers have also been used to study genetic diversity in quinoa. Anabalón Rodríguez and Thomet Isla (2009) used AFLP markers together with 20 morphological descriptors to characterize 14 accessions of quinoa located in the south of Chile, and also highland accessions. These local, ancestral varieties are usually conserved and selected by Mapuche communities and other smallholder farmers, passed down through the generations. Also included in the study are three varieties from the Tarapacá region and one enrolled variety (Regalona-Baer), with *C. album* and *Chenopodium ambrosioides* as the controls. The accessions clustered into two groups: (i) the coastal type, which included accessions from the north of Chile, and highland accessions, and (ii) the pre-cordillera accessions. The results are in accordance with previous morphological and isozyme studies, which separate quinoa into two types: a coastal type (Chile) and an Andean plateau type. This study indicates that the Chilean lowland germplasm could be more genetically diverse than previously thought. The authors explained that this level of genetic diversity could also be due to out-crossing with the weedy populations of *C. album* and *C. hircinum*, coupled with the ancestral seed exchange system and selection due to edaphoclimatic and photoperiod factors. This work also complements the work of Ruas (1999) in that the *C. album* and *C. ambrosioides* accessions were also differentiated from the *C. quinoa* accessions. However, with the use of AFLP markers, one *C. album* accession was grouped together with one *C. ambrosioides* accession (Anabalón Rodríguez and Thomet Isla 2009), whereas with the use of RAPD markers, the *C. album* and the *C. ambrosioides* accessions were placed in different groups (Ruas et al. 1999).

More recent studies have utilized microsatellites or SSR because these are frequently codominant, multi-allelic, highly reproducible, polymorphic, ubiquitous and widely distributed in the plant genome (Bhargava and Fuentes

2010). Christensen et al. (2007) used 35 of the first SSR markers developed to assess the level of genetic diversity in quinoa germplasm collections. A total of 152 accessions from the USDA and CIP-FAO collections were used in the study, including accessions from Peru, Bolivia, Ecuador, Argentina, and Chile. Results show that the accessions clustered into two main groups: one group included accessions from the lowlands of Chile together with a set of USDA accessions, whereas the other group consisted of accessions from the Andean highlands of Peru, Bolivia, Ecuador, Argentina, and extreme northeastern Chile.

Using multiplex fluorescent SSR markers, Fuentes et al. (2009) studied the genetic diversity patterns of 59 accessions from the northern and southern regions of Chile. These accessions were classified as Altiplano (28) and coastal (31) ecotypes. Both cluster analysis (UPGMA) and principal component analysis separated the accessions into two discrete groups. The first group comprised quinoa accessions from the north (Andean highlands), whereas the second group consisted of accessions from the south (lowland or coastal). Coastal quinoas showed more diversity than Chilean highland quinoas using both cluster and principal component analysis. The coastal group showed continuous variation between extreme subgroups represented. Their results are in contrast with those of del Castillo et al. (2007), which showed that Chilean accessions, as representatives of coastal quinoas, were not genetically diverse from the highland populations of Peru and Bolivia.

Fuentes et al. (2012) examined the correlation of seed exchanges on the genetic diversity of quinoa in Chile. In total, 34 quinoa accessions representative of Chile, 20 SSR markers, and 92 field interviews were used in their study. Results showed that a wide genetic diversity exists in quinoa grown along the main growing areas in Chile. The accessions were classified into two major groups, and further subdivided into five populations. Population I has nine accessions representative of the northern zone of Chile; Population II has seven accessions of the central zone; Population III has nine accessions from the southern zone and one from the central zone; Population IV has six accessions from the highlands of Peru, Bolivia, and Argentina; and Population V had two accessions, one each from Ecuador and Colombia. The genetic distances among populations were consistent with their geographic origin. Thus, quinoa populations with the lowest genetic distances between them were Populations I and IV (highlands) and II and III (lowlands). This grouping correlated well with the geographic origins of the accessions, together with edaphic and climatic conditions of their origin. This grouping also correlated with the expansion of quinoa and is a genetic reflection of the sociolinguistic context of ancient people inhabiting the Andes region.

Costa Tártara et al. (2012) studied the genetic structure of cultivated quinoa in Northwest Argentina, a region that has been underrepresented in germplasm collections and previous studies. Aside from being underrepresented, Northwest Argentina is also the southern extreme of quinoa distribution within the Central Andes. The authors used 22 SSR markers on 35 accessions, which showed a high level of genetic diversity. Cluster analysis separated the populations into four distinct groups; the groupings were consistent with the geographic origin of the accessions. The first group consisted of 5 accessions from the Transition area that is characterized by high altitudes, whereas the second group consisted of 12 accessions from Puna, the highland Plateau. The third group consisted of 8 accessions from Eastern humid valleys, and the fourth group consisted of 10 accessions from the dry valleys. Even though the Northwest Argentina is considered marginal in terms of quinoa cultivation, high levels of genetic diversity was still found in the accessions used. This genetic differentiation could be related to regional ecogeography and is also affected by the use of landraces.

SUMMARY

A better understanding of quinoa genetic diversity has been gleaned from all the studies discussed earlier. First, the wide range of environments in which quinoa can grow has a direct influence on its genetic diversity. Over a long period of time, the adaptation of quinoa to extremely dissimilar climatic conditions may have contributed to its wide genetic diversity (Costa Tártara et al. 2012). Phenotypic and genetic diversity studies have shown that quinoa accessions are most often clustered according to their geographic origin, indicating that genetic variability has a spatial structure and distribution (Risi and Galwey 1989a, 1989b; Ortiz et al. 1998; Rojas 2003; del Castillo et al. 2007; Costa Tártara et al. 2012; Curti et al. 2012). The effect of population isolation on genetic diversity was linked not only to geographical distance but also to climatic and orographic barriers (del Castillo et al. 2007). Furthermore, the congruence between genetic differentiation and ecogeography suggests that quinoa all over the southern Andes may be undergoing similar processes of genetic differentiation and that variation in quinoa ecotypes may be influenced significantly by gradients of frost and aridity (Curti et al. 2012).

The use of more informative molecular markers has made it possible to elucidate the genetic diversity of quinoa accessions. In their study, Christensen et al. (2007) suggested that the Argentinean accessions represent introductions from both the southern Bolivian highlands and the Chilean lowlands. A later study, using morphological data, also proposed that the main introductions of quinoa to Argentina came from southern Bolivia (Curti et al. 2012). On the other hand, instead of the Chilean lowlands, it is the Chilean highlands that could be the alternative route for quinoa introduction to Argentina (Curti et al. 2012). In Southern Chile, the high genetic diversity of coastal/lowland ecotypes has been investigated using dominant AFLP markers (Anabalón Rodríguez and Thomet Isla 2009) and codominant SSR markers (Fuentes et al. 2009; Fuentes et al. 2012). These studies showed that Chilean lowland germplasm is much more genetically diverse than previously thought. This high genetic diversity could be due to the continuous hybridization with relatives that coexist in the field (Fuentes et al. 2009), active crop/weed complexes having a monophyletic coevolving behavior (Rana et al. 2010). Christensen et al. (2007) also reported diversity data that were more consistent with an allogamous system rather than an autogamous one.

Genetic events, such as genetic drift, genetic bottlenecks, and founder effects, have been illustrated by genetic diversity studies in quinoa. Quinoa may have been domesticated twice – once in the High Andes and a second time in the Chilean lowlands (Christensen et al. 2007; Fuentes et al. 2009). This Chilean domestication event is a genetic bottleneck (Jellen et al. 2011). Fuentes et al. (2012) showed a fragmented pattern of diversity between the central zone and the southern zone of Chile, corresponding with the isolation of quinoa farmers. The average heterozygosity and proportion of polymorphic loci in the central zone of Chile were lower than that for the southern zone. This indicates a subpattern of geographic bottleneck within lowland/coastal quinoa. The reduction in genetic diversity levels in the northern highland and coastal lowland regions compared to the southern highland regions near Lake Titicaca could be a result of founder effects associated with the dispersal of the crop from its center of origin; alternatively, it could be a response to selective adaptation in the more uniform ecological regions of the northern highlands and coastal lowlands (Christensen et al. 2007). The diffusion of quinoa into Northwest Argentina from the centers of origin and diversity could also have been mediated by founding events associated with early dispersal from the central Andean region of Peru and Bolivia (Curti et al. 2012).

Human activities, specifically seed exchanges, have significantly affected the genetic diversity of quinoa. The influence of ancient exchange routes affecting germplasm distribution contributes to the genetic structure of quinoa (Costa Tartara 2012). As the earliest societies along the Andean range tested new soils and climates, the genetic diversity of quinoa probably evolved through cycles of seed exchange and the domestication process (Fuentes et al. 2012). It has been shown through genetic data and farmer interviews that genetic population structure can be reinforced by seed exchanges among farmers (del Castillo et al. 2007). Genetic relationships within quinoa populations have revealed the influence of various production systems on quinoa biodiversity. Thus, studies of genetic diversity in quinoa must be linked with sociological and agronomic studies (Fuentes et al. 2012).

The results of genetic diversity studies in quinoa has implications in our conservation efforts and, consequently, on our plant breeding programs. Since the 1960s, *ex situ* germplasm banks of quinoa have been established (Bonifacio 2003). However, some ecological zones are still inadequately represented in these gene banks, such as the northern Andes (especially Ecuador and Colombia), the Yungas (eastern slopes of the Bolivian and Peruvian Andes) (Christensen et al. 2007), and Argentina (Mujica and Jacobsen 2002). In areas such as Northwest Argentina, where it has been shown that phenotypic and genetic diversity is structured according to the site of origin (Curti et al. 2012; Costa Tártara et al. 2012), conservation programs should carefully consider the areas to be preserved. Also, some accessions in the USDA and the CIP-FAO collections could be heterogeneous lines of mixed genotypes as multiple alleles at marker loci have been detected. Accessions should be screened and purified before use in plant breeding programs (Christensen et al. 2007). Germplasm collections should also be replenished (Mujica and Jacobsen 2002). Meanwhile, *in situ* germplasm banks are found in specific ecological environments of the Andean region. The conservators of these quinoa landraces are farmers of the indigenous population (Bonifacio 2003). As the pattern of genetic diversity in *ex situ* collections does not necessarily reflect the extant genetic structure of *in situ* landraces, caution must be taken when extrapolating the results of studies using *ex situ* accessions to *in situ* collections (del Castillo et al. 2007). Armed with the knowledge of quinoa diversity in a particular region, the challenge then for plant breeders is to preserve the genetic diversity so that quinoa could still adapt to stressful environments, while achieving a level of homogeneity required for commercial production (Curti et al. 2012).

REFERENCES

Anabalón Rodríguez L, Thomet Isla M. Comparative analysis of genetic and morphologic diversity among quinoa accessions (*Chenopodium quinoa* Willd.) of the South of Chile and highland accessions. J Plant Breeding Crop Sci 2009;1:210–216.

Barbazuk WB, Emrich SJ, Chen HD, Li L, Schnable PS. SNP discovery via 454 transcriptome sequencing. Plant J 2007;51:910–918.

Batley J, Barker G, O'sullivan H, Edwards KJ, Edwards D. Mining for single nucleotide polymorphisms and insertions/deletions in maize expressed sequence tag data. Plant Physiol 2003;132:84–91.

Batley J, Edwards D. SNP applications in plants. In: Oraguzie NC, Rikkerink EHA, Gardeiner SE, de Silva HN, editors. Association mapping in plants. New York: Springer; 2007. p 95–102.

Bennett MD, Smith JB. Nuclear-DNA amounts in angiosperms. Phil Trans Royal Soc London Series B-Biol Sci 1991;334:309–345.

Bhargava A, Shukla S, Ohri D. Karyotypic studies on some cultivated and wild species of *Chenopodium* (Chenopodiaceae). Genet Resour Crop Evol 2006;57:1309–1320.

Bhargava A, Shukla S, Ohri D. Genome size variation in some cultivated and wild species of *Chenopodium* (Chenopodiaceae). Caryologia 2007a;60:245–250.

Bhargava A, Shukla S, Rajan S, Ohri S. Genetic diversity for morphological and quality traits in quinoa (*Chenopodium quinoa* Willd.) germplasm. Genet Res Crop Evol 2007b;54:167–173.

Bhargava A, Shukla S, Ohri D. Genetic variability and interrelationship among various morphological and quality traits in quinoa (*Chenopodium quinoa* Willd.). Field Crops Res 2007c;101:104–116.

Bhargava A, Fuentes FF. Mutational dynamics of microsatellites. Mol Biotech 2010;44:250–266.

Bonifacio A. 1995. Interspecific and intergeneric hybridization in chenopod species. MS Thesis, Brigham Young University, Provo, UT, USA.

Bonifacio A. *Chenopodium* sp.: Genetic resources, ethnobotany, and geographic distribution. Food Rev Int 2003;19:1–7.

Bundock PC, Eliott FG, Ablett G, Benson AD, Casu RE, Aitken KS, Henry RJ. Targeted single nucleotide polymorphism (SNP) discovery in a highly polyploid plant species using 454 sequencing. Plant Biotech J 2009;7:347–354.

Chawla HS. Introduction to Plant Biotechnology. Enfield, NH: Science Publishers Inc.; 2000.

Ching A, Caldwell K, Jung M, Dolan M, Smith O, Tingey S, Morgante M, Rafalski A. SNP frequency, haplotype structure and linkage disequilibrium in elite maize inbred lines. BMC Genet 2002;3:19.

Christensen SA, Pratt DB, Pratt C, Nelson PT, Stevens MR, Jellen EN, Coleman CE, Fairbanks DJ, Bonifacio A, Maughan PJ. Assessment of genetic diversity in the USDA and CIP-FAO international nursery collections of quinoa (*Chenopodium quinoa* Willd.) using microsatellite markers.

Plant Genetic Resources: Characterization and Utilization 2007;5:82–95.

Coles ND, Coleman CE, Christensen SA, Jellen EN, Stevens MR, Bonifacio A, Rojas-Beltran JA, Fairbanks DJ, Maughan PJ. Development and use of an expressed sequenced tag library in quinoa (*Chenopodium quinoa* Willd.) for the discovery of single nucleotide polymorphisms. Plant Sci 2005;168:439–447.

Costa Tártara SM, Manifesto MM, Bramardi SJ, Bertero HD. Genetic structure in cultivated quinoa (*Chenopodium quinoa* Willd.), a reflection of landscape structure in Northwest Argentina. Conserv Genet 2012;13:1027–1038.

Crossa J, Burgueno J, Dreisigacker S, Vargas M, Herrera-Foessel SA, Lillemo M, Singh RP, Trethowan R, Warburton M, Franco J, et al. Association analysis of historical bread wheat germplasm using additive genetic covariance of relatives and population structure. Genetics 2007;177:1889–1913.

Curti RN, Andrade AJ, Bramardi S, Velásquez B, Bertero HD. Ecogeographic structure of phenotypic diversity in cultivated populations of quinoa from Northwest Argentina. Ann Appl Biol 2012;160:114–125.

del Castillo C, Winkel T, Mahy G, Bizoux JP. Genetic structure of quinoa (*Chenopodium quinoa* Willd.) from the Bolivian altiplano as revealed by RAPD markers. Genet Res Crop Evol 2007;54:897–905.

Deschamps SP, La Rota M, Ratashak JP, Biddle P, Thureen D, Farmer A, Luck S, Beatty M, Nagasawa N, Michael L, et al. Rapid genome-wide single nucleotide polymorphism discovery in soybean and rice via deep resequencing of reduced representation libraries with the illumina genome analyzer. Plant Genome 2010;3:53–68.

Eathington SR, Crosbie TM, Edwards MD, Reiter R, Bull JK. Molecular markers in a commercial breeding program. Crop Sci 2007;47:S154–S163.

Elshire RJ, Glaubitz JC, Sun Q, Poland JA, Kawamoto K, Buckler ES, Mitchell SE. A robust, simple genotyping-by-sequencing (GBS) approach for high diversity species. PLoS One 2011;6:e19379.

Fairbanks DJ, Burgener KW, Robison LR, Andersen WR, Ballon E. Electrophoretic characterization of quinoa seed proteins. Plant Breeding 1990;104:190–195.

Fairbanks DJ, Waldrigues DF, Ruas CF, Ruas RM, Maughan PJ, Robinson LR, Andersen WR, Riede CR, Panley CS, Caetano LG, Arantes OMN, Fungaro MH, Vidotto MC, Jankevicius SE. Efficient characterization of biological diversity using field DNA extraction and RAPD markers. Braz J Genet 1993;16:11–33.

Fuentes FF, Martinez EA, Hinrichsen VP, Jellen EN, Maughan PJ. Assessment of genetic diversity patterns in Chilean quinoa (*Chenopodium quinoa* Willd.) germplasm using multiplex fluorescent microsatellite markers. Conserv Genet 2009;10:369–377.

Fuentes FF, Bhargava A. Morphological analysis of quinoa germplasm grown under lowland desert conditions. J Agron Crop Sci 2011;197:124–134.

Fuentes FF, Bazile D, Bhargava A, Martinez EA. Implications of farmers' seed exchanges for on-farm conservation of quinoa, as revealed by its genetic diversity in Chile. J Agric Sci 2012;150:702–716.

Gao D, Schmidt T, Jung C. Molecular characterization and chromosomal distribution of species-specific repetitive DNA sequences from *Beta corolliflora*, a wild relative of sugar beet. Genome 2000;43:1073–1080.

Garg K, Green P, Nickerson DA. Identification of candidate coding region single nucleotide polymorphisms in 165 human genes using assembled expressed sequence tags. Genome Res 1999;9:1087–1092.

Gore MA, Wright MH, Ersoz ES, Bouffard P, Szekeres ES, Jarvie TP, Hurwitz BL, Narechania A, Harkins TT, Grills GS, et al. Large-scale discovery of gene-enriched SNPs. Plant Genome 2009;2:121–133.

Gupta PK, Varshney RK. The development and use of microsatellite markers for genetic analysis and plant breeding with emphasis on bread wheat. Euphytica 2000;113:163–185.

Heiser CB, Nelson DC. On the origin of the cultivated chenopods (*Chenopodium*). Genetics 1974;78:503–505.

Jannink JL, Lorenz AJ, Iwata H. Genomic selection in plant breeding: From theory to practice. Brief Funct Genomics 2010;9:166–177.

Jarvis DE, Kopp OR, Jellen EN, Mallory MA, Pattee J, Bonifacio A, Coleman CE, Stevens MR, Fairbanks DJ, Maughan PJ. Simple sequence repeat marker development and genetic mapping in quinoa (*Chenopodium quinoa* willd.). J Genet 2008;87:39–51.

Jellen EN, Kolano BA, Sederberg MC, Bonifacio A, Maughan PJ. 2011. Chenopodium. In: Kole C, editor. Wild crop relatives: genomic and breeding resources. Berlin: Springer. pp. 35-61.

Joubes J, Chevalier C. Endoreduplication in higher plants. Plant Mol Biol 2000;43:735–745.

Kolano B, Gomez Pando L, Maluszynska J. Molecular cytogenetic studies in *Chenopodium quinoa* and *Amaranthus caudatus*. Acta Societatis Botanicorum Poloniae 2001;70:85–90.

Kolano B, Plucienniczak A, Kwasniewski M, Maluszynska J. Chromosomal localization of a novel repetitive sequence in the *Chenopodium quinoa* genome. J Applied Genet 2008a;49:313–320.

Kolano B, Siwinska D, Maluszynska J. Endopolyploidy patterns during development of *Chenopodium quinoa*. Acta Biol Cracov Bot 2008b;51:85–92.

Kolano B, Gardunia BW, Michalska M, Bonifacio A, Fairbanks DJ, Maughan PJ, Coleman CE, Stevens MR, Jellen EN, Maluszynska J. Chromosomal localization of two novel repetitive sequences isolated from the *Chenopodium quinoa* Willd. genome. Genome 2011;54:710–717.

Kolano B, Siwinska D, Gomez Pando L, Szymanowska-Pulka J, Maluszynska J. Genome size variation in *Chenopodium quinoa* (Chenopodiaceae). Plant Syst Evol 2012a;298:251–255.

Kolano B, Tomczak H, Molewska R, Jellen EN, Maluszynska J. Distribution of 5S and 35S rRNA gene sites in 34

Chenopodium species (Amaranthaceae). Bot J of the Linnean Soc 2012b;170:220–231.

Kolano B, Bednara E, Weiss-Schneeweiss H. 2013. Isolation and characterization of reverse transcriptase fragments of LTR retrotransposons from the genome of Chenopodium quinoa (Amaranthaceae). Plant Cell Reports 32: 1575–1588

Kraakman AT, Niks RE, Van Den Berg PM, Stam P, Van Eeuwijk FA. Linkage disequilibrium mapping of yield and yield stability in modern spring barley cultivars. Genetics 2004;168:435–446.

Mason SL, Stevens MR, Jellen EN, Bonifacio A, Fairbanks DJ, Coleman CE, Mccarty RR, Rasmussen AG, Maughan PJ. Development and use of microsatellite markers for germplasm characterization in quinoa (*Chenopodium quinoa* willd.). Crop Sci 2005;45:1618–1630.

Maughan PJ, Bonifacio A, Jellen EN, Stevens MR, Coleman CE, Ricks M, Mason SL, Jarvis DE, Gardunia BW, Fairbanks DJ. A genetic linkage map of quinoa (*Chenopodium quinoa*) based on AFLP, RAPD, and SSR markers. Theor Appl Genet 2004;109:1188–1195.

Maughan PJ, Kolano BA, Maluszynska J, Coles ND, Bonifacio A, Rojas J, Coleman CE, Stevens MR, Fairbanks DJ, Parkinson SE, Jellen EN. Molecular and cytological characterization of ribosomal RNA genes in *Chenopodium quinoa* and *Chenopodium berlandieri*. Genome 2006;49:825–839.

Maughan PJ, Turner TB, Coleman CE, Elzinga DB, Jellen EN, Morales JA, Udall JA, Fairbanks DJ, Bonifacio A. Characterization of *Salt Overly Sensitive 1* (SOS1) gene homoeologs in quinoa (*Chenopodium quinoa* willd.). Genome 2009a;52:647–657.

Maughan PJ, Yourstone SM, Jellen EN, Udall JA. SNP discovery via genomic reduction, barcoding and 454-pyrosequencing in amaranth. Plant Genome 2009b;2:260–270.

Maughan PJ, Yourstone SM, Byers RL, Smith SM, Udall JA. SNP genotyping in mapping populations via genomic reduction and next-generation sequencing: Proof-of-concept. Plant Genome 2010;3:166–178.

Maughan PJ, Smith SM, Rojas-Beltran JA, Elzinga D, Raney JA, Jellen EN, Bonifacio A, Udall JA, Fairbanks DJ. Single nucleotide polymorphism identification, characterization, and linkage mapping in quinoa. Plant Genome 2012;5:114–125.

Miller MR, Dunham JP, Amores A, Cresko WA, Johnson EA. Rapid and cost-effective polymorphism identification and genotyping using restriction site associated DNA (RAD) markers. Genome Res 2007;17:240–248.

Moose SP, Mumm RH. Molecular plant breeding as the foundation for 21st century crop improvement. Plant Physiol 2008;147:969–977.

Morton BR, Bi IV, Mcmullen MD, Gaut BS. Variation in mutation dynamics across the maize genome as a function of regional and flanking base composition. Genetics 2006;172:569–577.

Mujica A, Jacobsen S-E. Genetic resources and breeding of the Andean grain crop quinoa (Chenopodium quinoa Willd). PGR Newsl 2002;130:54–61.

Ochoa J, Frinking HD, Jacobs T. Postulation of virulence groups and resistance factors in the quinoa/downy mildew pathosystem using material from Ecuador. Plant Pathol 1999;48:425–430.

Ortiz R, Ruiz-Tapia EN, Mujica-Sanchez A. Sampling strategy for a core collection of Peruvian quinoa germplasm. Theor Appl Genet 1998;96:475–483.

Ossowski S, Schneeberger K, Clark RM, Lanz C, Warthmann N, Weigel D. Sequencing of natural strains of *Arabidopsis thaliana* with short reads. Genome Res 2008;18:2024–2033.

Palomino G, Hernandez LT, Torres ED. Nuclear genome size and chromosome analysis in *Chenopodium quinoa* and *C. berlandieri* subsp. *nuttalliae*. Euphytica 2008;164:221–230.

Powell W, Morgante M, Andre C, Hanafey M, Vogel J, Tingey S, Rafalski A. The comparison of RFLP, RAPD, AFLP, and SSR (microsatellite) markers for germplasm analysis. Mol Breeding 1996;2:225–238.

Rana TS, Narzary D, Ohri D. Genetic diversity and relationships among some wild and cultivated species of *Chenopodium* L. (Amaranthaceae) using RAPD and DAMD methods. Curr Sci 2010;98:840–846.

Risi JC, Galwey NW. The pattern of genetic diversity in the Andean grain crop quinoa (*Chenopodium quinoa* Willd.) I. Associations between characteristics. Euphytica 1989a;41:147–162.

Risi JC, Galwey NW. The pattern of genetic diversity in the Andean grain crop quinoa (*Chenopodium quinoa* Willd.) II. Multivariate methods. Euphytica 1989b;41:135–145.

Roa NK. Plant genetic resources: advancing conservation and use through biotechnology. African J Biotech 2004;3:136–145.

Rojas W, Barriga P, Figueroa H. Multivariate analysis of the genetic diversity of Bolivian quinua germplasm. Plant Genet Res Newsl 2000;122:16–23.

Rojas W. Multivariate analysis of genetic diversity of Bolivian quinoa germplasm. Food Rev Int 2003;19:9–23.

Ruas PM, Bonifacio A, Ruas CF, Fairbanks DJ, Andersen WR. Genetic relationship among 19 accessions of six species of *Chenopodium* L., by random amplified polymorphic DNA fragments (RAPD). Euphytica 1999;105:25–32.

Russell J, Booth A, Fuller J, Harrower B, Hedley P, Machray G, Powell W. A comparison of sequence-based polymorphism and haplotype content in transcribed and anonymous regions of the barley genome. Genome 2004;47:389–398.

Shen R, Fan JB, Campbell D, Chang WH, Chen J, Doucet D, Yeakley J, Bibikova M, Garcia EW, Mcbride C, et al. High-throughput SNP genotyping on universal bead arrays. Mutation Res-Fundamental Mol Mech Mutagenesis 2005;573:70–82.

Shi H, Ishitani M, Kim C, Zhu J-K. The *Arabidopsis thaliana* salt tolerance gene *SOS1* encodes a putative Na^+/H^+ antiporter. Proc Natl Acad Sci U S A 2000;97:6896–6901.

Smith BD, Yarnell RA. Initial formation of an indigenous crop complex in eastern North America at 3800 B.P. Proc Natl Acad Sci U S A 2009;106:6561–6566.

Stevens MR, Coleman CE, Parkinson SE, Maughan PJ, Zhang HB, Balzotti MR, Kooyman DL, Arumuganathan K, Bonifacio A, Fairbanks DJ, Jellen EN, Stevens JJ. Construction of a quinoa (*Chenopodium quinoa* Willd.) BAC library and its use in identifying genes encoding seed storage proteins. Theor Appl Genet 2006;112:1593–1600.

Tapia ME, Mujica SA, Canahua A. Orígen, distribución geográfica, y sistemas de producción en quinua. In: Primera reunion sobre genetica y fitomejoramiento de la quinua. Puno: Universidad Nacional Técnica del Altiplano (UNTA), Instituto Boliviano de Technologia Agropecuaria (IBTA), Instituto Interamericana de Ciencias Agricolas (IICA), Centro de Investigación Internacional para el Desarollo; 1980. p A1–A8.

Tautz D. Hypervariability of simple sequences as a general source for polymorphic DNA markers. Nucleic Acids Res 1989;17:6463–6471.

Van K, Hwang EY, Kim MY, Park HJ, Lee SH, Cregan PB. Discovery of SNPs in soybean genotypes frequently used as the parents of mapping populations in the United States and Korea. J Hered 2005;96:529–535.

Walsh BM, Adhikary D, Maughan PJ, Emshwiller E, Jellen EN. *Chenopodium* (Amaranthaceae) polyploidy inferences from SOS1 data. Manuscript in preparation.

Wang J, Lin M, Crenshaw A, Hutchinson A, Hicks B, Yeager M, Berndt S, Huang WY, Hayes RB, Chanock SJ, et al. 2009. High-throughput single nucleotide polymorphism genotyping using nanofluidic dynamic arrays. BMC Genomics 10:561. doi:10.1186/1471-2164-10-561. Available from http://www.biomedcentral.com/content/pdf/1471-2164-10-561.pdf

Weber J, May PE. Abundant class of human DNA polymorphisms which can be typed using the polymerase chain reaction. Am J Hum Genet 1989;44:388–396.

Wilson HD. Domesticated *Chenopodium* of the Ozark Bluff dwellers. Econ Bot 1980;35:233–239.

Wilson HD. Genetic variation among tetraploid Chenopodium populations of southern South America (sect. *Chenopodium* subsect. *Cellulata*). Syst Bot 1981;6:380–398.

Wilson HD. Allozyme variation and morphological relationships of *Chenopodium hircinum* (s.l.). Syst Bot 1988a;13: 215–228.

Wilson HD. Quinoa biosystematics I: domesticated populations. Econ Bot 1988b;42:461–477.

Wilson HD, Manhart J. Crop/weed gene flow: *Chenopodium quinoa* Willd. and *C. berlandieri* Moq. Theor Appl Genet 1993;86:642–648.

Windhausen VS, Atlin GN, Hickey JM, Crossa J, Jannink JL, Sorrells ME, Raman B, Cairns JE, Tarekegne A, Semagn K, et al. Effectiveness of genomic prediction of maize hybrid performance in different breeding populations and environments. G3 (Bethesda) 2012;2:1427–1436.

Zhang F, Zhao Z. The influence of neighboring-nucleotide composition on single nucleotide polymorphisms (SNPs) in the mouse genome and its comparison with human SNPs. Genomics 2004;84:785.

Chapter 8
Ex Situ Conservation of Quinoa: The Bolivian Experience

Wilfredo Rojas and Milton Pinto
PROINPA Foundation, Av. Elias Meneces km 4, El Paso, Cochabamba, Bolivia

INTRODUCTION

During the past four decades, germplasm collections maintained *ex situ* have grown in number and size as a result of the intense worldwide efforts to conserve plant genetic resources for food and agriculture (PGRFA). These collections are kept under very different conditions, depending on policies at the national or international level, the institutional environment, available expertise, facilities and budget, and the degree of national and international collaboration (Engels and Visser 2003). According to the second report on the state of the art of PGRFA (FAO 2010), the total number of samples stored *ex situ* worldwide has increased by approximately 20% (1.4 million) since 1996, reaching a total of 7.4 million samples. This growth in quantity and diversity of samples over a range of germplasm requires that collections be managed with the highest standards of conservation.

Genebanks are essential for the food security and sovereignty of every nation. They are part of a nation's ancestral and cultural heritage, and as such are a responsibility that should be assumed by society and the state. Toward this goal, conservation of genetic resources requires institutional support, including sustained financing, trained staff with specialized expertise and essential equipment needed to maintain germplasm collections and conduct conservation activities.

However, the mere creation of a genebank does not guarantee the conservation of plant genetic resources of a country, as shown by the first and second reports on the state of the art of PGRFA in the world (FAO 1996, 2010). Increasingly, more attention is paid as to how and when to regenerate germplasm from a collection, because of the increasing costs of germplasm maintenance and regeneration (Engels and Visser 2003). There is also the possibility that, in time, genetic erosion occurs in a bank that is improperly managed. The economics of genebank management and operation is implicit in all conservation efforts. It does not merely consist of assigning budgets to specific operations of the bank, but is related to internal decision making on the expenditures of the institution. These are crucial decisions, constituting a more important issue, because those who initially support its implementation continuously demand improvements in the registration of its operations.

In the absence of good planning, genebank management can develop and evolve in many ways. In addition, local germplasm conditions vary greatly, calling for several management approaches and producing a diversity of experiences. In developing countries, conservation issues and technological advancements are not

Quinoa: Improvement and Sustainable Production, First Edition. Edited by Kevin Murphy and Janet Matanguihan.

always a priority, and decisions regarding these often come from a political perspective. This has been the case regarding the germplasm collection of quinoa from Bolivia, where conditions have not always been easy. The Bolivian collection has the greatest diversity in the world, particularly in comparison to other countries in the Andean region. Quinoa is not merely an "Andean grain." More than any crop in the region, quinoa encapsulates the Bolivian culture. It is strongly rooted in our local customs, in indigenous consumption and production. Not only does Bolivia have the greatest genetic diversity of quinoa but it also has the highest acreage of quinoa grown in the world, making it the top exporter of quinoa. These aspects validate quinoa's strategic importance for Bolivia.

In this chapter, the center of origin and diversity of quinoa are discussed, together with the beginnings of the Bolivian germplasm collection. The operations and the management process followed for more than 45 years since the first quinoa accessions were collected, and critical times throughout the collection's existence, are also discussed. This chapter also provides a summary of the work of the Promotion and Research on Andean Products (PROINPA) Foundation during the period when it was in charge of genebank management and conservation by delegation of the Bolivian State, until the bank achieved recognition at both national and international levels.

CENTERS OF ORIGIN AND DIVERSITY OF QUINOA

According to the Russian scientist Vavilov, the center of origin of a cultivated plant is the region with the greatest diversity of plant types, both cultivated and wild progenitors. The Andean region is among the eight centers of origin of plants grown in the world (Vavilov 1951). It is also considered as the center of one of the most important American civilizations (Gandarillas et al. 2001). By consensus, all scientists who have conducted morphological, genetic, and systematic studies consider that quinoa originated in the Andes of Peru and Bolivia (Gandarillas 1979). The Andean region has the highest genetic diversity of both wild and cultivated quinoa occurring in natural conditions and in cultivated fields. Considering the agroecological conditions under which the quinoa species develops, subcenters of diversity have been identified. In these subcenters of diversity, quinoa ecotypes have developed and adapted, resulting in variability in botanical, agronomical, and species adaptation characteristics.

Five ecotypes of quinoa have been identified in the Andes region, based on genetic variability, adaptation, and several highly heritable morphological characteristics (Lescano 1989; Tapia 1990). Four of the five major ecotypes or groups of quinoa are found in Bolivia (inter-Andean valleys, highlands, salt flatlands, and Yungas). Only the sea level quinoas are not found in Bolivia but grow in Chile. The following are descriptions of the ecotypes:

1. *Quinoas of Inter-Andean Valleys* have adapted to grow between 2500 and 3,500 meters above sea level (masl) and grow up to 2.5 m or more in height. They produce many branches with lax inflorescence and are usually resistant to downy mildew (*Peronospora farinosa*). This group of quinoas are typically grown intercropped with maize in 5–6 transverse grooves, as edges of other crops, or scattered in the same field.
2. *Quinoas of the Highlands* are able to grow from 3,600 to 3,800 masl in the region of the Peruvian-Bolivian Altiplano, as pure or unique cultures, and in larger areas. The plants grow between 0.5 and 1.5 m, with the stem ending in a main panicle that is usually compact. It is in this Altiplano region where the greatest variability of the crop has been reported, and where grains with more specialized use are produced. This group includes the largest number of improved varieties, and is susceptible to mildew when grown in areas of high humidity.
3. *Quinoas of the Salt Flatlands* grow in the salt flat areas south of the Bolivian Altiplano. This is the driest area in the region, with 200–300 mm of rainfall. In this region, it is customary to plant quinoa as a single crop

at distances of 1 m × 1 m in holes to make better use of low humidity. These quinoas are known as "Royal Quinoa," with larger grain size (>2.2 mm in diameter), characterized by a thick pericarp and high saponin content.

4. *Quinoas of the Yungas or Edge of the Rainforest* consist of a small group of quinoas that have adapted to the conditions of the Yungas or edge of the Bolivian rainforest, mainly in the valleys of Cochabamba, at altitudes between 1,500 and 2,200 m. They are characterized by somewhat branched development, reaching heights of up to 2.20 m. The plants are green but the entire plant turns into a striking orange color when in full bloom.
5. *Sea-level quinoas* are found in the areas of Linares and Concepción, Chile to 36° South Latitude. Plants of this type are more or less robust, 1.0–1.4 m tall, with branched growth and cream-colored, transparent grains (Chullpi type). These quinoas share numerous similarities with *Chenopodium nuttalliae* (Huahzontle) grown in isolation in Mexico at 20° North Latitude.

The quinoa genetic diversity conserved and available in the different countries from the Andean Region are classified in five major groups. Quinoa from Bolivia is distributed in the inter-Andean valleys, highlands, salt flatlands, and Yungas. Quinoa from Colombia and Ecuador is distributed in the inter-Andean valleys, quinoa from Peru is found in the inter-Andean valleys and highlands, while quinoa from Chile is grouped in the salt flatlands and sea-level quinoa ecotypes (Fig. 8.1).

Rojas (2003) studied the genetic variability of the Bolivian germplasm collection and has identified seven subcenters of diversity in Bolivia: one subcenter in the salt flatlands of Potosi and Oruro, two sub-centers in the northern Altiplano in La Paz, one in the central Altiplano in La Paz and Oruro, one sub-center located geographically in the transition Salt Flatlands and Altiplano, and two subcenters in the Andean valleys of Cochabamba, Chuquisaca, and Potosi. These subcenters comprise a wide genetic diversity of the crop, expressed in the variability of plant, inflorescence, and seed color; variability of inflorescence types, culture cycle duration, nutritional (protein) and agroindustrial value; and variability in saponin and betacyanin content in leaves. This diversity of traits enables quinoa to adapt to different ecological and soil conditions such as salinity or acidity, rainfall, temperature, altitude, frost, and drought.

GEOGRAPHICAL DISTRIBUTION OF QUINOA

Given the wide distribution of its center of origin, quinoa can be considered as an oligocentric species, with the Andean region as its center of diversity with multiple routes of diversification (Mujica 1992). Quinoa is grown throughout the vast Andean region that was once ruled by the Incas (Lescano 1994). It is found from Colombia (Pasto), to northern Argentina (Jujuy and Salta) and Chile (Antofagasta and Concepción). It extends from 5° of North latitude in the south of Colombia, to 43° of South latitude in the Tenth Region of Chile. It ranges from sea level in Chile to 4,000 masl in the Peruvian and Bolivian Altiplano. Thus, there are quinoas of the coast, valleys, inter-Andean valleys, salt flatlands, and Altiplano (Fig. 8.2). A considerable part of the genetic variability of the species is associated with geographical distribution (Rojas 2003).

Specifically, in *Colombia*, quinoa is grown in the Department of Nariño, in the towns of Ipiales Puesrés, Contadero, Cordova, San Juan, Mocondino, and Pasto. A department is a political division, each with a governor and an assembly. Colombia has 32 departments. In *Ecuador*, quinoa is grown in the areas of Carchi, Imbabura, Pichincha, Cotopaxi, Chimborazo, Loja, Latacunga, Ambato, and Cuenca. In *Peru*, quinoa is grown in the areas of Cajamarca, Callejon de Huayllas, Mantaro Valley, Andahuayllas, Cusco, and Puno (Altiplano), which comprises 75% of the cultivated area. In *Bolivia*, quinoa is cultivated in the Altiplano of La Paz, Oruro, and Potosi and in the inter-Andean valleys of Cochabamba, Chuquisaca, Potosi, and Tarija (Rojas et al. 2010b). In *Chile*, the quinoa-growing areas are the Chilean

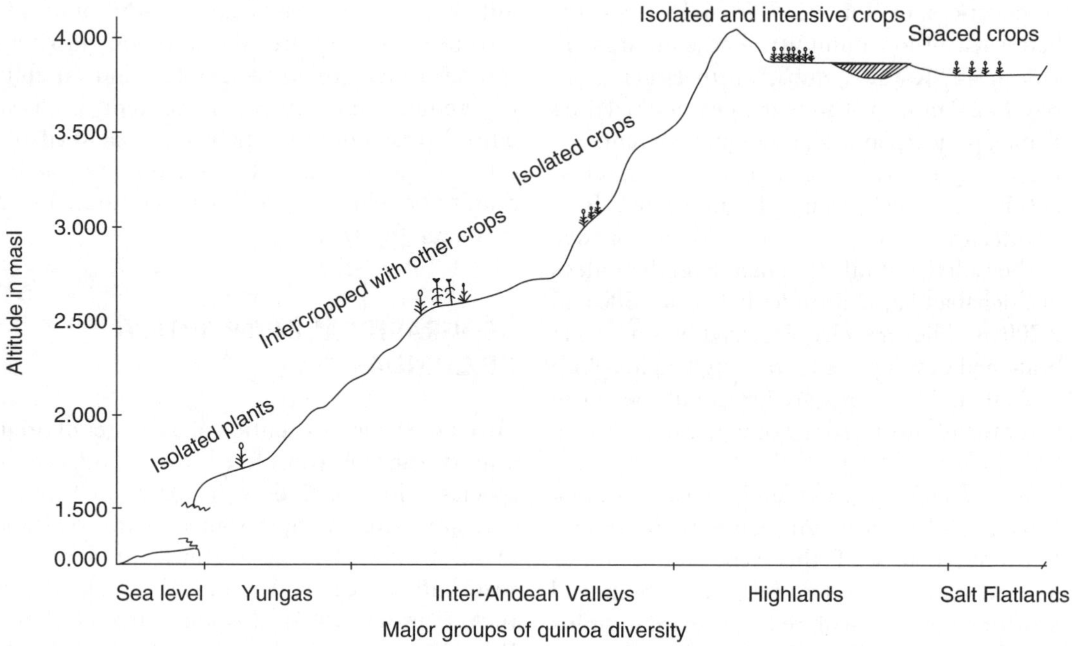

Fig. 8.1 Profile distribution of the five major quinoa ecotypes in the Andean Region.

Altiplano (Isluga and Iquique) and Concepción. There are also reports of the Ninth and Tenth Regions growing quinoa (Barriga et al. 1994). In *Argentina*, quinoa is produced in isolation in the highlands of Jujuy and Salta. Cultivation also expanded into Calchaquíes Valleys of Tucuman (Gallardo and Gonzalez 1992).

GENEBANKS OF THE ANDEAN REGION

To safeguard the phenotypic and genotypic variability of quinoa in the Andean region, gene banks have been established since the 1960s in several countries in the region. These genebanks are connected to the agricultural sector and universities of Argentina, Bolivia, Colombia, Chile, Ecuador, and Peru, and are responsible for their management and conservation.

According to Rojas et al. (2013), out of the 16,422 accessions conserved worldwide, 14,502 are conserved in genebanks from the Andean region, with Bolivia and Peru managing germplasm collections with the highest number of accessions (Fig. 8.3). In Bolivia, there are six genebanks which contains 6,721 accessions of quinoa. These genebanks are located at the Toralapa Center of the National Agricultural and Forestry Research Institute (INIAF), at the UMSA Experimental Station in Choquenaira at the Centre for Research in Biotechnology and Plant Genetic Resources of the UTO, at the Tiahuanacu Academic Unit of UCB, at the Kallutaca Experimental Center of UPEA, and at the Center for Community Research and Promotion. The genebank with the largest number of accessions is the one managed by INIAF, with 3,178 accessions, known in Bolivia as the national collection of quinoa germplasm. Next in line are the collections of UTO and UMSA with 1,780 and 1,370 accessions respectively (Rojas et al. 2010a, 2013).

In Peru, there are eight genebanks where 6,302 accessions of quinoa are conserved. The genebanks are located in the experimental stations of INIA in Illpa (Puno), Andenes (Cusco), Canaán (Ayacucho), Santa Ana (Huancayo), and Baños del Inca (Cajamarca). The following universities also conserve quinoa germplasm:

Fig. 8.2 Geographic distribution of quinoa in the Andean Region. (*See color insert for representation of this figure.*)

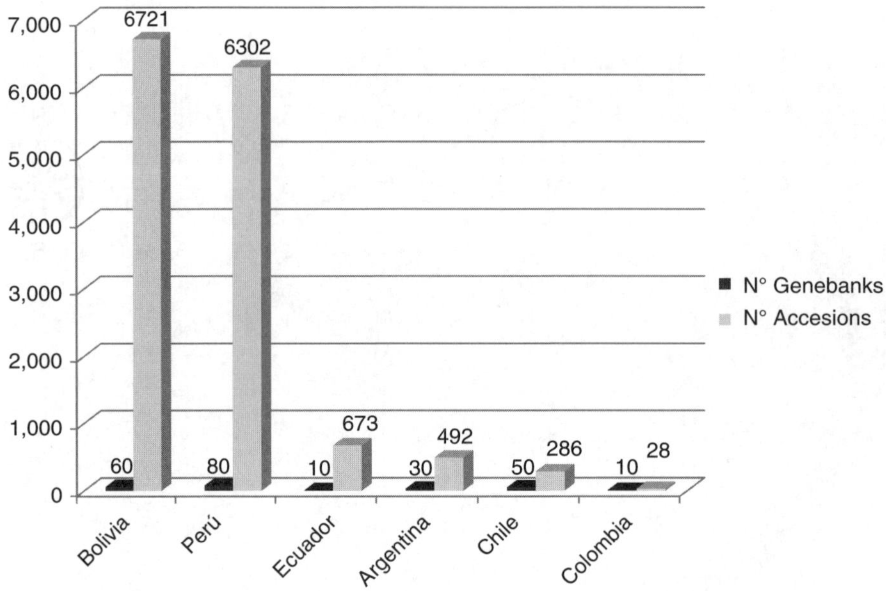

Fig. 8.3 Number of accessions and genebanks conserving quinoa germplasm in the Andean countries.

Universidad Agraria La Molina in Lima, Universidad Nacional de San Antonio Abad of Cusco, and the Universidad Nacional del Altiplano in Puno (Mujica 1992; Bonifacio et al. 2004; Bravo and Catacora 2010; Gómez and Eguiluz 2011). The most important collections of quinoa germplasm are those of Universidad Nacional Agraria La Molina, Universidad Nacional del Altiplano, and INIA Puno with 2,089, 1,910 and 1,029 accessions.

In Argentina, the national network of seed genebanks holds a total of 492 accessions of quinoa conserved in the Base Genebank of the National Institute of Agricultural Technology (INTA). These accessions are partially duplicated in the Active Genebank of the Argentinian Northwest and at "La Consulta" Genebank. The collection of 492 accessions is a result of the joint efforts of the Faculty of Agriculture of the Universidad de Buenos Aires (UBA) and the INTA (Rojas et al. 2013).

In Chile, 286 accessions are conserved, out of which 203 accessions are in the Base Genebank of Vicuña Experimental Center – from the National Institute of Agricultural Research (INIA), and the rest are in the genebank of the Faculty of Agrarian Sciences at UACH, in the Active Genebank of the Carillanca Regional Research Center – INIA, in the Universidad Arturo Prat (UNAP) and in the Baer Seed Bank. In Ecuador, there are 673 accessions of quinoa conserved by the National Department of Biotechnology and Phytogenetic Resources at the Santa Catarina Experimental Station from the National Institute of Agricultural and Livestock Research (INIAP). In Colombia, the genebank of the Colombian Corporation of Agricultural Research at Tibaitatá conserves 28 accessions (Rojas et al. 2013)

BOLIVIAN COLLECTION OF QUINOA GERMPLASM

History and management of the quinoa germplasm

According to Rojas et al (2010a), the first efforts to establish a germplasm collection of quinoa and other Andean crops date back to the early 1960s, at the initiative of Humberto Gandarillas who went on collection trips in the highlands, salt flatlands, and inter-Andean valleys of Bolivia. With the material Gandarillas collected, the first quinoa genebank was established. Initially, the genebank was under the responsibility of the Patacamaya Experiment Station but was later under the National Quinoa Program of the Bolivian Institute of Agricultural Technology (IBTA) that operated in Bolivia until 1998.

After the closure of IBTA in the late 1990s, the Patacamaya Experimental Station became dependent on the Prefecture of La Paz and the quinoa germplasm collection was in imminent danger of being lost. The Prefecture of La Paz was administered by the newly created Departmental Agricultural Services (SEDAG) – La Paz unit, and management of the quinoa germplasm collection also became one of its responsibilities. Unfortunately, during this time, germplasm collection had no financial support. In the absence of a clear policy regarding the collection, its conservation and management was discontinued.

At the initiative of quinoa researchers and executives of the Foundation for the PROINPA, every effort was made to maintain the plots of quinoa germplasm that had been sown in the northern Altiplano through the Durable Resistance Project in the Andean Region (PREDUZA). During the 1997–1998 crop year, the PREDUZA project was in full operation and managed by researchers from the former National Quinoa Program of IBTA. The material was planted by the lakeside of Lake Titicaca mainly for accessibility and ease in management. In addition, the environmental conditions in the northern Altiplano were conductive to assessment for mildew resistance. The germplasm had been divided into two lots, where 50% was assessed at the Belen Experiment Station (Omasuyos province) while the other 50% at the Choquenaira Experimental Station (Ingavi province), to replicate the quinoa germplasm collection at harvest. Both stations were managed by the Faculty of Agronomy, UMSA.

Under these circumstances, the authorities of the Ministry of Agriculture, Livestock and Rural Development (MAGDER), later the Ministry of Rural Development, Agriculture and Environment (MDRAyMA) currently known as the Ministry of Rural Development and Land (MDRyT) through letter PDTA-2216-BO-C No. 418/98, enabled the PROINPA Foundation to take over the conservation of the quinoa genebank. The PROINPA Foundation was entrusted to administer, manage, and conserve the quinoa collection within the framework of Bolivia's legislation. To fulfill its responsibilities, PROINPA on its own initiative raised funds and received support from several agencies, including the Danish Cooperation for Development (DANIDA), International Plant Genetic Resource Institute (IPGRI) currently known as Bioversity International, Program of Small Donations of the United Nations Development Program (PPD/PNUD), International Fund for Agricultural Development (FIDA), and the McKnight Foundation. This marked a new beginning concerning the conservation of plant genetic resources of Bolivia. The PROINPA Foundation has taken over the management and conservation of the quinoa genebank for more than 10 years.

Later, the quinoa genebank received financial support from the national government in February 2001 through agreements signed between MAGDER, Agricultural Services Program – Coordination Unit (UCPSA), and PROINPA. After the creation of the National System of Genetic Resources for Food and Agriculture (SINARGEAA) in 2003, six national genebanks came into operation and MDRAyMA endorsed PROINPA as the entity in charge of the National Genebank of Andean Grains, including the quinoa germplasm collection. During this period up to December 2008, PROINPA has managed and conserved these genetic resources under the SINARGEAA framework and with the support of projects under the Neglected and Underutilized Species (NUS) – IFAD and United Nations Environmental Program/Global Environmental Facility (UNEP/GEF) frameworks.

Subsequently, INIAF, an institution created through a national decree in 2008 and designated as the national competent authority in Genetic Resources, enabled funding for the SINARGEAA Transition Plan (2009–2010). Under this framework, INIAF and PROINPA signed an interagency cooperation agreement, through which the coordination and interaction mechanisms for the transfer of genetic material and relevant equipment to the new competent authority were established.

Current status of quinoa germplasm

In the crop year 2010, the quinoa germplasm collection and the Genebank of High Andean Grains were transferred to INIAF. The transfer was completed on July 23, 2010 and was notarized, recording in detail the conservation status of each accession, along with related documentation (databases, publications, and other protocols), equipment and materials basis for conservation. The transfer also included a period of at least 18 months of joint work between the two institutions, including staff training and capacity building to facilitate immediate management of the genebanks.

This moment closed a very important stage for PROINPA in the management of the National Genebank of Andean Grains, and, in particular, the administration of the Bolivian collection of quinoa germplasm which lasted for 12 years. It was a labor of dedication to increase the number of accessions of the germplasm collection, improve the quality of conservation, upgrade the collection's documentation, and generate a knowledge database that can be used in different fields, from genetic improvement to agribusiness. All of this has been made possible through complementary projects with which the genebank has been articulated with different users, from academic, scientific, and technical users to communities that work with *in situ* conservation. This work and dedication has been recognized both nationally and internationally. Conservation farmers, government agencies, international cooperation agencies based in Bolivia, and entities outside the country, have lauded the genebank for its high conservation quality and possession of unique materials (Rojas et al. 2010a).

At present, the Bolivian quinoa germplasm collection is part of the National Genebank of Andean Grains, preserved in the INIAF Toralapa Center. The collection has wide genetic variability and conserves 3,178 cultivated and wild accessions collected in highland communities, salt flatlands, inter-Andean valleys and along the edge of rainforests in the regions of La Paz, Oruro, Potosi, Cochabamba, Chuquisaca, and Tarija. In addition, it includes germplasm from Peru, Ecuador, Colombia, Argentina, Chile, Mexico, the United States, England, Holland, and Denmark.

STEPS FOR *EX SITU* MANAGEMENT AND CONSERVATION OF QUINOA

Ex situ conservation is considered complementary to *in situ* conservation as it is not possible to conserve all species *ex situ* or "off site." It consists of a set of activities towards the management of plant genetic resources. It is carried out by creating genebanks and germplasm collections, involving a series of steps and procedures that require trained personnel.

Seed genebanks are viable options for *ex situ* conservation of quinoa germplasm. Conditions in these seed genebanks promote maximum storage time with minimal physiological activity and negligible loss of viability and germination. Quinoa seeds have been classified as "orthodox" according to its behavior in storage (Ellis et al. 1988). In "orthodox" seeds, one can manage humidity and temperature in order to keep seeds viable for longer periods of time, whereas in "recalcitrant" seeds this is not possible. Because quinoa seeds are "orthodox," their viability can be maintained in a predictable manner within a range of environmental conditions through temperature and seed moisture reduction.

The "orthodox" behavior of quinoa seeds in storage makes it possible to devise a management strategy for the seed genebanks. The protocols developed and the results achieved with the Bolivian quinoa seed germplasm collection, during the time that it was managed by PROINPA, are discussed in this section. An effective protocol for *ex situ* management has been developed by several specialists in Plant Genetic Resources in Latin America (Jaramillo and Baena 2000). The protocol has been adapted to quinoa germplasm, including the collection, preliminary multiplication, storage, characterization, and evaluation of germplasm. It also includes the regeneration and multiplication of germplasm, plus their documentation and utilization.

Collection of quinoa germplasm

"Collection of germplasm" is the process of obtaining seed samples that represent wild plant populations or varieties of cultivated species. The collection of germplasm is performed for the conservation of species diversity, to obtain germplasm that can be used in breeding or to search for and find new populations not available previously in the genebanks (Sevilla and Holle 2004). Germplasm collections are necessary and fully justified in areas considered as centers of origin and diversity because cultivated varieties and wild relatives that coexist and evolve over time can be found in such places. Germplasm collection is the first stage or step that needs to be carried out in a management strategy for *ex situ* conservation. Utmost caution and care should be taken in technical and logistical planning and in job execution. For instance, seed samples collected should be viable to ensure that these will germinate and develop into plants similar to the mother plants if grown in comparable environments.

Since 1964, when germplasm collections of quinoa began in Bolivia, the *Centralized Collection* method has been used. This method is based on the work of a team of scientists and researchers who visit various sites in the highlands, salt flatlands, inter-Andean valleys, and areas of crop distribution in the country. The team collects seeds to form the main quinoa germplasm collection. Although the germplasm collection holds significant numbers of quinoa accessions, the variability stored does not represent all the existing diversity of quinoa in the country. For this reason, quinoa germplasm was collected in areas that had not been represented in the collection.

Since 2002, the *Decentralized Collection* method for quinoa germplasm collection has been implemented in the country. This method entails collaboration with local entities or groups such as extension services, farmer organizations, NGOs and universities, among others. It is based on itinerant collection of samples by a team of scientists throughout different ecological regions. This collection method also involves interaction with local experts and constitutes a practical alternative to the Centralized Collection Method, because local experts have extensive ecogeographic and cultural knowledge of the area. These local experts also know how to choose the best collection time and can collect during fruiting and later stages (Guarino et al. 1995).

Technical procedure for quinoa germplasm collection

For both the *Centralized Collection* and *Decentralized Collection* methods, it is critical that technical and logistical planning should be in sync to ensure the success of a quinoa germplasm collection trip. Technical planning involves defining what to collect and determining why, where, how, and when to collect. On the other hand, logistical planning means organizing the mission. Arrangements need to be made for the effective implementation of the technical planning aspects, such as forming the team or collection team, preparing the itinerary, arranging the transportation, acquiring permits, and assembling the necessary implements for collection, such as instruments, equipment, and supplies (Guarino et al. 1995).

A protocol for quinoa germplasm collection has been adapted, and it can be applied to both the centralized and decentralized collection methods. The protocol meets the minimum requirements needed for germplasm collection work. Some important points in the protocol are specified in the following sub-sections.

Collection form

The collection form adapted for quinoa germplasm collection is shown in Annex 8.1; it is based on the Germplasm Collection Form published by the International Plant Genetic Resources Institute or IPGRI (Jaramillo and Baena 2000). The form has the minimum necessary information about the origin of the collected populations.

Unavoidable variables were included to record morphological and ethnobotanical information, used to compare qualitative characteristics at the time of characterization and evaluation of quinoa accessions (Rojas 2002b).

Sources of collection

Sources of collection are those sites where samples of germplasm can be obtained. Farmers' fields and storage facilities constitute the main sources for quinoa collections because these provide more homogeneous seed samples. Moreover, the information needed on the collection form is more reliable, often coming from the farmers themselves. Nevertheless, rural fairs, markets, and other places of sale are also considered sources of collection, depending on available information regarding the origin of the sample so it can be recorded in the collection form. Finally, wildlife habitats are also considered sources of collection for the registration of wild populations, although wild populations of quinoa can also be found in farmers' fields.

Sampling strategy

The objective of a collection trip is to gather a group of accessions that would be a representative sample of the genetic diversity of a taxon. This requires adequate knowledge of the area and of the target species. For this reason, information on topography, geology, soil, climate, and vegetation of the area must be collected. In addition, information on the distribution, phenology, reproductive biology, genetic diversity, storage behavior, and ethnobotany of the species, that is, knowledge about their uses, must be gathered (Guarino et al. 1995).

The number of samples to be collected in a geographic area depends on the existing genetic variability found in the area. The number of sample plants to be taken per population would depend on the characteristics of the plot. Usually, a seed sample of 50 plants per population of quinoa is recommended. The plants should be randomly selected (random sampling) in case of homogeneous plots, and selected at small intervals (stratified sampling) in case of variable sites (Genebank Standard 1994). In the case of wild species, sampling should be done without compromising the natural preservation of plant populations (Querol 1988).

From each plant, it is recommended that the glomeruli to the average height of the main panicle be cut on the glomerular basal axis with a scissor. A sufficient quantity of seed can be produced with the material collected from at least 50 plants, which may vary from 20 to 100 g of seed, depending on the plant structure. For quinoa, 3 g of seed on average equals approximately 1,000 seeds. Therefore, the amount of seeds collected would exceed the recommended minimum amount of 1,500 to 2,000 seeds for autogamous and pollinated populations (Genebank Standard 1994; Hawkes 1980). Consequently, the seeds collected from an accession could be considered as representative samples.

Sample handling and documentation

Whenever possible, samples to be collected should be healthy, viable, and fresh. Plants should be collected in the mature stage so it can tolerate desiccation without losing viability. The moisture content of the seed sample should be between 10% and 12%. However, if the collection is done immediately after a rain, samples should be dried or exposed to the sun for the duration of the expedition.

For each sample of quinoa germplasm, the collection form described in Annex 8.1 must be completed. Information on origin, agromorphological characteristics, and uses is generally provided by farmers through informal interviews. Both the collected seed and collection form should be placed in paper envelopes identifying clearly the expedition name and the sample number.

History and evolution of quinoa germplasm collections

Rojas et al. (2001) published a catalog on the Bolivian quinoa germplasm collection and summarized the history and progress of quinoa collections. The first germplasm collection of quinoa and Andean crops organized in the Andean region was in the Patacamaya Station in 1966, at the initiative of Humberto Gandarillas. He went on collection trips throughout the highlands and inter-Andean valleys of Bolivia, with financial support from the Project Bolivia II Oxfam-FAO and later from the Institute of Andean Crops, Ministry of Agriculture of Bolivia (Tapia 1977).

Later, the germplasm collection was augmented with valuable donations received from the Technical University of Oruro (56 accessions) and IICA (239 accessions from Peru), expanding the collection to 1,375 accessions. The genetic material from the collection was evaluated at the Patacamaya Station, and as a result, 17 landraces of quinoa were developed and described (Gandarillas 1968). In the late 1960s and early 1970s, 446 accessions were received from Peru as donation and exchange, including 131 accessions ensuing from mass selections from the National University of the Altiplano in Puno. In the same period, a collection of 159 cultivated and wild accessions were received from the central Altiplano of Bolivia (La Paz and Oruro) and 65 accessions collected by the OAS, without registration data and collection date.

By the mid-1970s, the accessions of quinoa and Andean crops numbered to 2,045. However, loss of genetic material and separation of the quinoa collection from the potato, oca, and ullucus collections reduced the number to 1,458. Subsequently, Waldo Tellería made a collection in the Department of Oruro and increased the germplasm collection to 1,472. In 1978, the collection was expanded to 1,487 accessions through a donation of accessions from northern Argentina. In the same year, Humberto Gandarillas made collections in the Altiplano and inter-Andean valleys of the country and the inclusion of three accessions from Mexico increased the collection to 1,516 accessions. Later, in 1981, Humberto Gandarillas, Gualberto Espindola, and Florencio Zambrana undertook different collection trips in the country, increasing the collection to 1,752 accessions. In 1982, the entry of eight accessions from Ecuador (INIAP) and northern Chile were registered, increasing the collection to 1,761 accessions.

Between 1983 and 1985, Humberto Gandarillas, Gualberto Espindola, Raúl Saravia, Alejandro Bonifacio, Emigdio Ballon, German Nina, and Estanislao Quispe embarked on several collection trips in the country and enlarged the collection to 1,985 accessions. In the same period, quinoa varieties from Peru, Ecuador, Chile, and Mexico were incorporated in the collection. A single accession from northern Argentina was received in 1987. In 1989, 15 accessions were collected by Guillermo Prieto, Raúl Saravia and Alejandro Bonifacio in the central Altiplano, making the total number of accessions 2,001.

In 1992, the genebank recorded 2,012 germplasm accessions. In the same year, 20 accessions were incorporated from the southern and central Altiplano, through collections made by Gualberto Espindola, Genaro Aroni, and Juan Tupa. In 1993, 54 accessions were received from Cochabamba as a donation from the NGO Wiñay Siway, Comprehensive Services Cooperative Punata, Radio Esperanza, and Second Alandia. In addition, four accessions were received from INIAP Ecuador, for a total of 2,090 accessions. In the same year, the Mañica Substation of IBTA (Potosí), through Severino Bartolome, collected 147 cultivated and wild accessions. Furthermore, 182 accessions (wild material and advanced lines classified as bitter and sweet) from the breeding area of the Quinoa National Program were incorporated, increasing the collection to 2,419 accessions.

In 1994, Wilfredo Rojas, Nicholas Monasteries, and Gualberto Espindola collected nine accessions from the South Altiplano and nine accessions from the Pacajes province in La Paz. An exchange with the Caquiaviri Technical College resulted to 65 accessions incorporated in the collection. INIA Peru donated nine accessions, bringing the collection to a total of 2,511 accessions. In 1995, 24 accessions were added to the collection, mainly varieties and breeding lines. By the time of the closure of IBTA, the number of germplasm accessions managed at the Patacamaya Experiment Station numbered at 2,535.

Through the work executed by the PROINPA Foundation, complementary and planned collections were carried out. In 1998, 12 advanced lines from the breeding component were incorporated in the collection, and 56 accessions collected in the South Altiplano by Alejandro Bonifacio were registered and likewise included. In 1999, 13 accessions from the global quinoa test originating from Peru, Ecuador, England, Holland, and Denmark were included, along with 85 accessions collected in the northern, central, and southern Altiplano. By this time, the collection numbered a total of 2,701 accessions (Rojas et al. 1999).

Between 2000 and 2002, through decentralized collections, 135 accessions were incorporated, for a total of 2,836 accessions in the collection (Rojas 2002b). In the years 2002–2003, 113 accessions were collected and the quinoa collection numbered 2,949 accessions (Rojas et al. 2003a). In 2003–2004, 172 accessions were collected for a total of 3,121 accessions (Rojas and Pinto 2004). Finally, through complementary collection, 57 accessions were collected, bringing the total count to 3,178 accessions. Table 8.1 shows the origin and number of accessions in the quinoa germplasm collection in Bolivia.

Table 8.1 Origin and Number of Accessions in the Quinoa Germplasm Collection Preserved in the INIAF Genebank, Bolivia.

Country	Department/ Region	Number of Accessions	Subtotal
Bolivia	La Paz	1006	2,357
	Oruro	630	
	Potosi	470	
	Cochabamba	124	
	Chuquisaca	108	
	Tarija	19	
Peru	Ancash	5	675
	Junín	18	
	Ayacucho	40	
	Cusco	36	
	Puno	567	
	Ica	9	
Ecuador	Norte	11	28
	Centro	17	
Chile	Norte	1	18
	Sur	17	
Argentina	Jujuy	16	16
Mexico	Norte	3	6
	Centro	3	
USA	New Mexico	1	1
Denmark		2	2
Netherlands		2	2
England		2	2
OAS		60	60
NI[a]		11	11
Total			3,178

Source: Rojas et al. 2010a, 2013.
[a]NI, Non-identified.

Distribution of quinoa germplasm collection

According to the distribution studies conducted with the Bolivian collection of quinoa germplasm (Rojas 2002a; Rojas et al. 2009), the variability of quinoa is distributed from 15° 42′ of South Latitude in the Omasuyo province of the Department of La Paz, to 21° 57′ of South Latitude in the M. Omiste province of the Department of Potosi. Additionally, it extends from 64° 19′ of Western Latitude in the province Tomina of the department of Chuquisaca, to 69° 09′ of Western Longitude in the province Manco Kapac of the Department of La Paz. Its altitudinal distribution varies from 2,400 to 4,200 masl (Fig. 8.4).

Fig. 8.4 shows a greater variability of quinoa along the highlands (Altiplano), mainly in areas adjacent to the road that stretches from Lake Titicaca through La Paz, Oruro, Challapata, Sevaruyo and Uyuni, Salinas de Garci Mendoza, Daniel Campos, and Lipez. Equally important is the geographical distribution of the accessions through the Andean valleys of Cochabamba, Chuquisaca Potosí, and Tarija.

Preliminary multiplication of quinoa germplasm

During the initial stage of quinoa germplasm management, there is a need to verify compliance with minimum conditions to ensure that the seed samples meet optimal quantity and quality standards for storage and conservation. Preliminary multiplication of germplasm must be carried out when the sample does not meet these conditions. Preliminary multiplication consists of

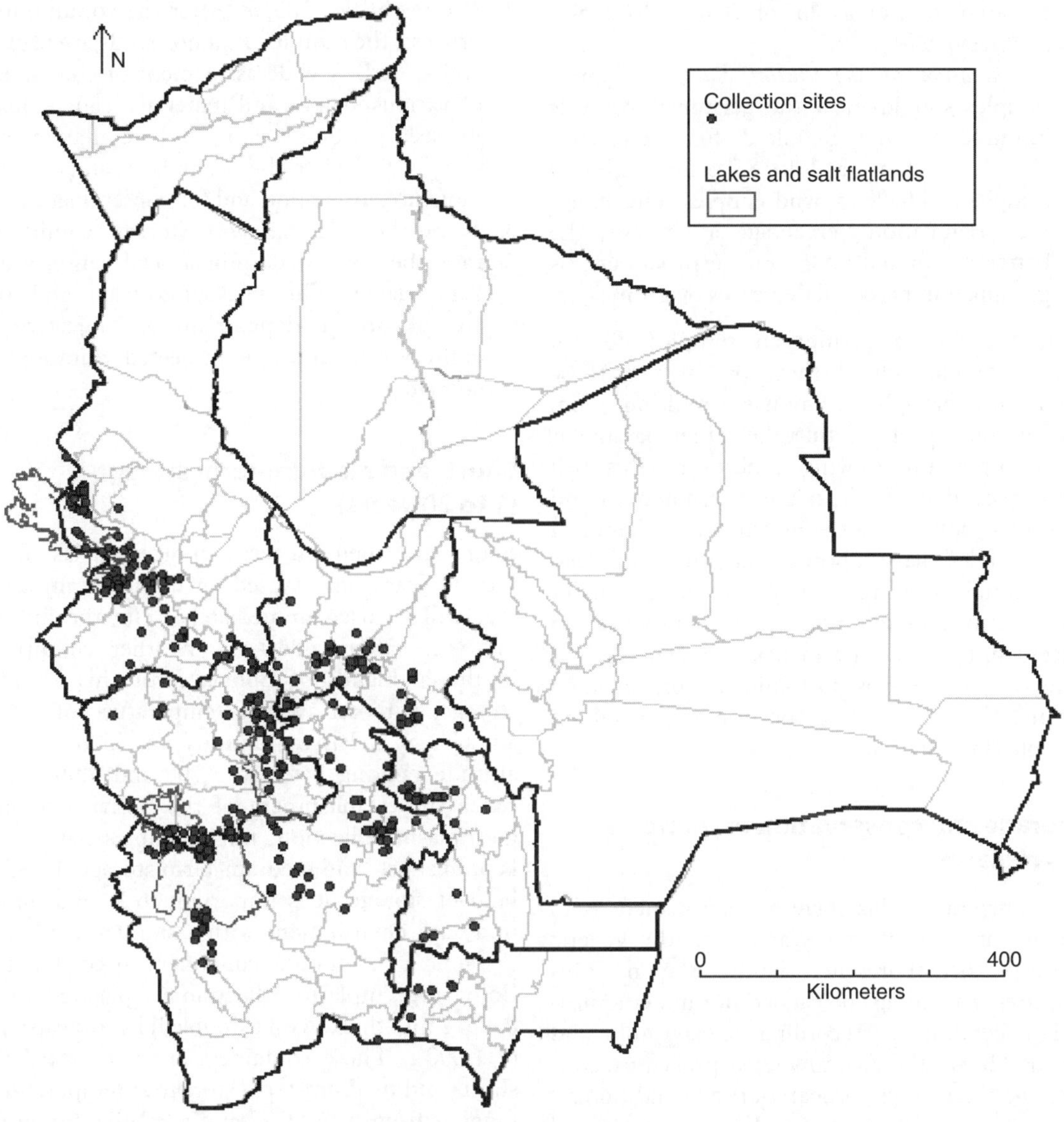

Fig. 8.4 Geographical distribution of the Bolivian quinoa germplasm collection.

increasing the sample size of germplasm under optimal culture conditions to ensure sufficient and viable samples that have maintained their original genetic identity (Jaramillo and Baena 2000).

There are two parameters that need to be taken into consideration for the preliminary multiplication of quinoa germplasm samples:

1. *Number of Seeds Per Sample*. According to the Standard for Genebanks (1994), the minimum amount required is from 1,500 to 2,000 seeds, which weigh an average of 4.5–6 g. However, at least 60 g of quinoa seed should be retained under short-term storage in order to have the minimum required amount while at the same time having enough seed needed for field

research tests or a plot of 70 m^2 (Rojas and Bonifacio 2001).

2. *Germination of the Initial Sample.* Quinoa samples should meet the germination levels required by the Standard for Genebanks (1994), which are at least 85% for cultivated samples and 65% for wild samples. The higher the germination percentage at storage, the better it would be for long term storage, as germination potential decreases over time.

If the quinoa germplasm samples do not meet one or both of these parameters, these must go through preliminary multiplication, which consists of meeting the target parameter under optimum growing conditions (this will be explained in detail in the "regeneration and multiplication" stage). On the other hand, if the samples have optimal quantities of seed and germination rate, they are placed under short-term storage. During this process, an accession number is permanently assigned to the sample, and the passport information obtained during the collection of the quinoa germplasm sample is registered.

Storage and conservation of quinoa germplasm

The purpose of this stage of management is to keep quinoa seeds viable and with the genetic identity that represents the population. This requires that storage be carried out under appropriate conditions. According to Jaramillo and Baena (2000), the recommended procedure consists of three steps: preparation or conditioning of samples, packing of samples, and storage of samples.

1. *Conditioning of the Sample.* This procedure is performed to obtain a clean sample of quinoa seeds, free of physical impurities such as perigone residues, seed stalks of the panicle, broken, infected, or foreign seeds. The conditioning process also includes the initial measurement of moisture content of the sample to ensure longevity in storage. This measurement can be determined by direct or indirect quantification of moisture content.
2. *Packing of the Sample.* After the conditioning process, the sample of quinoa seeds are packed and stored. A wide assortment of containers, of various shapes and materials, can be used for seed packing as long as the containers are airtight to isolate the germplasm and prevent moisture absorption and/or contamination.
3. *Storage of the Sample.* Storage conditions must be such that quinoa seed samples are kept viable. The packaging used and the site of storage depend on the objective of conservation and the projected duration of storage.

Short- and medium-term storage (1 to 20 years)

Short- and medium-term storage can last from 1 to 20 years, and is used for quinoa germplasm that will be used immediately or in the following years after collection. Weather conditions in the highlands of Bolivia favor this kind of storage, with an average temperature of 10°C, an average relative humidity of 45% and an altitude ranging between 3,700 and 3,900 masl. Since the establishment of the Bolivian quinoa germplasm collection, most samples have been kept in short- and medium-term storage. For this kind of storage, it is important to keep samples in a dark environment with a simple ventilation system. The storage containers used for the Bolivian germplasm collection are plastic bottles 0.4 to 2 mm thick, with a double lid and a capacity of 1,000 g. These containers are well suited for short- and medium-term storage at temperatures ranging from 8 to 20°C and a relative humidity of 15% to 60% (IPGRI 1996). Under these conditions, samples can be stored and maintained from 1 to 20 years, depending on the genetic material.

Long-term storage (80 to 100 years)

Seeds conserved under these conditions are maintained for approximately 80–100 years. According to the Standard for Genebanks (1994), most orthodox seed species can be kept indefinitely at temperatures between −10 and −20°C, with a seed moisture content of 3–7% and a viability not lower than 85%. In order to implement long-term storage of the quinoa collection, research on the use of silica gel and borax as seed drying agents was conducted. However, these drying agents were ineffective, and moisture levels failed to meet the requirements of the Standard for Genebanks (Rojas and Camargo 2002). In the subsequent year, a protocol for the implementation of long-term storage of quinoa germplasm was established (Rojas and Camargo 2003) and described as follows.

Protocol for the implementation of long-term storage

a. *Analysis of Germination*. The methodology established by the ISTA (1993) is especially applicable for the analysis of quinoa seed germination, meeting the standards in terms of duration, number of seeds, drying levels, and incubation temperature. The formula for calculating the germination percentage is

$$G\% = \text{No. germinated seeds} / \text{No. total seeds} \times 100$$

b. *Initial Seed Moisture*. Initial seed moisture can be quantified directly or indirectly. Electronic analyzers (moisture meters) can be used for direct quantification of seed moisture, while methods described in the manual on seed technology for genebanks (Ellis et al. 1988) can be used for indirect quantification. Using the differences in initial seed weight (wet weight) and final seed weight (dry weight), the initial humidity of the seeds can be calculated using the following formula:

$$H\,\% = \text{Wet weight} - \text{Dry weight} / \text{Wet weight} \times 100$$

c. *Reduction of the Moisture Content of Seeds*. To reduce the initial moisture content of seeds, a procedure using a dehumidifier has been standardized. The dehumidifier is programmed to 20°C and seed moisture reaches a range of 3–7% after exposure for 24.5 h (Leon et al. 2007), thus meeting the humidity levels required by the Genebank Management Standard (Table 8.2).

d. *Seed Packing*. Once moisture levels reach the required levels (Genebank Management Standard 1994), the quinoa seed must be packed immediately and vacuum sealed in trilaminated aluminum packs, which come in different sizes. These envelopes are airtight and well suited for long-term storage, where temperatures range

Table 8.2 Initial Seed Moisture Content and Final Moisture Content in 14 Accessions of Quinoa Exposed to 20°C for a Period of 24.5 h.

Number	Quinoa Accessions	Initial Moisture, %	Achieved Moisture, %
1	2350	12.40	4.90
2	2511	11.05	4.90
3	2857	11.10	3.65
4	2417	10.90	3.55
5	2401	12.00	4.30
6	1608	12.55	5.30
7	2840	12.40	5.10
8	1600	11.80	3.70
9	1462	10.90	4.65
10	1289	11.00	4.20
11	0550	10.40	4.55
12	0577	11.20	4.20
13	2237	9.70	4.75
14	2374	10.05	4.85

from 8 to −20°C and the relative humidity is from 10% to 20% (IPGRI 1996).

e. *Storage*. The aluminum packages should be stored in freezers at −20°C, as indicated in the Standard for Genebanks (FAO/IPGRI 1994).
f. *Germination Monitoring of Stored Samples*. To monitor the germination rate of stored samples, the following steps are recommended: (i) open the envelopes to remove seeds, and immediately vacuum seal the package to preserve the remaining seeds; (ii) rehydrate seed in a boiling water bath for 50–60 min (Rojas and Camargo 2003); and (iii) perform a germination test following the procedure established by ISTA (1993).

Following this procedure, 247 germplasm accessions of quinoa that belong to the "core collection" were conserved for long term storage in 2003. This is the first experience with long-term conservation of the Bolivian quinoa germplasm. After 5 years in storage, the first monitoring activity was performed in 2008 on the 247 accessions that had been put on long-term storage. Results were encouraging because the germination percentage remained stable between 90% and 98% when compared to the initial germination percentage.

Characterization and evaluation of the quinoa germplasm

Characterization and evaluation of accessions in a germplasm collection are complementary activities that involve describing the qualitative and quantitative attributes of the accessions. These activities are undertaken to differentiate accessions, determine their usefulness, structure, genetic variability, and relatedness, and find genes of value in the production or breeding of quinoa (Jaramillo and Baena 2000). Germplasm conservation goes hand in hand with its use, as experienced with the Bolivian collection of quinoa. This is only possible if the characteristics, attributes, and potential uses of the accessions are known. Information on the germplasm and its usefulness come from taking and analyzing a set of data on the germplasm sample in various stages of its management, but mainly during the characterization and evaluation stage.

Theoretically, the quantity of data that can be taken during characterization and evaluation is infinite. However, a good and useful description of plants is not determined by the number of variables used, but by its practical utility and accuracy (Querol 1988). Therefore, before describing an accession, the data to be registered must be well defined, although it also depends on the purpose and/or stage that is being followed in the process.

Stages of germplasm characterization and evaluation

It is important to differentiate the stages involved in characterization and evaluation of germplasm in order to carry out the work. These stages are complementary, based on data registration, and can be performed simultaneously. The stages are (a) correct identification; (b) characterization; (c) preliminary assessment; and (d) evaluation.

(a) *Correct Identification*. Before describing a germplasm accession, the sample must have accurate botanical identification because work is done through differentiations at the intraspecific level using the taxonomic classification system. Furthermore, compliance with passport data must be verified, a useful procedure for identifying duplicates among accessions that are often reintroduced to the genebank (Valls 1992).
(b) *Characterization*. In a strict sense, this stage involves the systematic description of accessions using a set of highly heritable and persistent qualitative characters. Characters such as growth habit, branching, plant color, morphology, etc., in addition to aiding the description, help differentiate the accessions.
(c) *Preliminary Assessment*. This stage involves the description of agronomic traits of accessions in a particular environment (space or time). The descriptors used are usually quantitative characters affected by the environment, such as yield and resistance to biotic and abiotic factors, among others. It is recommended that a preliminary assessment be conducted

when evaluating the entire collection of germplasm accessions or significant amounts of the collection.

(d) *Evaluation* – Further evaluation of germplasm consists of the description of agronomic or other traits of interest in the maximum number of possible environments (space or time). These traits are usually quantitative and affected by the environment, such as yield and resistance to biotic and abiotic factors, among others. The evaluation stage is carried out after the preliminary assessment and with those accessions that require further evaluation to assess their use. Unlike the preliminary assessment stage, the evaluation stage is similar to research trials and is established using statistical experimental designs.

The characterization and preliminary evaluation of quinoa germplasm can be performed simultaneously. A representative sample of the accessions is used, together with a list of crop descriptors and the necessary tools to register the information. The genetic material must be established in properly identified plots, and whenever possible, under uniform management conditions. Depending on the number of accessions, it will require between 3 and 5 days of work per week from planting to physiological maturity. The data should be recorded systematically and consistently to facilitate subsequent statistical analysis.

The procedure that must be followed to register each one of the target variables is in the list of "crop descriptors." However, when conducting preliminary characterization and evaluation of quinoa germplasm, there are important things to consider. First, sites with ecologies similar to the places of origin of the quinoa accessions must be chosen. Planting dates must match the natural periods of quinoa crop sowing, taking into account the area or place of origin of the accession. To achieve a representative population, plots should have four or six rows per quinoa accession and information should be gathered from the central rows (registration of information). Recording of qualitative variables and phenological phases should be performed according to the total number of plants established from the accession. For the recording or registration of quantitative variables, 10 plants randomly selected from the central rows during the "tipping panicle" stage are used. Moreover, it is recommended that information throughout the crop cycle be recorded.

In the over 40 years of its existence, the Bolivian quinoa germplasm has been characterized and evaluated with a focus on agromorphological information. The first catalog of the Bolivian collection of quinoa germplasm was published in 2001 (Rojas et al. 2001). The catalog describes the genetic variability of 2,701 quinoa accessions using 59 qualitative and quantitative variables. Although the information recorded was based on the "Quinoa Descriptor" (IBPGR 1981), the catalog includes more variables identified in the various characterization and evaluation processes carried out since the 1980s. Later, a new version of "Descriptor for Quinoa" was prepared and subsequently validated by researchers from Ecuador, Peru, and Bolivia (Rojas et al. 2003b). This document was the basis for the publication of the new list of "Descriptors for Quinoa and its Wild Relatives" edition spearheaded by Bioversity International and PROINPA (Bioversity International et al. 2013).

Assessment of nutritional value of the quinoa accessions began in 2001 and assessment of agroindustrial variables began in 2006. Information on 555 accessions of quinoa was recorded to guide the use of germplasm in the development of products from processed quinoa. In addition, most quinoa germplasm accessions were characterized at the molecular level.

Agromorphological variables

Starting in the 1980s, several characterization and evaluation tasks were performed on the quinoa germplasm from Bolivia. As a result, the genetic, morphological, and agronomic variability observed during the crop cycle of the quinoa germplasm was studied. Although this information is published in the germplasm catalogs, the parameters of some variables of interest were subsequently presented (Rojas et al. 2001, 2009;

Rojas 2003; Aroni et al. 2003; Rojas and Pinto 2013):

1. *Growth Habit*. Although branching and growth habit are influenced by planting density, four different clearly defined growth habits were identified in the quinoa germplasm collection (Fig. 8.5).
2. *Plant Color*. Between the stages of "tipping panicle" and "first flower," four colors typical of the quinoa crop are expressed – green, purple, red, and a mixture of colors. However, as plants form grain and reach physiological maturity, quinoa plants turn into different colors and combinations of colors including white, cream, yellow, orange, pink, red, purple, brown, gray, black, wild green, and mixtures.
3. *Shape and Density of the Panicle*. In terms of shape, there are three forms of panicles. The "amaranth-form" panicles are those with the glomeruli embedded directly in the secondary axis and have an elongated shape. "Glomerulated" panicles are those with the glomeruli inserted into the glomerulated axes and have a globular shape. "Intermediate" panicles express the features of the "amaranth-form" and "glomerulated" (Fig. 8.6). Likewise, there are two types of quinoa panicle in terms of density – lax (loose) or compact. The panicle density is determined by the length of the secondary axes and pedicels, being compact when both are short.
4. *Grain Color and Shape*. When quinoa grains reach physiological maturity they express a wide diversity of colors, including white, cream, yellow, orange, pink, red, purple, light brown, dark brown, greenish brown, and black. In the quinoa germplasm, 66 grain colors had been characterized (Cayoja 1996). Four grain shapes also exist in the germplasm (Fig. 8.7).
5. *Grain Diameter*. Grain diameter varies from 1.36 to 2.66 mm. The famous "Royal Quinoa" falls within this range of variation. Its large grains (2.20–2.66 mm) make it highly appreciated in the international market. On the other hand, the weight of 100 grains ranges from 0.12 to 0.60 g, and this variable is associated with grain size.
6. *Vegetative Cycle*. It is possible to find in the germplasm collection accessions that reach physiological maturity in 110 days and other accessions that reach maturity in 220 days. This feature is strongly dependent on the genotype. Quinoas from the inter-Andean

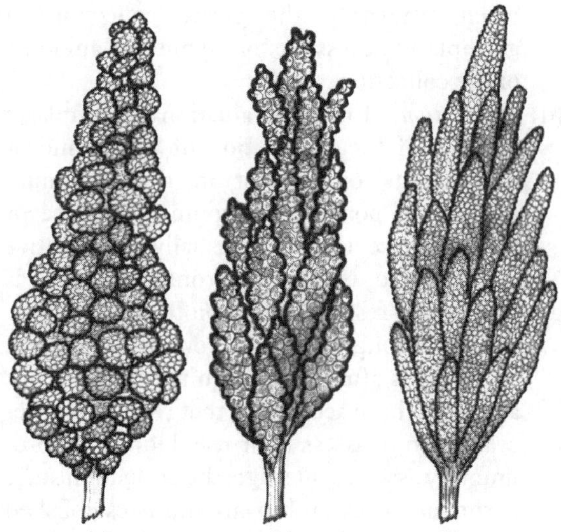

Fig. 8.6 Panicle shape: (a) glomerulated; (b) intermediate; and (c) amaranth-form.

Fig. 8.7 Quinoa grain shapes: (a) lenticular; (b) cylindrical; (c) ellipsoidal; and (d) conical.

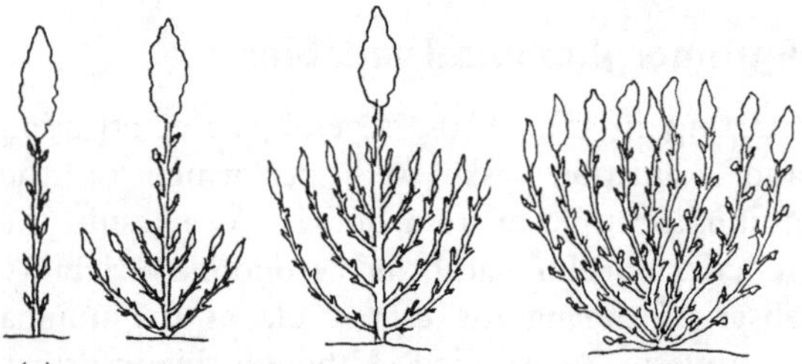

Fig. 8.5 Quinoa Growth habit: (a) simple; (b) branched to the lower third; (c) branched to the second third; and (d) branched with undifferentiated main panicle.

valleys mature later than those from the Altiplano. This wide range of variation of the vegetative cycle is promising as adaptation to climate variability and climate change and can be exploited further in quinoa breeding programs.

7. *Grain Yield Per Plant*. Yields per plant were recorded from 48 to 250 g. Yield is strongly dependent on both genotype and on yield component variables such as stem diameter, plant height, length and diameter panicle, grain diameter, etc.

Agro-food and nutritional value variables

A summary of the estimated statistical parameters (expressed on a dry basis) for each nutritional and agro-food value feature of quinoa germplasm is presented in Table 8.3 (Rojas and Pinto 2006, 2008; Rojas et al. 2007). The 555 accessions evaluated show wide variability for most of the traits studied, and is another indication of the genetic potential of quinoa germplasm.

As seen in Table 8.3, the amount of protein ranged from 10.21% to 18.39%. These values are broader than the range of 11.6–14.96% reported by βo (1991) and Moron (1999) cited by Jacobsen and Sherwood (2002). While the amount of protein is a basic aspect of nutritional value, its quality is also important and depends on the content of essential amino acids. The protein quality of quinoa is higher than that of cereal protein. Fig. 8.8 shows the variation and protein content distribution among the 555 accessions studied. It can be seen that majority of the accessions have a protein content ranging from 12% to 16.9%, while 42 accessions had a higher range of protein content, 17% to 18.9%. The latter group of accessions constitutes an important source of genes to drive the development of products with high protein content.

The fat content ranged from 2.05% to 10.88% with an average of 6.39% (Table 8.3). The upper range of these results is greater than the range of 1.8–9.3% reported by βo in 1991 and Morón in 1999 (Jacobsen and Sherwood 2002). Quinoa grain has high fat content owing to its high percentage of unsaturated fatty acids (Jacobsen and Sherwood 2002). The fat content of quinoa can be utilized to produce fine vegetable oils for cosmetic and culinary use.

The size of starch granule ranged from 1 to 28 μ (Table 8.3). It is very important that the starch granule is small to facilitate the process of texturing and insufflation because the spaces between granules allow more air to enter for the exchange and formation of air bubbles (Rojas et al. 2007). The size of starch granules is an important variable in establishing the functional character of quinoa, as a component of different mixtures with cereals and legumes.

The content of inverted sugars ranged from 10% to 35%. This variable indicates the quantity of sugar that starts fermentation by splitting or inversion, which is the parameter for determining the quality of carbohydrates. In addition, it is also an important parameter through which quinoa can be classified as food suitable for diabetics.

Table 8.3 Nutritional and Agro-Food Characteristics and Statistical Parameters for 555 Germplasm Accessions of Quinoa from Bolivia.

Component	Minimum	Maximum	Mean	SD
Protein, %	10.21	18.39	14.33	1.69
Fat, %	2.05	10.88	6.46	1.05
Fiber, %	3.46	9.68	7.01	1.19
Ash, %	2.12	5.21	3.63	0.50
Carbohydrates, %	52.31	72.98	58.96	3.40
Energy, Kcal/100 g	312.92	401.27	353.36	13.11
Starch granule, μ	1.00	28.00	4.47	3.25
Inverted sugars, %	10.00	35.00	16.89	3.69
Water filling, %	16.00	66.00	28.92	7.34

SD, standard deviation. Analyzed by LAYSAA, Cochabamba, Bolivia.

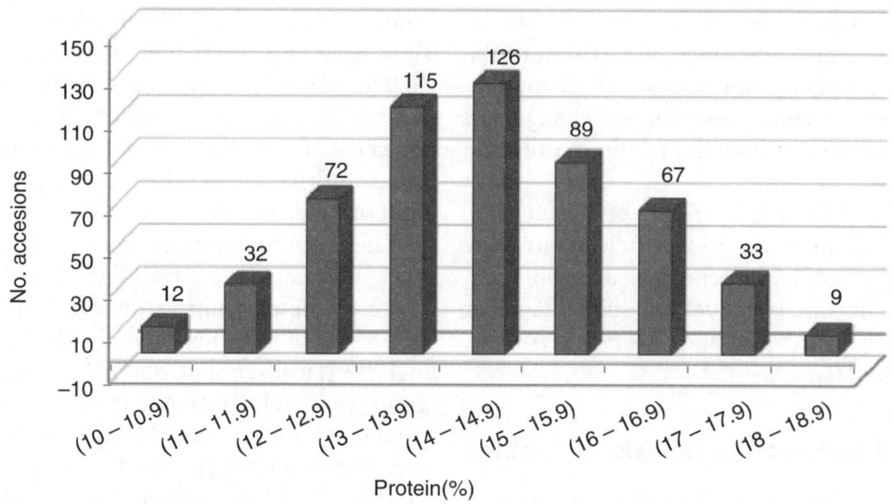

Fig. 8.8 Variation in protein content of 555 quinoa germplasm accessions.

The optimum percentage of "inverted sugar" content is ≥25%, and quinoa accessions having this optimum percentage can be used in flour blends to be used for breads, cereals, and other flour-based products. As long as the saponin is removed from the outside of the grain, flour blends with quinoa have a pleasant mouthfeel.

The percentage of "water filling" shows a variation range of 16–66%. This variable measures the water absorption capacity of starch for pasta elaboration and making of bread and pastries. The ideal value of this parameter for industrial application is ≥50%. There are quinoa germplasm in the collection which meet this parameter and can be an important source of genes to develop these products.

The quinoa in the Bolivian germplasm collection has a wide range of diversity as measured by agro-food and nutritive variables. This diversity is valuable when incorporating quinoa in processed products to make appropriate use of the genetic potential of quinoa. It is possible to select varieties with higher protein percentages (≥18%) and obtain more attractive products. Varieties with small starch granule diameters (≤3 μ) can be used for expanded and homogeneous popping. Varieties with stable percentages of amylose and amylopectin can be used for making puddings, gelatinised baby food, instant creams, and noodles, among others. Using quinoa in processed products is congruent with the goals of conservation and preservation of genetic diversity.

Molecular characterization

Two types of markers (SSR and ISSR) were used to characterize 86% (2,701 accessions) of the Bolivian collection. The markers revealed 3 to 13 polymorphic alleles, generating a genetic fingerprint for each quinoa accession. Furthermore, using the information generated, similar accessions can be grouped and related at a molecular level. In addition, the DNA extraction method for quinoa was standardized.

Multiplication and regeneration of quinoa germplasm

Although the seeds from the quinoa germplasm are stored in optimum conditions, with the passage of time they decrease in both quantity (because of use and distribution) and germination percentage. According to Jaramillo and Baena (2000), if the objective is to bring seed samples to an optimal number, the process is termed "multiplication"; if the goal is to recover the initial germination percentage, the process is called "regeneration" or "rejuvenation." In any case, samples obtained from the multiplication and/or regeneration process must be viable, healthy, of optimum size for storage, and genetically identical to the original sample.

Monitoring of seed quantity and percentage of seed germination

A seed sample is in optimum condition when it is viable in terms of germination and in sufficient quantity. If the sample does not meet any of these requirements, it must be multiplied and/or regenerated, a decision that comes from monitoring both the size and viability of accessions. Monitoring is governed by rules and procedures according to the Standard for Genebanks (FAO/IPGRI 1994).

Monitoring for sample size

The jars used to store the Bolivian collection of quinoa have a capacity of 1,000 g. It is best to fill the container with seed to its full capacity. However, this amount of seed decreases with the use of germplasm for research activities, crop breeding, introduction or reintroduction to communities, and also with the distribution of seeds to meet requests received by the genebank. Thus, monitoring for sample size should be continuous to keep the amount of seed from reaching a level below 60 g. It has been determined that 60 g of seed is the minimum acceptable level while at the same time keeping a sufficient quantity of seed readily available for a research test or a plot of 70 m^2 (Rojas and Bonifacio 2001).

The monitoring process must record the use and distribution of germplasm, and the information generated corresponds to the quinoa germplasm management data. Once the sample quinoa reaches a level less than 60 g, multiplication of the sample must be programmed, whereby the seeds are increased under optimal culture conditions.

Monitoring germination of the sample

According to the Standard for Genebanks (FAO/IPGRI 1994), the interval between germination tests depends on the species and storage conditions. For seeds that have been stored under short and medium term conditions, a germination test every 5 years is sufficient. Another important aspect to consider is the number of accessions held in the germplasm collection. In the case of quinoa, if there are 100–200 accessions, it is best to conduct annual monitoring of germination. However, for a greater number of accessions, a germination control test is recommended every 3–5 years, although this will also depend on the nature of the genetic material.

For germination control tests, a seed sample is subjected to germination tests according to ISTA guidelines (1993). The results obtained should be compared with the initial germination percentage of the same sample collected during the "preliminary multiplication" period. If the germination percentage is less than 85%, the sample must be regenerated. The decision to regenerate and/or multiply a sample should not depend on the amount of the sample, whether or not it is nearing the minimum quantity allowed. Germination percentage takes priority over the amount of seed. It will be more urgent to regenerate a large sample whose germination is low than to regenerate a small sample whose germination is optimal. However, regeneration should not be done often either because it is expensive and may compromise the genetic integrity of the germplasm if the sample is contaminated (Jaramillo and Baena 2000).

Technical procedure for multiplication and/or regeneration

Once the need to carry out regeneration and/or multiplication work is determined, the quinoa germplasm must be established in field plots or greenhouses and managed in the most optimal conditions for good crop development. This will ensure that the quinoa samples obtained are viable, healthy, in sufficient quantities, and genetically identical to the original sample.

Field plots/greenhouses

Depending on the place of origin of the quinoa accession and on the conditions of the site intended for multiplication and/or regeneration, it is recommended that this activity be done under field conditions so quinoa plants can express their genetic potential in natural environments. In field plots, there should be six rows per accession, with a furrow 5 m long. Once quinoa plants reach physiological maturity, the four central rows should be harvested, leaving 50 cm headers at both ends of the plot. If there is less available land, plots with four rows per accession can also be planted. To facilitate management, the rows should be spaced uniformly. Crop management practices usually carried out on the quinoa crop should be performed in a timely manner. However, if an accession has a small amount of seed or very low levels of germination, it requires more care to regenerate or multiply it. Therefore, the sample can be grown in greenhouses or under controlled environments, where suitable substrates for sowing can be prepared, soil moisture controlled with appropriate irrigation, and temperature regulated with simple aeration systems.

Prevention of pollen exchange

Great care should be taken to prevent gene flow and contamination by pollen exchange between two accessions, because quinoa is a partially allogamous species (10–15%). To overcome this challenge, accessions with short phenological cycle can be planted alternately with accessions that have long phenological cycle, so that their flowering phases do not coincide. Panicles can also be bagged using paper envelopes immediately before the quinoa plants enter the flowering phase. Another way to reduce allogamy and prevent contamination is to plant other crops in alternate rows with quinoa, preferably those similar in height or architecture. It is also equally important to avoid mechanical mixing of accessions, which is common when handling quinoa germplasm, especially during threshing and venting. Careful and correct management of the identification of every accession is recommended, from its establishment in the field to harvest and postharvest operations.

Finally, once the quinoa germplasm sample is multiplied and/or regenerated, the germination test must be redone. After fulfilling the requirements established by the Standard for Genebanks (FAO/IPGRI 1994) and for quinoa germplasm, the sample must be conditioned for the corresponding storage duration. The percentage of germination and the quantity of seed both serve as points of reference for the next monitoring cycle.

Regeneration schedule

When a germplasm collection consists of thousands of accessions, such as the case of the Bolivian quinoa germplasm collection, this stage in the genebank management requires considerable attention to technical details. Because of the wide genetic variability of the accessions, different growth traits and behaviors are expected when evaluating germination percentage over time.

To develop a regeneration schedule of the germplasm collection, the "monitoring of germination behavior" has been implemented since 2004 in a continuous manner in different groups of cultivated and wild quinoa. This work began with the formation of nine groups of cultivated quinoa and five groups of wild quinoa with similar germination percentages. Representative accessions were selected from each group, while taking into consideration the agromorphological variability and passport data of these accessions.

Fig. 8.9 shows half-year results for the period 2004–2010 from the four groups of cultivated quinoa (Q-1, Q-2, Q-3, and Q-4) and two groups of wild quinoa (QS-1 and QS-2), stored under short and medium term conditions. As expected, the results indicate that there is a variation in the levels of maintenance of seed germination. In general, an average reduction of germination from 8% to 40% was observed for the study period. From these results, it can be seen that Q-1 seeds need to be regenerated after 2 years of storage, the Q-2 seeds after 4 years, and the Q-3 seeds after 6 years, as germination levels are below the level required by the Standard for Genebanks (FAO/IPGRI 1994).

While this schedule of quinoa germplasm regeneration is still being developed, it clearly indicates that generalizations cannot be made

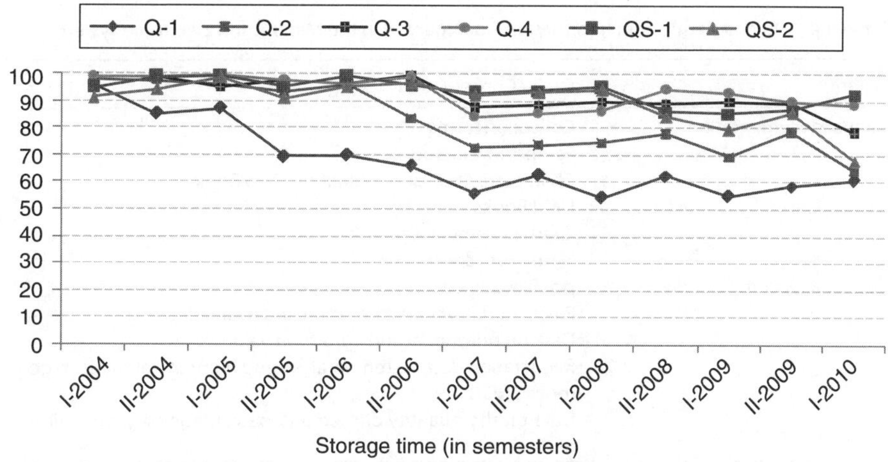

Fig. 8.9 Germination behavior in four groups of cultivated quinoa and two groups of wild quinoa. Q-1, Q-2, Q-3 and Q-4 are the cultivated quinoa groups, while QS-1 and QS-2 are the wild quinoa groups.

when managing a germplasm collection, particularly at the stage of seed regeneration. Without doubt, the wide genetic variability of the quinoa germplasm collection, as well as storage conditions, determines the internal moisture content of the seeds and the fat content of the seeds, among others.

Documentation and information on quinoa germplasm

The management and conservation of quinoa germplasm in its various stages comprises a series of activities for which information is needed or where information is derived. Recording, organizing, and analyzing data all comprise the "documentation" stage of germplasm management. Documentation is fundamental to the exploration and characterization of the germplasm, and in making decisions about its management. The value of a germplasm collection increases as it is characterized and studied; hence, the importance of good documentation (Jaramillo and Baena 2000). The likelihood that a particular quinoa accession will be used increases to the extent that information on its characteristics and genetic potential is available. A quinoa accession cannot be considered for use without pertinent information, so it is essential to document information as systematically as possible, and with as much detail.

The information on the Bolivian quinoa germplasm is managed under a manual and electronic documentation system. Manual documentation consists of recording information in field books or record books, which are practical to handle when gathering information during collection trips. These manually documented data are transferred to electronic documentation systems on Excel, pcGRIN, and SIRGEN databases (Table 8.4).

Information on the germplasm collection is organized into three data sets consisting of passport data collection, characterization and evaluation data, and management data. The passport and collection data consists of 31 variables recorded at the time of collection and/or reception of quinoa accessions. In the manual system, there is a book of "passport data" and a file of "collection forms" for the germplasm collection. More information on the 31 variables is presented in Annex 8.1.

For the characterization and evaluation data, agromorphological information is organized into 59 qualitative and quantitative variables, both in the manual and in the electronic system (Rojas et al. 2001; Pinto et al. 2009). The nutritional and agro-food value information of the quinoa grain is organized into six variables of nutritional value (protein, fat, ash, crude fiber, carbohydrates, moisture, and caloric energy) and three agro-food

Table 8.4 Quinoa Germplasm Information Documented in the Manual and Electronic Systems.

System	*Ex situ* Conservation
Manual documentation	Collection forms
	Passport Data Books
	Characterization and evaluation Data Book
	Field Books
	Seed movement logbook
	Herbarium specimens
Electronic documentation	pcGRIN quinoa
	Passport data, characterization and evaluation
	Data on nutritional and agro-food value
	Registration data on temperature and humidity of the storage room BNGA
	Data on the quantity of seed and percentage of germination of the accessions

variables (size of starch granule, percentage of inverted sugars, and water filling). Molecular information is organized in the electronic documentation system and consists of DNA fragments of different sizes (base pairs) revealed with SSR and ISSR markers.

The quinoa germplasm management data corresponds to the information generated from monitoring of the percent germination, sample size, initial seed moisture, number of regenerations and/or multiplications, temperature and humidity recording in the storage chamber, and distribution of seeds. Analysis of the information that has been generated and documented helps in the decision making process to improve and optimize the management of quinoa germplasm. In addition, using the genebank information, documents were produced and disseminated to promote the use of Bolivia quinoa germplasm. The following are some publications produced: document on the management strategy of the National Genebank of Andean grains, germplasm catalogs, technical bulletins, brochures, booklets, calendars, banners, reading for secondary school biology books (conservation of our plant genetic resources – genebanks), graduate and post graduate theses, scientific articles, and press publications.

Utilization of quinoa germplasm

The increasing population growth and reduction of land available for agriculture make it critical to increase food production and attain a more equitable distribution (FAO 1996). In order to do this, plant genetic resources need to be utilized judiciously (Cadima et al. 2009). The use of germplasm depends on knowing where it comes from, its characteristics and usefulness, and on keeping it viable and available (Jaramillo and Baena 2000). The strategy developed by PROINPA for the management of plant genetic resources involves three steps: direct use, indirect use, and social use.

Direct use

The direct use of germplasm consists of identifying promising quinoa accessions with desirable characteristics, to introduce or reintroduce these in their original form into other regions or communities. In the Bolivian quinoa genebank, direct use of germplasm began with the analysis of the results of agromorphological characterization and evaluation. Afterwards, quinoa accessions with desirable traits (high yield, large grains, plant architecture, and place of origin) were identified and given to farmers for direct use. Training courses on the varied uses of quinoa for food processing were also developed by the genebank for direct use by farmers and traditional agro-food processors.

Procedure to apply participatory evaluation techniques for quinoa

"Participatory evaluation techniques" are tools that allow farmers to be involved in the decision-making process regarding one or several new technologies. These techniques were applied to the quinoa accessions. This form of quinoa utilization requires direct work with farmers in communities where quinoa is produced (Pinto et al. 2010). Participatory evaluations were performed during the flowering stage and harvest.

The "participatory evaluation techniques" were implemented from first contact with farmers to delivery of seeds to participating families. First, quinoa accessions were identified on the basis of the analysis of results from the characterization and evaluation of the germplasm collection. Then, communities where these accessions would be produced were identified. Traditionally conservative communities and/or those with the potential to produce the crop were selected. The farmers' level of motivation and interest in participating in the process was also considered during the selection process. Afterwards, while respecting the customs of each place, contact was made with local communal authorities and farmers in the selected communities and planning commenced. With the active involvement of farmers, field evaluation plots were established and quinoa accessions planted. Throughout the course of the crop cycle, farmers and technicians conducted crop management activities such as tilling, weeding, fertilizing, roguing, and applying phytosanitary controls. At the flowering stage of the crop, a participatory evaluation was conducted involving 10 farmers (men and women) who used the "absolute evaluation" method that evaluates each alternative on a fixed or absolute scale. Afterwards, farmers from the local communities helped harvest and thresh the quinoa accessions. Harvest from each individual accession was placed in bags of the same size, for comparison of quantities obtained. Participatory evaluation of the grain was then conducted, using the "order of preferences" method that facilitates a comparison between the different alternatives. Again, 10 hardworking and communicative farmers (men and women) who had superior knowledge of the quinoa crop were chosen for this activity. A participatory evaluation of food products was also conducted to determine the culinary potential of the accessions, and the "order of preference" technique was again used, where farmers compare each technological alternative against the others. Finally, seeds of the quinoa accessions that were preferred and selected by the farmers were distributed to the farmers in the presence of technical staff and local community authorities.

Participatory assessments with quinoa germplasm

From the agricultural year 2002–2003 to 2007–2008, participatory assessments were conducted on 29 germplasm accessions and 4 local varieties (Local, Wila Jupha, Wila Cayuni and Acujuira) in 22 communities in the central and northern Altiplano of Bolivia (Table 8.5).

Two participatory evaluation techniques were applied, namely, the "absolute evaluation" technique in the flowering phase and the "order of preference" technique during harvest (grain). The most frequent selection criteria used for quinoa accessions were tall plants, large panicles, resistance to frost, large grain size, white grain, good yield, and rapid maturation. Fourteen accessions, namely, accessions 2527, 3130, 2522, 2529, 2394, L320, 2943, 2031, 2857, 1667, 2516, 1560, 1474, and 2401, were selected by farmers (60% male and 40% female). The seeds of the selected accessions were distributed in each community to the farmers who participated in the whole process, so they could cultivate these in their land and introduce these in their traditional crop systems for the next year.

The contribution of the farmer participatory process is reflected in the introduction and reintroduction of quinoa germplasm to communities. The flow of genetic material has strengthened the *in situ* conservation of crop diversity and its use in communities. Equally important, farmers have become aware of the existence of the Bolivian quinoa collection and the role of the genebank in case they deplete their seed stock.

Training courses on the varied uses of quinoa

From 2005 to 2006, training courses on food preparation based on quinoa germplasm accessions were conducted in communities surrounding Lake Titicaca. These courses were

Table 8.5 Quinoa Accessions Evaluated Using Participatory Techniques Throughout Six Farming Years (2002–2008) and the Participating Communities.

Agricultural Year	Communities	Quinoa Accessions and Local Landraces
2002–2003	Jalsuri	0533
	San Pedro – San Pablo	0547
	Kalla Arriba	1750
	Chahuira Chico	1667
	Vitu Calacachi	2031
	Pomposillo	2390
	Tacaca	2394
		2411
		2522
		2516
		2527
		L-26
2003–2004	Salviani	1667
		2031
		2390
		2394
		2516
		2522
		2529
		3130
		Local
2004–2005	Antarani	0027
	Patarani	0575
	Rosapata	1641
	Erbenkalla	1655
	Coromata Media	1659
		1667
		1927
		2031
		2390
		2394
		2516
		2522
		2527
		2529
		2561
		3130
		Local
2005–2006	Cachilaya	1713
	Cutusuma	0027
	Cutusuma alta	1667
	Titijoni	2031
	Cariquina grande	2394
	Jutilaya	2511
	Huancarama	2516
	Llanga	2527
		2857
		2943

Table 8.5 *(Continued)*

Agricultural Year	Communities	Quinoa Accessions and Local Landraces
		3130 Wila jupa Wila coyuni
2006–2007	Patarani Cariquina Grande	1474 1560 1641 2390 2401 2516 2527 2857 L320 Acujuira
2007–2008	Santiago de Okola	1474 2511 2689 2857 2943 Local

developed to promote the varied uses of quinoa. For these courses, accessions 2943, 2637, 0081, 0381, 1667, 2511, and 0027 were used. The quinoa accessions were evaluated by farmers in a participatory process to determine their suitability for various culinary products. The following food preparations were promoted: quinoa cake, quinoa cookies, quinoa and apple juice blend, boiled quinoa Valencia style, quinoa bread, quinoa fried buns, quinoa tamales, and quinoa pancake. The "order of preference" technique was used to determine that accessions 1667 and 2943 were good for making quinoa and apple juice blend, while accessions 2511 and 0027 were good for quinoa bread and cake.

The training courses were conducted in communities where participatory evaluations of the quinoa crop had been conducted previously. This was done in order to promote the use of quinoa in food preparations in communities where selected accessions of quinoa had been distributed to farmers. A total of 18 training courses in eight communities of five provinces from the department of La Paz were conducted, attended by 397 male and female farmers. There were more women who participated than men, because of the role of women in home food preparation (Table 8.6).

Indirect use

The indirect use of quinoa is linked to crop breeding. As with any crop improvement program, quinoa breeding objectives are to (i) increase production through yield, agronomic traits and, resistance to pests and diseases; and (ii) increase the quality of products through nutritional and agribusiness characteristics, such as grain shape, grain color, and storage life attributes.

In Bolivia, research on quinoa breeding began in the early 1960s in the Patacamaya Experimental Station, initially under the administration of the Institute of Andean Crops, which was under the Ministry of Agriculture, and then under the administration of the IBTA. During this period, Bolivia consolidated a quinoa breeding program, which was a pioneer in the Andean region, to develop not only quinoa varieties but also the technical skills of its workforce. Later, PROINPA continued the quinoa breeding program, making crosses and monitoring generations to obtain new quinoa varieties.

Table 8.6 Number of Communities, Courses and Participants in Training Events on New Ways of Preparing Quinoa Food Products (2005–2008).

Agricultural Year	N°	Community	Province	Number of Courses	Participants
2005–2006	1	Antarani	Pacajes	1	15
	2	Cariquina Grande	Camacho	1	23
	3	Titijoni	Ingavi	1	39
	4	Cachilaya Zone A	Los Andes	1	62
	5	Jutilaya	Camacho	1	37
	6	Cachilaya Zone B	Los Andes	1	48
	7	Cutusuma	Los Andes	1	47
	8	Coromata Media	Omasuyos	1	40
2006–2007	1	Titijoni	Ingavi	2	27
	2	Cachilaya	Los Andes	2	22
	3	Cariquina Grande	Camacho	2	37
2007–2008	1	Titijoni	Ingavi	1	11
	2	Cachilaya	Los Andes	2	34
	3	Cariquina Grande	Camacho	1	11
Total				18	453

In Bolivia, the initial focus of the quinoa breeding program was to develop high-performing varieties with large grain size, white color, and saponin free. In the course of time, changes in the quinoa market and weather patterns have adjusted priorities for crop improvement, without neglecting productivity. In the late 1990s, brown and black quinoa grains, known in the international market as "red quinoa" and "black quinoa," were included in the breeding process. At the end of the 1990s, the earliness trait was also considered in the breeding process to cope with delayed rains, thus introducing the possibility of planting crops until November and harvesting within a shorter growing cycle. There are 24 varieties of quinoa in Bolivia obtained by breeding through hybridization and/or selection (Table 8.7). There is also a complex of at least 54 known bitter varieties with the name "Royal Quinoa" (Bonifacio et al. 2012), of which the varieties Real Blanca, Toledo, Pandela, K'ellu, and Black Pisankalla dominate the export market.

Quinoa core collection

When germplasm collections are composed of a large number of accessions, such as the case of the Bolivian quinoa germplasm, the selection of a core collection is necessary to facilitate management and encourage the use of quinoa germplasm. Core collections are a subset of the complete collection, representing between 10 and 15% of the accessions, and have the highest possible genetic variability (70–80%) in the germplasm collection. It is important to highlight that core collections do not replace the total collection, yet are accepted as effective tools to improve the conservation and use of germplasm collections (Rojas 2010).

The quinoa core collection was selected through statistical analysis of 2,514 accessions using 18 quantitative variables. The core collection consists of 267 accessions, representing 10.6% of the accessions analyzed (Rojas 2010). The quinoa core collection has guided the work of the crop breeding program and has facilitated the selection of progenitors. The accessions that comprise the core collection are currently in hybridization and selection processes to develop varieties tolerant and/or resistant to biotic and abiotic factors.

Social use

Genetic resources are the basis for the survival of humanity. They are vital for developing countries, to be utilized for the benefit of current

Table 8.7 Bolivian quinoa varieties obtained through breeding.

Number	Variety	Material of origin		Year
1	Sajama	0547	0559	1967
2	Samaranti	Individual selection		1982
3	Huaranga	Selection S-67		1982
4	Kamiri	S-67	0005	1986
5	Chucapaca	0086	0005	1986
6	Sayaña	Sajama	1513	1992
7	Ratuqui	1489	Kamiri	1993
8	Robura	Individual selection		1994
9	Jiskitu	Individual selection		1994
10	Amilda	Individual selection		1994
11	Santa Maria	1489	Huaranga	1996
12	Intinayra	Kamiri	F4(28)xH	1996
13	Surumi	Sajama	Ch'iara	1996
14	Jilata	L-350	1493	1996
15	Jumataqui	Kallcha	26(85)	1996
16	Patacamaya	Samaranti	Kaslala	1996
17	Mañiqueña	Selection 1489		1999
18	Caraquimeña	Selection DC		2003
19	Jacha Grano	1489	Huaranga	2003
20	Kosuña	1489	L-349	2005
21	Kurmi	1489	Marangani	2005
22	Horizontes	1489	L-349	2007
23	Aynoq'a	Selection L-118		2007
24	Blanquita	Selection L-320		2007

Source: Personal elaboration based on Espindola and Bonifacio (1996), Bonifacio et al. (2006), and Rojas-Beltran et al. (2010).

and future communities. Therefore, there is great value in increasing awareness among the general population, especially the younger generation, about the importance of conservation and sustainable use of genetic resources. Based on the experience of PROINPA, the social use of quinoa are in distinguished advocacy and dissemination campaigns, participation in formal and informal education, and empowerment of local populations and authorities (Cadima et al. 2009).

Promotion and dissemination

As a key element in supporting the conservation of genetic resources, promotion and dissemination activities are aimed at encouraging the consumption and use of native species and crops. In this sense, information gathered from the genetic material in the quinoa germplasm collection can be the basis for advocacy and dissemination.

A key promotion activity of the genebank is participation in biodiversity fairs, as well as dissemination of information through mass media. To promote the benefits of quinoa, the genebank participated in urban and rural fairs with banners containing information on the *ex situ* conservation of quinoa germplasm. Germplasm samples were also exhibited, together with samples of quinoa biscuits and cakes, to highlight quinoa diversity (Table 8.8). Publications about the management and conservation of quinoa germplasm were also distributed to booth visitors. Information about the genetic wealth of quinoa and its nutritional value was promoted and disseminated through the newspapers "La Prensa" and "Nuestros Pueblos," in concert with genebank activities. The germplasm collection was also promoted through mass media such as Radio San Gabriel, which has urban and rural outreach. Radio scripts were also developed on topics such as

Table 8.8 Promotion and Dissemination of Quinoa Germplasm Through Urban and Rural Fairs and Mass Media.

Area	Promotion	2003–2004	2004–2005	2005–2006	2006–2007	2007–2008
Town	Fairs	IV Ecological Fair and Sustainable Development	1st Fair and Contest on Andean crops – Andean Agricultural Binational Project Peru-Bolivia. World food day II Fair Expo Puno 2004	World Food Day III FEXPO QUINUA 2006	IV FEXPO QUINUA 2007	V FEXPO Quinua International 2008 Bolivia Gourmet Food Festival Festiquinua 2008 Puno – Perú
	Publication in newspapers		Nuestros pueblos	La Prensa	La Prensa	La Prensa
	Radio programs			San Gabriel Radio	San Gabriel Radio	San Gabriel Radio
	TV program				Bolivia TV	Bolivia TV
Rural	Fairs		Chachacomani (Omasuyos Province) Guaqui (Ingavi Province)	Palcoco (Los Andes Province) Batallas (Los Andes Province), Desaguadero (Ingavi Province)	The 1st' Cultural &Productive Expo Fair Guaqui The 1st' Integral Productive Expo Fair Batallas	The 2nd' Integral Productive Fair Batallas The 1st' Agricultural and Tourism Fair Carabuco (Camacho Province)
	Radio programs			San Gabriel Radio	San Gabriel Radio	San Gabriel Radio

"Quinoa Germplasm Bank," "First Meeting of Farmers," "Diversification of Quinoa-based Food Uses," "Participatory Evaluations," "Crop Cultivation of Quinoa," "Quinoa Seed Flows in Fairs," Diversity Contests," "Testimonials on Diversity Contests," "Biodiversity Fairs," "Quinoa Planting," "Quinoa Pest and Disease Control," "Quinoa Harvest,""Medicinal Value of Quinoa," Quinoa Nutritional Value," and "Cultural Value of Quinoa," for a total of 28 scripts. The booklets were disseminated through the "Agricultural" program in Radio San Gabriel. Additionally, through the program "Palabra Rural" aired by Bolivia TV, documentaries entitled "Quinoa germplasm bank and its role in the conservation of strategic genetic resources for the country" and "Diffusion of quinoa varieties and accessions" were broadcast (Table 8.8).

Formal and informal education

More than 30 lectures and talks were given to farmers; students and teachers from schools, colleges and, public and private universities; as well as technicians and researchers from different research institutions dedicated to Andean grains. Depending on the target audience, the lectures covered the following topics: "Origin and Distribution of Quinoa," "Quinoa of Nutritional, Economic, Medical and Adaptive Importance," "Training Quinoa Germplasm Bank," "Conditions of Conservation of Genetic Material," "Conservation *ex situ,*" "*In situ* Conservation," and "Use of Genetic Material and Its Nutritional Value" (Table 8.9).

At conferences and lectures, seed samples of the quinoa germplasm collection were used to illustrate variation in shape, size, and color of grains. Quinoa panicles and plants were displayed to show differences in panicle shape and color, plant architecture, and other morphological traits. Toward the middle part and end of the talk, quinoa-based products such as biscuits, juice, pancakes, and nougats were tasted and appreciated by the audience.

CONCLUSIONS

Plant genetic resources are essential for food security and sovereignty of people groups worldwide, and contribute substantially to humanity's basic needs. These genetic resources are part of the ancestral and cultural heritage of countries. Their conservation and use is a responsibility that must be assumed by society in general, with the leadership of governments and the state. Those times when the responsibility of managing genetic resources fell solely on particular institutions or organizations are over. Now, all segments of the Bolivian society must assume the responsibility, because it concerns everyone. We must start by valuing this genetic heritage, recognizing that it is part of our cultural identity, and be proud of the genetic diversity under our stewardship. We need to teach our children the value and role played by plant genetic resources for the benefit of society as a whole.

The conservation of plant genetic resources requires institutional support through solid infrastructures backed by clear policies and headed by governments. State governments have to provide sustained financial resources for vital conservation activities. According to studies, the conservation of a sample of germplasm costs between 5 and 10 USD per year. No effort should be spared when allocating annual budgets because these are valuable resources that should be passed on from generation to generation to support our very existence. Governments must also play a leading role in the specialized training of personnel while ensuring work stability. At the same time, governments must establish infrastructures that meet the minimum equipment conditions required to manage and conserve germplasm collections.

Management of the Bolivian quinoa germplasm collection has allowed us to experience varied technical, social and financial situations. Vocation and commitment to work were the key elements that enabled us to make the Bolivian quinoa genebank the most important quinoa genebank in the world, with the greatest diversity and largest number of accessions (FAO 2010). This is an exceptional achievement, especially in a

Table 8.9 Conferences, Lectures, and Talks on the Quinoa Germplasm Collection Held at Schools, Colleges, Universities, Institutions, and for Farmers.

Area	Promotion	2004–2005	2005–2006	2006–2007	2007–2008
Town	Conferences in Schools, Universities, and Institutions	Villamil School San Calixto School Faculty of Nutrition UMSA Committee 336 IBNORCA	National Academy of Sciences (World Food Day) Tambo Quirquincho	Loyola University Public University of El Alto (UPEA) Bolivian – German Integral Senior Center (CISBA)	Loyola University Public University of El Alto (UPEA) Military College of Engineering (EMI) World Earth Day (Chancellery)
Rural	Lectures and talks to farmers, schools, and colleges		School Unit Marcelo Quiroga Santa Cruz and Farmers (Cachilaya) Humanistic Technical College Rene Barrientos and Farmers (Coromata Media)	Education Unit and Farmers (Titijoni) School Unit Marcelo Quiroga Santa Cruz and Farmers (Cachilaya)	Don Bosco School and Farmers (Santiago de Okola)

ANNEX 8.1
QUINOA GERMPLASM COLLECTION FORM

Fundación PROINPA

ANNEX 8.1: QUINOA GERMPLASM COLLECTION FORM

EXPEDITION:		
1.NAME(S) OF COLLECTOR(S):		
2. NAME OF FARMER:		
3. SAMPLE NUMBER:	4. ACCESSION NUMBER:	
5. DATE OF COLLECTION (DD/MM/YYYY):	6. REGISTRATION DATE:	
7. GENUS:		
8. SPECIES:		
9. LOCAL NAME(S):		
10. ETHNIC GROUP OF THE RESPONDENT:		
11. COUNTRY:	12.DEPARTMENT:	
13. PROVINCE:	14. MUNICIPALITY:	
15. CANTON:	16. COMMUNITY:	
17. LOCATION: at Km, fromin direction (North / South / East / West)		
18. LAT (° ' ") N/S	19. LONG (° ' ") E/O	20. ALT (m.a.s.l.)
21. TYPE OF SAMPLE		
1. Wild 4. Breeding line	2. Undergrowth 5. Advanced crop (variety)	3. Native crop 6. Other (specify)
22. SITE / SOURCE OF COLLECTION		
1. Habitat: Wild Prairie Desert Salt Flatlands	2. Farm:Field plot Huerto / Garden Crop Bordure Intercropping Storage	3. Market/Fair:City Town Community
4. Research Institute	5. Other (specify)	
23. USES OF THE PLANT / PARTS		
1. Food: Grains Leaf Stem / branches Root	2. Medicinal: Grains Leaf Stem / branches Root	3. Beverages: Grains Leaf Stem / branches Root
4. Crafts: Grains Leaf Stem / branches Root	5. Foliage: Grains Leaf Stem / branches Root	6. Ornamental: Grains Leaf Stem / branches Root
7. Other (specify)		
24. INFLORESCENCE COLOR (PANICLE):		25. STEM COLOR:
26. INFLORESCENCE SHAPE (PANICLE):		
1. Glomerulated	2.Amaranth-form	3. Intermediate
27. GROWTH HABIT		
1. Simple 2. Short Branches	3. Long Branches	4. Undefined main panicle
28. SAMPLE POPULATION:	1. Uniform	2. Mixed
29. NUMBER OF PLANTS SAMPLED:		
30. WERE PICTURES TAKEN?	Yes No	How many?
31. OBSERVATIONS:		

developing country such as Bolivia, where conservation of genetic resources and technological advancements are not always a priority. We have learned many valuable lessons from the whole process of managing the quinoa collection. There are still many goals to be accomplished, and we should be motivated to continue studying quinoa and other plant species, especially underutilized species, such as cañahua, amaranth, and lupine. These species, owing to particular characteristics, can be grown for food in places of extreme poverty and adverse weather conditions. Thus, utilization of these plant species should be a priority for all countries worldwide.

REFERENCES

Aroni JC, Aroni G, Quispe R, Bonifacio A. 2003. Catálogo de Quinua Real. Cochabamba, Bolivia: Fundación PROINPA, 51p.

Barriga P, Pessot R, Scaff R. 1994. Análisis de la diversidad genética en el germoplasma de quinua (*Chenopodium quinoa* Willd.) recolectado en el sur de Chile. Agro Sur 22 (No. Esp.): 4.

Bioversity International, FAO, PROINPA, INIAF, FIDA. 2013. Descriptores para quinua (*Chenopodium quinoa Willd.*) y sus parientes silvestres. Bioversity International, Roma, Italia; Organización de las Naciones Unidas para la Agricultura y la Alimentación, Roma, Italia; Fundación PROINPA, La Paz, Bolivia; Instituto Nacional de Innovación Agropecuaria y Forestal, La Paz, Bolivia; Fondo Internacional de Desarrollo Agrícola, Roma, Italia.

Bonifacio A, Mujica A, Alvarez A, Roca W. 2004. Mejoramiento genético, germoplasma y producción de semilla. Mujica A, Jacobsen S, Izquierdo J, Marathee JP, Quinua: ancestral cultivo andino, alimento del presente y futuro. FAO, UNA, CIP: Santiago, Chile. 125–187.

Bonifacio A, Rojas W, Saravia R, Aroni G, Gandarillas A. 2006. PROINPA consolida un programa de mejoramiento genético y difusión de semilla de quinua. Informe Compendio 2005–2006. Cochabamba, Bolivia: Fundación PROINPA. 65–70.

Bonifacio A, Aroni G, Villca M. 2012. Catálogo etnobotánico de la Quinua Real. Cochabamba, Bolivia: Fundación PROINPA. 1–123.

Bravo R, Catacora P. 2010. Situación actual de los bancos nacionales de germoplasma. Bravo R, Valdivia R, Andrade K, Padulosi S, Jagger M. Granos Andinos: avances, logros y experiencias desarrolladas en quinua, cañihua y kiwicha en Perú. Roma, Italia: Bioversity International. 15–18.

Cadima X, Gabriel J, Terrazas F, Rojas W. 2009. MODULO 3: Usos de los recursos genéticos. En: Cursos semipresencial sobre "Recursos Fitogenéticos, Riqueza Estratégica para el Desarrollo del País". COSUDE – DANIDA, Bioversity International, INIAF, Fundación PROINPA, pp. 1–35.

Cayoja MR. 1996. Caracterización de variables contínuas y discretas del grano de quinua (*Chenopodium quinoa* Willd.) del banco de germoplasma de la Estación Experimental Patacamaya. Tesis de Lic. en Agronomía. Oruro, Bolivia, Universidad Técnica Oruro, Facultad de Agronomía, pp. 1–129.

Ellis RH, Hong TD, Roberts EH. 1988. A low-moisture-content limit to logarithmic relations between seed moisture content and longevity. Ann Botany 61: 405–408.

Engels JMM, Visser L 2003. A guide to effective management of germplasm collections. IPGRI Handbooks for Genebanks. 6. IPGRI, Rome, Italy.

Espindola G, Bonifacio A. 1996. Catálogo de variedades mejoradas de quinua y recomendaciones para producción y uso de semilla certificada. Publicación conjunta IBTA/DNS: Boletín No. 2, La Paz, Bolivia, 1–76.

[FAO/IPGRI] Food and Agriculture Organization of the United Nations/International Plant Genetic Resources Institute. 1994. Genebanks standards. United Nations Food and Agriculture Organization & International Plant Genetic Resources Institute, Rome, 1–13.

[FAO]Food and Agriculture Organization of the United Nations. 1996. Informe sobre el estado de los recursos fitogenéticos en el mundo. Organización de las Naciones Unidas para la Agricultura y la Alimentación, Italia. 1–75.

[FAO] Food and Agriculture Organization of the United Nations. 2010. The second report on the state of the world's plant genetic resources for food and agriculture. FAO, Rome, Italy. 1–370.

Gallardo MG, Gonzalez JA. 1992. Efecto de algunos factores ambientales sobre la germinación de Chenopodium quinoa W. y sus posibilidades de cultivo en algunas zonas de la Provincia de Tucumán (Argentina). LILLOA XXXVIII. 55–64.

Gandarillas H. 1968. Razas de quinua. Bolivia, Ministerio de Agricultura. División de Investigaciones Agrícolas. Boletín Experimental No. 4, 1–53.

Gandarillas H. 1979. Genética y origen. Tapia M. Quinua y kañiwa, cultivos andinos. Bogotá, Colombia: CIID, Oficina Regional para América Latina, 45–64.

Gandarillas H, Gandarillas C, Gandarillas A. 2001. Humberto Gandarillas. Historia de la investigación para el desarrollo agropecuario en Bolivia. Memorias de un investigador. Cochabamba, Bolivia. 139–142.

Genebank Standards. 1994. United Nations Food and Agriculture Organization and International Plant Genetic Resources Institute [FAO/IPGRI], Rome, 1–13.

Guarino L, Ramanatha Rao V, Reid R. 1995. Collecting plant genetic diversity. Technical Guidelines. CAB International, Wallingford, Reino Unido, a nombre del Instituto Internacional de Recursos Fitogenéticos (IPGRI) y en colaboración con la Organización de las Naciones Unidas para la Agricultura y la Alimentación (FAO), la Unión para la Conservación Mundial (IUCN) y el Programa de las Naciones Unidas para el Medio Ambiente (PNUMA). Unidad 8.1.1 – Introducción a la colecta de germoplasma y Unidad 8.2.1 – Planificación de una colecta de germoplasma.

Gómez L, Eguiluz A. 2011. Catálogo del banco de germoplasma de quinua (*Chenopodium quinoa* Willd.). Universidad Nacional Agraria La Molina, Lima, Peru. 183.

Hawkes JG. 1980. Crop genetic resources field collection manual. IBPGR and EUCARPIA. University of Birmingham. Birmingham, England.

[IBPGR]International Board for Plant Genetic Resources. 1981. Descriptores de Quinua. Roma, Italy: IBPGR. 1–18.

[IPGRI]International Plant Genetic Resources Institute. 1996. Evaluation of seed storage containers used in genebanks. Report of a survey. Roma, Italia: IPGRI. 1–25.

[ISTA]International Seed Testing Association. 1993. International rules for seed testing. Seed Sci Technol, 21(Supplement. Suiza: ISTA): 1–288.

Jacobsen SE, Sherwood S. 2002. Cultivo de granos Andinos en Ecuador. Informe sobre los rubros quinua, chocho y amaranto. Organización de las Naciones Unidas para la Agricultura y la Alimentación (FAO), Centro Internacional de la Papa (CIP) y Catholic Relief Services (CRS). Quito, Ecuador. 1–89.

Jaramillo S, Baena M. 2000. Material de apoyo a la capacitación en conservación ex situ de recursos fitogenéticos. Instituto Internacional de Recursos Fitogenéticos, Cali, Colombia. 1–210.

Leon R, Pinto M, Rojas W. 2007. Evaluación de parámetros de temperatura y humedad para la conservación de largo plazo de quinua. Rojas W. Manejo, conservación y uso sostenible de los recursos genéticos de granos altoandinos, en el marco del SINARGEAA. Informe Anual 2006/2007. Proyecto SIBTA-SINARGEAA. MDRAMA - Fundación PROINPA, pp. 42–51.

Lescano JL. 1989. Avances sobre los recursos fitogenéticos altoandinos. En: Curso: cultivos altoandinos. Potosí, Bolivia. 17–21 de abril de 1989, pp. 19–35.

Lescano JL. 1994. Genética y mejoramiento de cultivos altoandinos: quinua, kañihua, tarwi, kiwicha, papa amarga, olluco, mashua y oca. Programa Interinstitucional de Waru Waru, Convenio INADE/PELT - COTESU. Puno, Perú, pp. 1–459.

Mujica A. 1992. Granos y leguminosas andinas. Hernandez J, Bermejo J, Leon J. Cultivos marginados: otra perspectiva de 1492. Organización de la Naciones Unidas para la Agricultura y la Alimentación. Roma: FAO. 129–146.

Pinto M, Poma S, Aroni G, Rojas W. 2009. Sistematización de la información de manejo de la colección de quinua y cañahua para su actualización en el SIRGEN. Pinto M. Proyecto manejo, conservación y uso sostenible de los recursos genéticos de granos Altoandinos, en el marco del SINARGEAA. Informe Final 2008–2009. La Paz, Bolivia: Fundación PROINPA. 30–36.

Pinto M, Polar V, Soto JL, Rojas W. 2010. Cerrando la brecha entre las prioridades de los productores y la de los investigadores: selección participativa de granos andinos. Rojas W, Pinto M, Soto JL, Jagger M, Padulosi S. Granos Andinos: avances, logros y experiencias desarrolladas en quinua, cañahua y amaranto en Bolivia. Roma, Italia: Bioversity International. 94–111.

Querol D. 1988. Recursos genéticos, nuestro tesoro olvidado: aproximación técnica y socio-económica. Lima, Perú: Industrial Gráfica S. A. 1–218.

Rojas W, Bonifacio A, Aroni G, Aroni JC. 1999. Recolección de nuevas accesiones de quinua y otras Chenopodiaceas. Informe Anual 1998–99. Fundación PROINPA, pp. 1–5.

Rojas W, Bonifacio A. 2001. Multiplicación de accesiones tardías de quinua bajo condiciones de invernadero. Informe Anual 2000/2001. Fundación PROINPA, pp.. 1–4.

Rojas W, Cayoja M, Espindola G. 2001. Catálogo de colección de quinua conservada en el Banco Nacional de Granos Altoandinos. La Paz, Bolivia: Fundación PROINPA, MAGDER, PPD-PNUD, SIBTA-UCEPSA, IPGRI, IFAD. 1–129.

Rojas W. 2002a. Distribución geográfica de la colección de germoplasma de quinua. Informe Anual 2001/2002. Proyecto Mcknight. Fundación PROINPA, pp. 1–5.

Rojas W. 2002b. Recolección de germoplasma de cañahua y quinua. Informe Técnico Anual 2001–2002. Año 1. Proyecto "Elevar la contribución que hacen las especies olvidadas y subutilizadas a la seguridad alimentaria y a los ingresos de la población rural de escasos recursos". IPGRI – IFAD. La Paz, Bolivia: Fundación PROINPA, pp. 13–21.

Rojas W, Camargo A. 2002. Reducción de la humedad del grano de quinua para almacenamiento a largo plazo. Informe anual 2001–2002 Proyecto Mcknight. Fundación PROINPA, pp. 1–6.

Rojas W. 2003. Multivariate analysis of genetic diversity of Bolivian quinoa germplasm. Food Reviews International 19:9–23.

Rojas W, Camargo A. 2003. Establecimiento de un método de reducción del contenido de la humedad del grano de quinua. Informe Anual 2002/2003. Proyecto Mcknight. Fundación PROINPA, pp. 1–4.

Rojas W, Pinto M, Camargo A. 2003a. Recolección descentralizada de germoplasma de cañahua y quinua. Informe Técnico Anual 2002–2003. Año 2. Proyecto IPGRI - IFAD "Elevar la contribución que hacen las especies olvidadas y subutilizadas a la seguridad alimentaria y a los ingresos de la población rural de escasos recursos." La Paz, Bolivia: Fundación PROINPA. 9–18.

Rojas W, Pinto M, Camargo A. 2003b. Estandarización de listas de descriptores de quinua y cañahua. En: Informe Técnico Anual 2002–2003. Año 2. Proyecto IPGRI-FAD "Elevar la contribución que hacen las especies olvidadas y subutilizadas a la seguridad alimentaria y a los ingresos de la población rural de escasos recursos." La Paz, Bolivia: Fundación PROINPA. 59–94.

Rojas W, Pinto M. 2004. Colecta descentralizada de quinua y cañahua. Informe Anual 2003/2004. Proyecto

SIBTA-SINARGEAA "Manejo, conservación y uso sostenible de los recursos genéticos de granos altoandinos, en el marco del SINARGEAA." Fundación PROINPA, pp. 11–19.

Rojas W, Pinto M. 2006. Evaluación del valor nutritivo y agroindustrial de accesiones de quinua y cañahua. Rojas W. Proyecto manejo, conservación y uso sostenible de los recursos genéticos de granos Altoandinos, en el marco del SINARGEAA. Informe Final 2005–2006. La Paz, Bolivia: Fundación PROINPA. 32–42.

Rojas W, Pinto M, Alcocer E. 2007. Diversidad genética del valor nutritivo y agroindustrial del germoplasma de quinua. Revista de Agricultura – Año 59 Nro. 41. Cochabamba, diciembre de 2007, pp. 33–37.

Rojas W, Pinto M. 2008. Evaluación del valor nutritivo de accesiones de quinua y cañahua silvestre. Pinto M. Proyecto implementation of the UNEP-GEF project, "In situ conservation of crop wild relatives through enhanced information management and field application." Informe de Fase 2005–2008. La Paz, Bolivia: Fundación PROINPA. 54–60.

Rojas W, Pinto M, Mamani E. 2009. Logros e impactos del Subsistema Granos Altoandinos, periodo 2003 – 2008. En Encuentro Nacional de Innovación Tecnológica, Agropecuaria y Forestal. INIAF. Cochabamba, 29 y 30 de junio de 2009, pp. 58–65.

Rojas W. 2010. Colección núcleo de granos Andinos. Rojas W, Pinto M, Soto JL, Jagger M, Padulosi S, Granos Andinos: avances, logros y experiencias desarrolladas en quinua, cañahua y amaranto en Bolivia. Roma, Italia: Bioversity International. 54–72.

Rojas W, Pinto M, Bonifacio A, Gandarillas A. 2010a. Banco de germoplasma de granos Andinos. Rojas W, Pinto M, Soto JL, Jagger M, Padulosi S, Granos Andinos: avances, logros y experiencias desarrolladas en quinua, cañahua y amaranto en Bolivia. Roma, Italia: Bioversity International. 24–38.

Rojas W, Pinto M, Soto JL. 2010b. Distribución geográfica y variabilidad genética de los granos andinos. Rojas W, Pinto M, Soto JL, Jagger M, Padulosi S, Granos Andinos: avances, logros y experiencias desarrolladas en quinua, cañahua y amaranto en Bolivia. Roma, Italia: Bioversity International. p.11- 23.

Rojas W, Pinto M. 2013. La diversidad genética de quinua de Bolivia. Vargas M. Congreso Científico de la Quinua (Memorias). Ministerio de Desarrollo Rural y Tierras - MDRyT, Viceministerio de Desarrollo Rural y Agropecuario - VDRA, Instituto Nacional de Innovación Agropecuaria y Forestal - INIAF, Instituto Interamericano de Cooperación para la Agricultura - IICA 14 y 15 de junio de 2013. La Paz, Bolivia. 77–92.

Rojas W, Pinto M, Alanoca C, Gómez L, León-Lobos P, Alercia A, Diulguerof S, Padulosi S. (2013). Estado de la conservación ex situ de los recursos genéticos de quinua. Bazile D, Bertero HD, Nieto C. Estado del arte de la quinua en el mundo en 2013. Santiago, Chile: Publicación FAO-RLC/Montpellier, Francia: CIRAD.

Rojas-Beltran J, Bonifacio A, Botani G, Maugham PJ. 2010. Obtención de nuevas variedades de quinua frente a los efectos del cambio climático. Informe Compendio 2007–2010. Cochabamba, Bolivia: Fundación PROINPA. 67–69.

Sevilla R, Holle M. 2004. Recursos genéticos vegetales. Luis León Asociados S.R.L. Editores, Lima, Perú. 445 p.

Tapia M. 1977. Investigaciones en el banco de germoplasma de quinua. En: Universidad Nacional Técnica del Altiplano. Curso de Quinua. Fondo Simón Bolívar. Puno, Perú: IICA - UNTA. 66–70.

Tapia M. 1990. Cultivos Andinos subexplotados y su aporte a la alimentación. Instituto Nacional de Investigación Agraria y Agroindustrial INIAA – FAO, Oficina para América Latina y l Caribe, Santiago de Chile.

Valls JFM. 1992. Caracterización morfológica, reproductiva y bioquímica del germoplasma vegetal. Capelo W. Memorias del curso internacional sobre recolección y evaluación de germoplasma forrajero andinodictado en. Riobambadel, 13al–23 de agosto de 1990. REPAAN. Riobamba, Ecuador. pp. 105–122.

Vavilov NI. 1951. Phytogeographic basis of plant breeding. The origin, variation, immunity and breeding of cultivated plants. Chronica Bot 13:1–316.

Chapter 9

Quinoa Breeding in Africa: History, Goals, and Progress

Moses F.A. Maliro[1] and Veronica Guwela[2]

[1]*Department of Crop and Soil Sciences, Bunda College Campus, Lilongwe University of Agriculture and Natural Resources, P.O. Box 219, Lilongwe, Malawi*

[2]*International Crops Research Institute for the Semi-Arid Tropics, P.O Box 1098, Lilongwe, Malawi*

INTRODUCTION

Origin of quinoa

Quinoa (*Chenopodium quinoa* Willd.) is an Andean crop belonging to the family *Chenopodiaceae*. It is a highly nutritious grain-like crop that has been a staple for centuries in South America, among pre-Columbian Andean farming communities from Colombia to Ecuador (Wilson 1990; Schlick and Bubenheim 1993; Bhargava et al. 2007). It is native to several countries of the Andean region, from Colombia to the north of Argentina and the south of Chile. History shows that the crop has been cultivated for at least 5,000 years in Latin America and was the staple food of the Inca Empire for many centuries (Schlick and Bubenheim 1993). In the 1500s, the Spanish conquerors banned quinoa cultivation in South America (Cusack 1984) and, hence, it became a minor crop, grown only by small-scale farmers for local consumption in remote areas of Bolivia, Peru, and Colombia (Jacobsen and Stølen 1993).

At present, the major producers of quinoa are Bolivia, Peru, and the United States. Quinoa cultivation has today transcended continental boundaries; thus, it is now grown in France, England, Sweden, Denmark, Holland, and Italy. In the United States, it is grown in Colorado and the Pacific Northwest, and in Canada, it is grown in Saskatchewan. Quinoa has shown high yields in Kenya and has also been grown successfully in the Himalayas and the plains of northern India (Jacobsen et al. 2003a, b).

Introducing quinoa in Africa

Food security in Africa

The nutritional superiority of quinoa compared to other crops has been recognized in recent years (Jacobsen et al. 2003b), and demand for quinoa has soared in developed countries where there is more consciousness about wellness through healthy diets. The increasing popularity of quinoa has triggered intensive research and breeding to promote its production and meet the growing market demand. In developing countries, particularly in Africa, the introduction of quinoa into the diet has the potential to contribute to food and nutritional security.

While food insecurity in some African countries may be attributed to political instabilities and civil wars, in the majority of African countries,

Quinoa: Improvement and Sustainable Production, First Edition. Edited by Kevin Murphy and Janet Matanguihan.

food insecurity is due to natural disasters. Recently, droughts and unpredictable rainfall patterns have been exacerbated by climate change effects. Even in times when food production has been adequate, many children and adults are still malnourished because maize is the predominant staple food and the main source of energy in the African diet. Protein-rich crops such as common bean and cowpea are grown as minor crops by most of the small-scale farmers who dominate African agriculture. Consequently, malnutrition has been one of major causes of infant mortality among children under 5 years of age (Table 9.1, FAO 2012a; Babatunde et al. 2011) and stuntedness among adults.

Prevalence of malnutrition

Quinoa production in Africa has the potential to contribute to reducing malnutrition, which is a problem among both rural and urban people in Malawi and other African countries. On the basis of three anthropometric indices, results of the 2010 Malawi Demographic and Health Survey (NSO 2010) showed that malnutrition is still high among children under 5 years of age. Height-for-age is the measure of linear growth and used as a measure of the nutritional status of a child. Children who are below −2 standard deviations from the reference mean for height-for-age are considered stunted or short for their age. Stuntedness is a reflection of the cumulative effect of chronic malnutrition. In 2010, the percentage of children who were stunted (below −2 SD) was 47%. In rural areas, 48% of children were stunted, whereas in urban areas, 41% of children were stunted.

Another index of malnutrition is the weight-for-height ratio, wherein children who are below −2 standard deviations from the reference mean are considered too thin for their heights or considered "wasted," a condition that reflects acute or recent nutritional deficit. The results of NSO 2010 showed that 4% of the children were wasted and half of them severely wasted. Rural areas had 4% of children wasted, whereas urban areas had 2% of children wasted.

The weight-for-age ratio is an overall indicator of a population's nutritional health; 13% of all children were underweight and 3% of these were

Table 9.1 Prevalence of undernourishment in selected African countries and in comparison with world overall levels (FAO 2012a).

Country	Proportion of undernourished in total population, %					
	1990–1992	1999–2001	2004–2006	2007–2009	2010–2012	Change so far
World	18.6	15.0	13.8	12.9	12.5	−32.8
Developed countries	1.9	1.6	1.2	1.3	1.4	NA
Developing countries	23.2	18.0	16.8	15.5	14.9	−35.8
Morocco	7.1	6.2	5.2	5.2	5.5	−22.5
Angola	63.9	47.5	35.1	30.7	27.4	−57.1
Benin	22.4	16.4	13.1	10.8	8.1	−63.8
Burundi	49.0	63.0	67.9	72.4	73.4	49.8
Cameroon	38.7	29.1	19.5	15.6	15.7	−59.4
Ethiopia	68.0	55.3	47.7	43.8	40.2	−40.9
Kenya	35.6	32.8	32.9	32.4	30.4	−14.6
Madagascar	24.8	32.4	28.1	29.1	33.4	34.7
Malawi	44.8	26.8	24.7	23.0	23.1	−48.4
Mali	25.3	21.5	14.7	9.5	7.9	−68.8
Mozambique	57.1	45.3	40.3	39.9	39.2	−31.3
Namibia	37.5	24.9	26.8	32.7	33.9	-9.6
South Africa	< 5	< 5	< 5	< 5	< 5	NA
Togo	32.8	25.2	20.4	19.8	16.5	−49.7
Zambia	34.3	43.9	48.3	47.5	47.4	38.2
Zimbabwe	44.1	43.1	38.2	33.9	32.8	−25.6

severely underweight, and 13% of rural children were underweight as compared with 10% of urban children. In addition, the impact of weaning could be seen in younger children where data on all three indices showed that the nutritional status of children deteriorated after 6 months of age, when children were being weaned. This case of malnutrition is prevalent across Africa as shown in Table 9.1 (FAO 2012a).

Malnutrition compromises the body's immunity, rendering children, pregnant women, and lactating mothers, as well as the elderly, vulnerable to diseases. Malnutrition contributes heavily to the high mortality rates in Africa, especially in children. Therefore, introducing quinoa and promoting its production and consumption in Africa would be one of the interventions to solve the malnutrition crisis.

Nutritional value of quinoa

Quinoa grain has an excellent balance of oil, fat, and protein content and has a unique composition of amino acids. The protein in quinoa contains all eight essential amino acids needed for human growth and development. It is exceptionally high in lysine, cystine, and methionine, amino acids that are typically low in other cereals (Schlick and Bubenheim 1993). Such nutritive value makes quinoa a relatively cheap yet excellent alternative grain used in people's diets and to fortify various food products. Quinoa grain ground into flour can be used for biscuits and cakes, added directly into soups, and can also be fermented with millet to make beverages (Schlick and Bubenheim 1993). New promising products from the grain include milk, protein concentrates, and natural colorants. The fresh leaves and tender shoots are eaten raw in salads or cooked and eaten as a vegetable in African dishes.

Quinoa as alternative crop in Malawi

Quinoa can grow in a wide range of climatic conditions and potentially can be grown almost anywhere in the world. It has been reported that some quinoa varieties can tolerate salinity levels as high as those present in sea water, with electrical conductivity (EC) 40 dS% to 400 mM NaCl high salt tolerance (Jacobsen et al. 2001, 2003a; Koyro and Eisa 2008; Hariadi et al. 2011; Adolf et al. 2012) and can grow under extremely dry conditions. It is a potential alternative crop in drought-prone areas of Africa, as it can grow in areas with as low as 200 mm annual rainfall in pure sand. In areas where frost occurs, quinoa can survive night frost (−8°C for 2–4 h) (Jacobsen et al. 2007). With the growing demand for quinoa grain in the United States, Europe, and Asia, the crop is a potential alternative export crop for many African countries such as Malawi. For so many years, Malawi has depended on tobacco as its main source of foreign income, but lately, international demand for tobacco has dwindled due to anti-smoking lobbies in many importing countries. Thus, Malawi needs an alternative export crop and quinoa could well be it.

Increasing awareness about quinoa

In recognition of its excellent nutritional properties and high adaptability, FAO selected quinoa as one of several promising crops with the potential to sustain food security in the next century (Jacobsen et al. 2003b). The FAO and United Nations also declared 2013 as the Year of Quinoa (FAO 2012b; UN 2012) to recognize the Andean indigenous peoples who have maintained, protected, and preserved quinoa as food for present and future generations. Owing to the recognition given by the FAO and United Nations to quinoa, more African agricultural and nutritional scientists have become aware of its possible role in alleviating African food insecurity.

Ecological adaptation of quinoa

The high and balanced nutritive value of quinoa, in addition to other desirable attributes such as tolerance to adverse growing conditions (Jacobsen et al. 2007), sparked a growing interest to introduce this crop into other regions of the world. According to Jacobsen et al. (2003b), there is a good potential of growing quinoa where environmental conditions are similar to those of the Andean region, specifically in parts of southern Europe, the United States, Africa, and Asia. Studies on quinoa adaptability conducted in Europe produced promising results (Geerts et al. 2008; Jacobsen et al. 1996; Jacobsen 1997, 2003) such that in Denmark and certain areas in Europe, quinoa had been considered a future food and fodder crop (Sigsgaard et al. 2008). Quinoa also has a growing popularity in Canada (AAFRD 2005). Africa can therefore take advantage of the growing world demand to produce quinoa for export in addition to contributing to its own food security. The prospects of introducing quinoa into African farming systems are high as the crop is adapted to a wide range of ecological zones in the Andean region where it originates. Quinoa can produce grain yields even in unfavorable soil and climatic conditions (Garcia et al. 2003). Quinoa's tolerance to arid conditions makes it an attractive crop for farmers in drought-prone areas; hence, its cultivation has been expanding to Kenya, India, North America, and Europe. Most quinoa growers are small farmers and the crop holds the promise of improved income. Some of the poorest Andean indigenous farming communities have already benefited greatly from the rising prices driven by demand for quinoa in world markets.

GOALS OF QUINOA BREEDING IN AFRICA

Quinoa was first introduced to Africa in the late 1990s in Kenya and recently in Malawi in 2012. Quinoa cultivars from the Andean region and cultivars bred in the United States, Canada, and Denmark were supplied by the International Potato Centre (CIP) and Washington State University (WSU) for adaptability studies. The goal of these initial breeding experiments was to identify cultivars and varieties of quinoa that could grow well and yield grain for production and consumption by African communities. The ultimate goal of these experiments is to contribute to efforts in reducing malnutrition problems in Africa. From the breeding work, it is also expected that quinoa varieties will be developed and grown for the domestic and export market, thereby presenting alternative cash crops, especially for small-scale farming communities of Africa. The breeding experiments that have been carried out so far have focused on evaluation of introduced varieties for adaptation to local ecological and climatic conditions of Africa, specifically in Malawi and Kenya.

Quinoa studies under Malawi conditions

Cultivars introduced

Quinoa was first introduced in Malawi in 2012 with 13 cultivars that originated from South American countries and those that have been bred for US and Denmark conditions (Table 9.2) and for which seeds were supplied by WSU. Testing cultivars of diverse origin increases the likelihood of selecting cultivars that can perform well under local conditions.

General climate conditions

Malawi has a diverse range of agroecological zones ranging from cool, highland areas to warm, low-lying areas. Its climate is described as subtropical, which is relatively dry and strongly seasonal. The warm-wet season runs from November to April, during which 95% of the annual precipitation occurs. Annual average rainfall varies from 725 to 2,500 mm with Lilongwe in the Central region having an average of 900 mm, Blantyre in the Southern region with 1,127 mm, Mzuzu in the Northern region with 1,289 mm, and Zomba in the Eastern region with 1,433 mm rainfall.

A cool, dry winter season runs from May to August with mean temperatures varying between 17 and 27°C, with temperatures falling between 4 and 10°C. Frost occurs in isolated areas in June and July. A warm, dry season lasts from September to October with average temperatures varying between 25 and 37°C. Relative humidity ranges from 50% for the drier months of September and October to 87% for the wetter months of January and February (MoNREE 2013) (Fig. 9.1).

Malawi has landforms that include highlands, escarpments, plateau and low-lying areas of the lakeshore, and the lower Shire valley. The highlands consist of isolated mountainous areas with

Table 9.2 Quinoa varieties introduced in Malawi in 2012 for testing and their background information.

No.	Variety	Origin	Background information
1	Ecuadorian	Ecuador	Not provided
2	Black-seeded	Colorado, USA	Developed from cross between *Chenopodium quinoa* and *Chenopodium berlandieri*, very tall variety (>2 m tall)
3	Inca Red (a.k.a. Pasankalla)	Bolivia	Member of the "Salares" ecotype of quinoa
4	Brightest Brilliant Rainbow	Oregon, USA	Not provided
5	Bio-bio	Chile	Not provided
6	Cherry Vanilla	Oregon, USA	Not provided
7	Multi-Hued	British Columbia, Canada	Not provided
8	Red Head	Oregon, USA	Not provided
9	QQ74	Chile	Chilean landrace
10	Puno	Denmark	Bred by Sven-Erik Jacobsen
11	Titicaca	Denmark	Bred by Sven-Erik Jacobsen
12	QQ065	Chile	Originally from extremely rainy region of southern Chile (>2,500 mm annual precipitation); shortest variety observed (~0.8 m) and has shown high resistance to postharvest sprouting in Malawi trials
13	Rosa Junin	Peru	Not provided

elevations between 1,320 and 3,000 masl such as the Nyika, Viphya, and Mulanje, whereas Dedza and Zomba are between 960 and 1,600 masl. The escarpments are areas that are around the highland plateaus and mountains. The plateaus are at elevations of 750 to 1,300 masl. The low-lying areas along the lakeshore are at 465–600 masl where the land is flat to gently undulating from the north to the south along the main outlet of Lake Malawi. The lowest lying areas are in the Shire valley at the southernmost part of the country with most elevations at less than 180 masl. The climate for each specific agroecological area of the country is moderated by variation in the landform, thereby giving a wide range of conditions across the country, from cool to warm to hot areas and from wet to drier areas (Fig. 9.1).

Fig. 9.1 Map of Malawi showing temperature variation across the country and these mainly influenced by mountains and low-lying areas that characterize the country.

Plant growth performance

A wide range of crop species can be grown in a country with a diversity of geographical areas, precipitation, and temperature. Thus, it should be possible to grow quinoa quite well in Malawi. For instance, there are varieties of maize, a staple food crop for the country, that have been bred and perform well in each specific agroecological zone. Adaptation trials that include quinoa cultivars and genotypes collected from diverse geographical sources should generate information on which specific quinoa varieties will do well in various agroecological zones of Malawi.

Eleven varieties/cultivars were evaluated for their adaptation in two ecological areas of the country using indices such as plant growth, flowering, and grain yield. The first site was Bunda (1,200 masl) in Lilongwe district, representing the mid-altitude agroecological zones. The second site was Bembeke (1,560 masl) in Dedza district, representing higher altitude agroecological zones. The experiment was conducted from July through the warmer months of August, September, and October under irrigation conditions. The cultivars were also evaluated under warm rain-fed conditions during the December to April 2012/2013 cropping season at the Bunda site.

Results in the Bunda and Bembeke sites and the rain-fed site at Bunda showed promise for quinoa introduction in Malawi. In the mid-altitude site (Bunda), all cultivars grew slowly for the first 21 days after sowing but grew rapidly when the weather started to warm up in mid-August. In the Bembeke highland site, quinoa grew extremely slowly for the first 2 months, probably due to the prolonged cold period experienced during the trials. By the time the weather warmed up, the plants were close to physiological maturity. Consequently, quinoa plants at the highland site were short, reaching only 50 cm in height, while all the 11 cultivars at the mid-altitude site grew to heights of 90–100 cm. Plant height significantly affected panicle length that in turn affected the number of inflorescence, ultimately determining grain yield. The results suggested that there will be severely reduced grain yield if quinoa is grown in the highland areas of Malawi during the winter season. The results also highlight the need to continue evaluating a diverse number of cultivars to select for genotypes adapted to specific agroecological areas across seasons in Malawi.

The slow plant growth attributed to very low temperatures also affected the period to maturity. At the mid-altitude or warmer site, the plants had 100% flowering by the end of 30 days, whereas at the highland or cooler site, the plants took 51 days to attain 100% flowering. The different quinoa varieties at the cooler site (Bembeke) took longer to reach maturity (102–119 days), whereas the quinoa genotypes at the warmer site (Bunda) reached maturity earlier (90 days) corresponding to earliness to flowering.

Genotype performance

The performance of the different genotypes showed variation in plant height at harvest, number of days to harvest (maturity period), panicle length, grain yield (kg/ha), and harvest index (HI) (Fig. 9.2). Promising high yielding varieties and genotypes that yielded up to or about 3 tons per hectare (t/ha) were "Brightest Brilliant" and "Cherry Vanilla" (both bred in Oregon, USA), "Multi-Hued" (bred in British Columbia, Canada), "Titicaca" (bred in Denmark), and QQ74 (from Chile). These promising quinoa cultivars could be recommended for commercial production under the climatic conditions of Malawi. The 3 t/ha quinoa grain yield was higher or comparable to the maize yields that small-scale farmers get (2 t/ha) under rain-fed conditions in Malawi. Moreover, the quinoa yields in Malawi did not differ from those reported in other regions of the world (Bertero et al. 2004). This implies that if farmers allocate a piece of land to quinoa production in place of maize, they are likely to harvest the same or higher quantity of quinoa grain, with the added benefit that quinoa is more nutritious than maize.

Quinoa studies in Kenya

Experimental sites and climatic conditions

Quinoa breeding studies in Kenya were first conducted as part of a multienvironment trial that involved a diverse set of 24 cultivars tested in 14 sites under irrigation conditions from 1999 to 2000 (Bertero et al. 2004; Oyoo et al. 2010). The other 13 sites were in Peru, Brazil, Bolivia, and Vietnam. In Kenya, the cultivars were planted at the Kabete Field Station of the University of Nairobi to determine their adaptability and yield performance. The site is located at an altitude of 1,820 masl, with temperatures ranging from 13 to 23°C, and a bimodal rainfall pattern with a mean annual rainfall of 970 mm. The soils have an average pH of 5.0–6.8, classified as humic nitisols (Siderius and Muchena 1977). The quinoa accessions were evaluated under both the long rainy seasons of 1999 (March–June) and the short rainy seasons of 1999 to 2000.

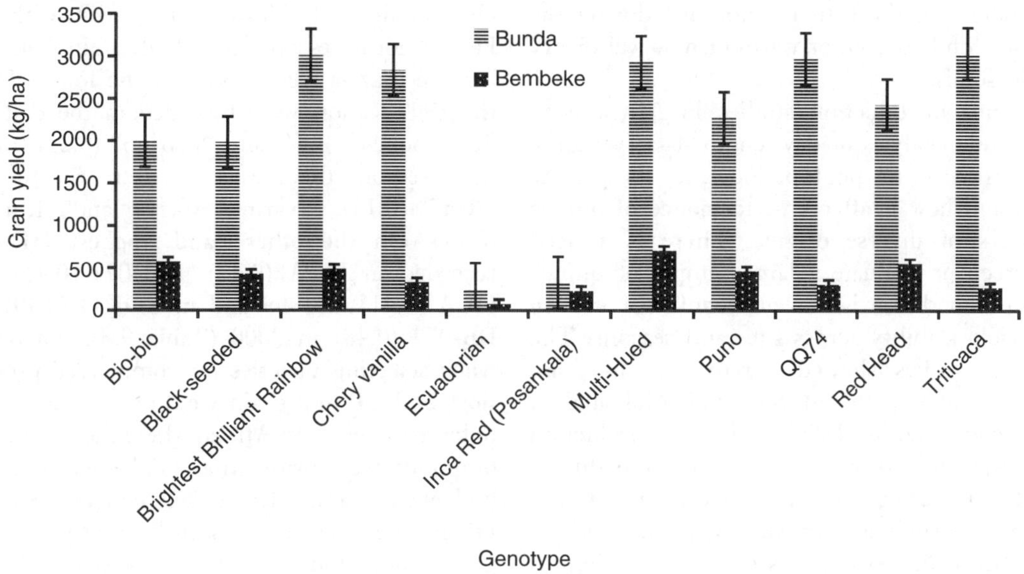

Fig. 9.2 Yield (kg/ha) of 11 quinoa genotypes and cultivars grown under irrigation from July to October 2012 at the Bunda and Bembeke sites in Malawi.

Plant growth and grain yield

The cultivars tested in Kenya showed varying performance in terms of plant growth and grain yield. The cultivars had shorter growth and maturity periods at 65–98 days during the 1999 season and 72–123 days in the 2000 growing season (Oyoo et al. 2010) compared to the maturity periods ranging from 110 to 190 days when quinoa is grown under Andean climatic conditions (Jacobsen and Stølen 1993; Bertero et al. 2004). The earlier maturity periods of quinoa in Kenya has been attributed to temperature effects imposed by the lower altitude of 1,820 masl compared to altitudes of over 3,000 masl in the Andean region where quinoa originated.

The overall grain yields of the cultivars evaluated (1.583 to 2.097 t/ha) were comparable to the yield obtained in the Andean and other regions of the world, and some cultivars even had higher yield. Some genotypes had yields up to 4 t/ha in the Nairobi sites, presenting one of the highest yields of quinoa obtained from quinoa introduction studies around the world (Jacobsen 2003). These results strongly suggest that quinoa production under Kenyan conditions is possible. However, these results cannot be generalized for the entire country as plant growth and yield are influenced by several environmental factors.

Latitude is one of the factors that influence crop introduction in a new environment. Latitude determines day length, which is critical for crops that are either long-day or short-day sensitive. The photoperiod sensitivity of quinoa for seed filling plays an important role in the crop's adjustment to new environments (Bertero et al. 2004), as it permits an accelerated seed filling when photoperiod shortens. However, this characteristic limits the adaptation of quinoa to higher latitudes. South Chilean cultivars are known to have less sensitivity to photo period in seed filling. Thus, adaptation of quinoa to high latitudes has been used as a selection index for the lack of or for less sensitivity to photoperiod in seed filling.

Cultivar performance also varied significantly from year to year even when grown during the same cropping season, as influenced by variation in temperature and precipitation. The growth range among cultivars has been reported to be almost the same, but the actual length of the growth period (longer or shorter) and the speed of plant growth (faster or slower) are highly dependent on the actual conditions during the season – on how cold or warm or how wet or dry the season is.

Quinoa introduction studies in Africa need to include analysis of day length data (variation in sensitivity to photoperiod and its genetic base) and how it affects performance of quinoa cultivars of diverse origins. There is a need to screen or evaluate a large sample of quinoa germplasm that exists in genebanks for growth and yield stability across sites and seasons. The Kenyan studies showed a reduction in grain yields of quinoa for all genotypes evaluated in 2000 compared to 1999. This yield reduction was attributed to lower rainfall received during the latter season compared to the first. Other parameters such as plant height, panicle length, and HI, which is strongly associated with biomass and grain yield, varied among cultivars tested and were also significantly affected by the cropping season. From the initial trials, three cultivars have performed best in terms of yield and show promise of commercial cultivation under Kenyan conditions – "Narino" (from Colombia, adapted to the altiplano), CICA-127 (from Peru, adapted to the valley areas), and ECU-420 (from Ecuador, adapted to the valley areas).

Harvest indices are useful in selecting genotypes with high grain production. The capacity of cultivars for high grain production as measured by HI varied from 0.24 to 0.47 and 0.15 to 0.49 for the 1999 and 2000 seasons, respectively. Such HI showed that the genotypes and cultivars tested were as productive in Kenya as in the areas where quinoa is traditionally grown (Bertero et al. 2004; Oyoo et al. 2010). However, genotypes with high HI did not necessarily have high grain yields per hectare (kg/ha). For instance, the lowest HI in the 1999 season were recorded in the cultivars "Canchones" (0.24) and "Narino" (0.25). In the 2000 season, the lowest HI were recorded in "Real" (0.15), "Narino" (0.17), and "Kamiri" (0.17). On the other hand, highest HI were recorded in ILPA (0.47), RM-072 (0.41), and HUA (0.41) in 1999 and in EDK-4 (0.49) and BB-079 (0.48) in 2000 (Table 9.3). Therefore, when selecting varieties for commercial production, both HI and grain yield per unit area need to be considered. In Africa, where the fresh leaves of quinoa are considered as a delicious vegetable, both as salad and when cooked, cultivars with low HI may be selected for both leafy vegetable and grain production, whereas those with high HI can be selected for grain production only (Rojas et al. 2000).

International quinoa trials comprising 14 sites (Kenya included) across three continents investigated the implication of genotype (G) × environment (E) interaction for genetic improvement of grain yield and grain size in low altitude areas. The studies focused on examining the relative size of components of variance for G and G × E interaction for grain yield, time to physiological maturity, above-ground biomass, HI, and grain size to group quinoa cultivars according their relative responses to testing environments (Bertero et al. 2004). Results from these studies showed a strong across-cultivar and across-environment variation for grain yield and its physiological determinants and grain yield.

Table 9.3 An extract of the mean performance of 24 cultivars of quinoa in Nairobi, Kenya, evaluated under the 1999 international quinoa trials.

Location	Latitude	Altitude, masl	Mean temperature, °C	Mean photoperiod, h	Sowing month	Days to crop maturity	Grain yield, gm^{-2}	Grain size, mm	Above-ground biomass, gm^{-2}	Harvest index
Nairobi L	1°15′S	1,819	18.9	12.4	October	77.1	209.7	2.1	698.7	0.31
Nairobi S	1°15′S	1,819	18.2	11.7	March	84.5	158.3	2.0	806.5	0.25

Source: Bertero et al. (2004).

Note: Nairobi L and Nairobi S are crops grown during the long (L) and short (S) rainy seasons in that environment (Maurice 2001). Crop duration is expressed as days from sowing to maturity. Temperature and photoperiod data are means for that period.

Both yield determinants, that is, HI and biomass also showed large G × E interaction to the G component of variance ratio. From this study, the investigators classified the sites into four groups of environments based on grain yield and Kenya was in the group of warmer and low-altitude environments. The Kenya sites were also reported to be in the shortest crop duration areas. In terms of cultivar performance, it was reported that all the valley type of cultivars showed their best performance in the valley sites, that is, Kenya sites, above the mean yields in the Atiplano sites and below mean yields in other environments. These results suggest that these cultivars are better adapted to the warmer and low altitude areas and can thus be adaptable in most of the tropics in Africa.

Pests and diseases

The varieties tested in Malawi did not have serious infestation of pests and diseases. Insect pests observed during the trials were sucking insects such as aphids and shield or stinkbugs, but at levels that did not warrant spraying of chemicals to control them. The aphids seemed to be more prevalent under dry weather as is the case in other crops, though the infestation was light and did not seem to attract new waves of infesting aphids. It may be too early to conclude that quinoa does not have serious pests and diseases under Malawian conditions until adaptation studies have been conducted over several agroecological zones and seasons. Future trials need to include studies on quinoa pests and diseases to substantiate these initial observations.

CHALLENGES AND CONSIDERATIONS FOR FUTURE RESEARCH

Plant lodging

Both irrigated and the rain-fed experiments under Malawi conditions showed that most of the quinoa varieties tested are susceptible to lodging, resulting to grain yield losses. African climates are characterized by the occurrence of heavy rain storms that may make the lodging worse. Therefore, selecting and breeding for lodging-resistant varieties should be one of the goals of quinoa breeding in Africa.

Acceptability

Acceptability of quinoa into African diets is a key factor for the successful introduction of the crop. Maize has been the staple food of Malawians for centuries. They do not consider that they have had a meal until maize-based food is eaten; other crop-based food products are considered merely as snacks. Such an attitude is reflected even at the national level, where the country is considered to have a food shortage whenever maize harvests are lean, even if other food crops are available. It is only in recent times that the urban populations are changing their eating habits and have begun to accept potato, rice, and wheat-based food products in their main meals. Malawians will likely have an attitude to quinoa similar to their attitudes toward other food crops. Hence, in conjunction with quinoa breeding and adaptability studies, there should be studies in the acceptability and incorporation of quinoa in the Malawian and African diet. Recipes and quinoa-based products should be developed to promote quinoa consumption in Africa.

The majority of African farmers are small-scale farmers who consider maize as the main food crop and allocate most or all of their resources in growing maize. The farmers' priority is to plant maize when the first rains come, as is the case for tobacco in regions where it is grown as the main cash crop. Under rain-fed cropping conditions, maize and other cash crops are planted first, followed by other important crops such as groundnut, soybean, and millet. Crops with less market value are planted last. However, without the application of inorganic fertilizers, crops planted late in the season usually do not have a vigorous take off compared to crops planted early with the first rains. In order to apply inorganic fertilizers to late-planted crops, farmers have to be convinced that the new crop has a potential of earning them income as well as being valuable in their diets.

Agronomic practices

The farming systems of Africa are dominated by small-scale farms characterized by land holdings of less than 2 ha, almost no mechanization, and with few input resources (Jayne et al. 2003). Successful promotion of quinoa as a new crop will need technologies adapted to small-scale farming conditions. For instance, most farm operations are manually done and sowing of quinoa seed at very close plant spacing of 10 cm would be time consuming. Farmers might opt to use the broadcast sowing method to save time, but this sowing method is impractical as quinoa has extremely small seeds. Investigations on optimal seed rates need to be conducted and recommended to farmers. Application of fertilizers in cereal crops is recommended to boost productivity, but most small-scale farmers can hardly afford inorganic fertilizers. Most of the inorganic fertilizers that they do manage to obtain are reserved for their staple crops such as maize. Alternative methods of improving soil fertility for quinoa production, such as application of animal manure or green manure, need to be investigated.

Rain-fed versus irrigated cropping systems

Crop production in small-scale farms is mostly under rain-fed conditions. In regions of Africa where rains are limited to one season and only for a few months, quinoa may need to be produced under irrigated conditions. Results from on-going experiments suggest that irrigated quinoa grown shortly after the end of the rainy season gives a better grain yield than under the rain-fed conditions in Malawi. However, irrigation as practiced by most small-scale farmers is with the use of watering cans. Irrigation technologies that would conserve soil moisture (such as mulching) would make quinoa production more feasible and affordable.

CONCLUSION

Results of initial experiments in Malawi and Kenya have shown that quinoa can grow well under varying agroecological zones, from warmer to cooler areas. However, more genotypes from the Andean region and those that have been bred elsewhere need to be evaluated under the various agroecological zones and seasons of Africa. Such evaluations will allow selection of more cultivars adapted to specific ecological zones and cropping seasons that prevail in the region. Studies have shown that cultivars from South America and those bred in North America can give the same or higher grain yields (3–4 t/ha) when grown in Malawi and Kenya, comparable to the yields of the same cultivars when grown in areas where quinoa is traditionally grown. Such varieties can be adopted in Africa and promoted to alleviate food insecurity common in many African countries.

To facilitate the introduction and adoption of quinoa in Africa, there should be more research on cropping technologies relevant to small-scale farms that characterize a large section of African agriculture. Aside from agronomic factors, socioeconomic factors that affect the adoption of new crops also need to be considered. For instance, the acceptability of a new grain crop can be influenced by grain color, taste, or the ease of processing of the grain for local dishes, even though the crop may be high yielding and nutritious. Acceptability of quinoa into the African diet, and the factors that influence the adoption of change, should also be studied. Plant breeders and agronomists should work hand-in-hand with farmers, extension agents, nutrition experts, government agencies, and relevant stakeholders to introduce quinoa to Africa, taking into account not only the agronomic challenges but also the social, market, and political challenges of introducing a new crop into the African landscape and psyche.

REFERENCES

[AAFRD]Alberta Agriculture, Food and Rural Development. 2005. Quinoa – the next cinderella crop for Alberta? A report by Alberta Agriculture Food and Rural Development (AAFRD). Available from: http://www1.agric.gov.ab.ca/$department/deptdocs.nsf/all/afu9961

[FAO]Food and Agriculture Organization of the United Nations. 2012a. Economic growth is necessary but not sufficient to accelerate reduction of hunger and malnutrition. Rome: FAO.

[FAO]Food and Agriculture Organization of the United Nations. 2012b. Master plan for the International Year of Quinoa – a future sown thousands of years ago. Available from http://www.fao.org/fileadmin/templates/aiq2013/res/en/master_plan.pdf

[MoNREE]Ministry of Natural Resources, Energy and Environment. 2013. Climate of Malawi. Department of Climate Change and Meteorological Services (DCCMS) of the Ministry of Natural Resources, Energy and Environment (MoNREE). Accessed from http://www.metmalawi.com/climate/climate.php

[NSO]National Statistical Office. 2010. Malawi Demographic and Health Survey 2010. National Statistical Office Zomba, Malawi and MEASURE DHS, ICF Macro Calverton, Maryland, USA, pp. 1–51.

[UN]United Nations. 2012. United Nations Resolution on the declaration of the IYQ 2013. A resolution adopted by General Assembly – A/RES/66/221 in March 2012. Accessed from http://www.un-ngls.org/IMG/pdf/IYQ.pdf

Adolf VI, Shabala S, Andersen MN, Razzaghi F and Jacobsen SE. 2012. Varietal differences of quinoa's tolerance to saline conditions. Plant Soil 357:117–129.

Babatunde RO, Olungunju IF, Sola O. 2011. Prevalence and determinants of malnutrition among under five children of farming households in Kwara State, Nigeria. 10.5539/jas.v3n3p173

Bertero HD, de la Vega AJ, Correa G, Jacobsen SE, Mujica A. 2004. Genotype and genotype-by-environment interaction effects for grain yield and grain size of quinoa (*Chenopodium quinoa* Willd.) as revealed by pattern analysis of international multi-environment trials. Field Crops Res 89:299–318.

Bhargava A, Shukla S, Rajan S, Ohri D. 2007. Genetic diversity for morphological and quality traits in quinoa (*Chenopodium quinoa* Willd.) germplasm. Genet Res Crop Evol 54:167–173. 10.1007/s10722-005-3011-0

Cusack DF. 1984. Quinoa: grain of the Incas. Ecologist 14:21–23.

Garcia M, Raes D, Jacobsen SE. 2003. Evapotranspiration analysis and irrigation requirements of quinoa (*Chenopodium quinoa*) in the Bolivian highlands. Agr Water Manage 60: 119–134.

Geerts S, Dirk Raes D, Garcia M, Vacher J, Mamani R, Mendoza J, Huanca R, Morales B, Miranda R, Cusicanqui J, Taboada C. 2008. Introducing deficit irrigation to stabilize yields of quinoa (*Chenopodium quinoa* Willd.). Eur J Agron 28:427–436.

Hariadi Y, Marandon K, Tian Y, Jacobsen SE, Shabala S. 2011. Ionic and osmotic relations in quinoa (*Chenopodium quinoa* Willd.) plants grown at various salinity levels. J Exp Bot 62:185–193.

Jacobsen SE. 1997. Adaptation of quinoa (*Chenopodium quinoa*) to Northern European agriculture: studies on developmental pattern. Euphytica 96: 41–48.

Jacobsen SE. 2003. The worldwide potential for quinoa (*Chenopodium quinoa* Willd.). Food Rev Int 19:167–177.

Jacobsen SE, Stølen O. 1993. Quinoa morphology, phenology and prospects for its production as a new crop in Europe. Eur J Agron 2:19–29.

Jacobsen SE, Hill J, Stølen O. 1996. Stability of quantitative traits in quinoa (*Chenopodium quinoa*). Theor Appl Genet 93:110–116.

Jacobsen SE, Quispe H, Mujica A. 2001. Quinoa: an alternative crop for saline soils in the Andes. In: CIP Program Report 1999–2000. Lima, Peru: CIP. 403–408.

Jacobsen SE, Mujica A, Jensen CR. 2003a. The resistance of quinoa (*Chenopodium quinoa* Willd.) to adverse abiotic factors. Food Rev Int 19:99–109.

Jacobsen SE, Mujica A, Ortiz R. 2003b. The global potential for quinoa and other Andean crops. Food Rev Int 19:139–148.

Jacobsen SE, Monteros C, Corcuera LJ, Bravo LA, Christiansen JL, Mujica A. 2007. Frost resistance mechanisms in quinoa (*Chenopodium quinoa* Willd.). Eur J Agron 26:471–475.

Jayne TS, Yamano T, Weber MT, Tschirley D, Benfica R, Antony Chapoto A. 2003. Smallholder income and land distribution in Africa: implications for poverty reduction strategies. 10.1016/S0306-9192(03)00046-0

Koyro HW, Eisa SS. 2008. Effect of salinity on composition, viability and germination of seeds of *Chenopodium quinoa* Willd. Plant Soil 302:79–90.

Maurice OE. 2001. Adaptability of some quinoa genotypes to Kenian highland conditions with special reference to Kiambu district. MSc Thesis. Faculty of Agronomy, University of Nairobi, Kenia, pp. 1–79.

Oyoo ME, Githiri SM, Ayiecho PO. 2010. Performance of some quinoa (*Chenopodium quinoa* Willd.) genotypes in Kenya. S Afr J Plant Soil 27:2.

Rojas W, Barriga P, Figueroa H. 2000. Multivariate analysis of the genetic diversity of Bolivian quinoa germplasm. PGR Newsl 122:16–23.

Schlick G, Bubenheim DL. 1993. Quinoa: an emerging "new" crop with potential for CELSS. NASA Technical Paper 3422. 1–12.

Siderius W, Muchena FN. 1977. Soils and environmental conditions of agricultural research stations in Kenya. Miscellaneous Soil Paper M5. Kenya Soil Survey. National Agricultural Research Laboratory, Nairobi.

Sigsgaard L, Jacobsen SE, Christiansen JL. 2008. Quinoa, *Chenopodium quinoa*, provides a new host for native herbivores in Northern Europe: case studies of the moth, *Scrobipalpa atriplicella*, and the Tortoise Beetle, *Cassida nebulosa*. J Insect Sci 8: 1–4.

Wilson HD. 1990. Quinoa and relatives (*Chenopodium* sect. *chenopodium* subsect. *cellulata*) Econ Bot 44Supplement 3:92–110.

Chapter 10

Quinoa Cultivation for Temperate North America: Considerations and Areas for Investigation

Adam J. Peterson and Kevin M. Murphy

Department of Crop and Soil Sciences, Washington State University, Pullman, WA, USA

INTRODUCTION

Quinoa has a relatively recent history of commercial cultivation in North America. The first major efforts at commercial cultivation were begun in 1983 by a partnership between Colorado State University and Sierra Blanca Associates, focusing on growing quinoa in high altitude locations of Colorado (Johnson 1990). In addition, private efforts to test and grow quinoa varieties in North America were begun by John Marcille in northern Washington state and Emigdio Ballón in northern New Mexico (Ballón 1990; Wilson and Manhart 1993). Quinoa production expanded outside Colorado to Wyoming and Northern New Mexico (Ward 1994). By the late 1980s, efforts to grow quinoa in Canada were underway (National Research Council 1989; Tewari and Boyetchko 1990; Small 1999).

Cultivation of quinoa in North America is currently centered in the San Luis Valley of Colorado and in the Canadian Prairies through the Northern Quinoa Corporation, with headquarters in Saskatoon, Saskatchewan. Despite efforts to cultivate quinoa in North America and other parts of the world, the majority of worldwide quinoa cultivation is still concentrated in South America. In recent years, the large gap between the supply of quinoa and the growing demand for it has led to high prices and mounting concern about the social and environmental impacts of quinoa production in the major quinoa-producing countries (Jacobsen 2011; Romero and Shahriari 2011). These concerns, along with the desire in many areas for regionally sourced food and interest in quinoa as a crop for improving food security, have led to efforts to adapt and grow quinoa in various parts of the world (Jacobsen 2003).

TOLERANCE TO ABIOTIC STRESSES

Heat tolerance

Quinoa production in North America has so far been limited to regions with cool summers, where maximum temperatures do not exceed 35°C (95°F). Initial trials in Colorado showed that quinoa was unsuccessful at elevations lower than 2,100 m (7,000 ft) due to pollen sterility and plant dormancy caused by high temperatures (Johnson and Croissant 1985; Johnson 1990).

Quinoa: Improvement and Sustainable Production, First Edition. Edited by Kevin Murphy and Janet Matanguihan.

Multiple reports exist indicating that high temperatures pose a constraint to quinoa production in North America, namely in Minnesota (Oelke et al. 1992), New York (Dyck 2012), Virginia (Bhardwaj et al. 1996), and Alberta (AAFRD 2005). Oelke et al. (1992) suggested that heat susceptibility would limit the expansion of quinoa cultivation to coastal areas in central California, high altitude locations in central and northern Washington state, and parts of the Canadian prairies. High temperatures have been noted as having a detrimental effect on yield and seed filling in locations outside of North America such as Morocco (Benlhabib et al. 2004), Chile (Fuentes and Bhargava 2011), Greece (Iliadis et al. 2001), and Italy (Pulvento et al. 2010).

Chronic heat stress can also affect quinoa, in contrast to the acute heat stress events that can often occur under field conditions. In an early study of a variety of unspecified origin, Aguilar (1968) found that quinoa subjected to a constant temperature of 32°C (90°F) had greatly reduced height, biomass, and seed yield compared to more mild temperature treatments. Overall development time was shortened, and red stem color present in the tested variety was enhanced at hotter temperatures.

Current work at Washington State University (WSU) has found maximum summer temperatures to be a major limiting factor for successful seed set in quinoa. While temperatures did not exceed 35°C (95°F) during anthesis in a 2011 field experiment, high temperatures during the seed fill stage may have caused significant reductions in yield. Many of the inflorescences lacked seeds or contained empty seeds (Peterson 2013). Despite the large reduction in yield in 2011, several varieties, namely Colorado 407D, QQ74, and Kaslaea, exhibited greater heat tolerance than others.

Variation in adaptation to hot, dry conditions has also been observed among Chilean highland varieties grown in a field experiment in Atacama Desert of northern Chile (Fuentes and Bhargava 2011). The effect of heat stress on quinoa was reviewed in detail by Bonifacio (1995), who noted that heat stress could cause reabsorption of the endosperm, a phenomenon recognized by Bolivian farmers as "phuna" or "puna." As the endosperm of the quinoa seed is of minimal size and the perisperm and embryo are the two largest components of the seed (Prego et al. 1998), this reabsorption likely extends to more seed components than the endosperm. Bonifacio also observed that high temperatures inhibited anther dehiscence in outdoor plantings in Provo, Utah in 1993 and found varietal differences in heat tolerance.

Irrigation may be an important tool in quinoa cultivation by ameliorating the effects of heat stress. Preliminary results from a WSU irrigation trial showed that irrigation substantially increased yields under heat stressed growing conditions in eastern Washington state (Hannah Walters, personal communication, 2013). Further investigation into the role of heat stress, water availability, and the interaction of these factors in quinoa is warranted.

As most areas of temperate North America have summer temperatures in excess of quinoa's threshold of tolerance (NCDC 2011), the development of heat tolerant varieties could greatly expand the area suitable for quinoa production and increase harvest security in areas with damaging heat waves. Despite the heat susceptibility of quinoa, closely related *Chenopodium* spp. inhabit parts of North America that experience intense summer heat (Jellen et al. 2011). In addition, archeological evidence indicates the existence of a domesticated chenopod pseudograin, analogous to quinoa, in the prehistoric Eastern North American agricultural complex (Smith 1985). This implies the possibility of developing a heat-tolerant chenopod pseudocereal for North America outside of high altitude or cool maritime locations. If the existing variation for heat tolerance within quinoa proves insufficient, related *Chenopodium* spp. such as *C. berlandieri* and *C. bushianum* to which quinoa can hybridize may provide a promising source for introgression of genes for heat tolerance.

Drought tolerance

Quinoa is considered to have remarkable drought tolerance. It has been reported to grow with as little as 200 mm (8 in) annual precipitation in pure sand (Aguilar and Jacobsen 2003). Yields exceeding 1,000 kg/ha (890 lbs/acre) have been reported with as little as 50 mm (2 in) irrigation in the Atacama Desert of northern Chile. However, yields are much improved in arid regions under irrigation (Martínez et al. 2009). Initial research in Colorado found that quinoa had highest yields with 208 mm (8.2 in) of combined irrigation and rainfall on a sandy loam soil (Flynn 1990). Later recommendations to growers in the San Luis Valley were 25–38 cm (10–15 in) of combined irrigation and rainfall (Johnson and Croissant 1985). The addition of organic matter has been shown to be effective in increasing quinoa yield under arid conditions (Martínez et al. 2009).

The effect of drought on yield varies depending on the stage of plant development. Geerts et al. (2008) found that water use efficiency improved when there was adequate water supply during the initial germination phase and during flowering and seed fill, even though drought occurred during vegetative growth. In contrast, there were severe negative effects on yield when drought occurred during flowering and during seed fill stages. Jensen et al. (2000) also found yield decreases when drought was simulated during the flowering stage and during the seed fill stage. When drought was instead simulated during the early vegetative phase, yield increased compared to the control.

Sensitivity to drought in later developmental stages may be responsible for the reported lack of seed set in cool growing locations. Quinoa trials in Sweden showed a total failure of seed set during a year with a near absence of precipitation in July (Ohlsson 2000). This contrasts with favorable performance the previous year, when rainfall was more equally distributed (Ohlsson and Dahlstedt 1999). Some Chilean lowland varieties grown in cool locations at sites in maritime western Washington state failed to set seed in 2013 (unpublished data). As these sites are characterized by a mild oceanic Mediterranean climate with seasonal summer drought that would coincide with later growth stages of quinoa, the impact of drought on seed set deserves further investigation, as well as the variability for seed set among Chilean lowland varieties.

In contrast, some reports indicate greater drought tolerance in later growth stages. Jacobsen and Stølen (1993) noted that in Denmark, the greatest impact from drought occurs during the vegetative stage. Likewise, Darwinkel and Stølen (1997) reported greater drought tolerance in later growth stages. Razzaghi et al. (2012) found that yield did not significantly decrease when simulated drought was applied during the seed filling stage.

Cold tolerance

Cold temperatures are an important limiting factor for quinoa cultivation in many areas of North America. The occurrence of frosts places limits on the planting date of spring planted quinoa, which is of particular importance where the long time to maturity for quinoa is an issue. At high altitude locations, cooler temperatures favor quinoa cultivation, although the frost-free growing period is shorter. This limits cultivation to the earliest maturing varieties and can place plants in the field at risk to early frosts during sensitive growth periods (Johnson and Croissant 1985).

One untested possibility is the cultivation of quinoa as a winter crop in extreme southern locations of the United States. Although heat would not be an issue over the relatively cool winters, the potential occurrence of damaging frosts in many of these locations would likely limit any cultivation of quinoa to locations that do not experience temperatures lower than the reported limits for quinoa, such as areas of southern Florida that experience a tropical climate.

There are varying reports on the frost tolerance of quinoa under field conditions. Risi and Galwey (1984) reviewed contradictory reports regarding the frost tolerance of quinoa in South America, with some sources indicating little or no frost resistance and others reporting high frost resistance.

In a study by Jacobsen et al. (2007), frost tolerance measured under controlled conditions varied between an Altiplano and a Valley-type variety. In addition, frost tolerance was found to correlate with the ability of quinoa to produce more soluble sugars and, to a lesser extent, proline, with exposure to cooler temperatures (Jacobsen et al. 2007). Growth stage also affects tolerance to frost. Exposure to frost after the initiation of flowering proved far more damaging than exposure during earlier growth stages. A 66% reduction in yield was seen when plants in anthesis were exposed to −4°C (25°F), whereas seedlings at the two-leaf growth stage exposed to the same conditions only had a 9% reduction in yield. Humidity also interacted with the impact of frost on quinoa, with drier conditions resulting in lower percentage of plant survival (Jacobsen et al. 2005).

There are a few accounts of frost tolerance in Chilean lowland varieties. In England, Risi and Galwey (1984) found that several Chilean varieties tolerated several frosts in spring, one of which reached −5°C (23°F). In Colorado, reports indicate that quinoa could withstand light frosts of −1 to 0°C (30–32°F). In accordance with Jacobsen's findings, a heavy frost of −4.4°C (24°F) during flowering caused losses exceeding 70% (Johnson and Croissant 1985). Oelke et al. (1992) suggested that temperatures below −2°C (28°F) during flowering would cause significant losses. However, once seeds are in the soft dough stage, frost resistance increases and plants were reported to withstand temperatures down to −7°C (20°F). Darwinkel and Stølen (1997) reported varietal differences in frost tolerance and noted −3°C (27°F) as the threshold for quinoa.

Salinity tolerance

Soil salinity is a significant agricultural problem for large parts of temperate North America. In the United States, there are 2.2 million hectare (Mha) (5.4 million acres (Mac)) of saline agricultural land, with a further 30.8 Mha (76.1 Mac) of agricultural land under threat of salinization (USDA 2011). The majority of these saline-affected soils are confined to arid and semi-arid regions of the West (USDA 1989). In Canada, salinity is a significant problem in the Prairie provinces of Alberta, Saskatchewan, and Manitoba, affecting an estimated 2.19 Mha (5.4 Mac) (ARD 2004).

Quinoa is generally recognized as one of the most saline-tolerant crops known (Jacobsen 2007). In its native range, quinoa is cultivated in areas with highly saline soils. For example, varieties of the Salares ecotype are grown around salt flats in Southern Bolivia (Risi and Galwey 1984).

Quinoa has exhibited unparalleled levels of salinity tolerance under controlled conditions. Koyro and Eisa (2007) and Hariadi et al. (2011) demonstrated the ability of quinoa to survive and produce seed under 500 mM NaCl, an equivalent concentration to that of seawater. Jacobsen et al. (2003) found that seeds of the variety "Kanckolla" were able to germinate at 57 mS/cm, a level exceeding that of seawater.

The salt tolerance of quinoa has been demonstrated to vary significantly within and between ecotypes. In an analysis of Peruvian quinoa germplasm, Gómez-Pando et al. (2010) found significant variation for saline tolerance at both the seedling and the adult stages. Tolerance at the seedling stage was not necessarily correlated with tolerance during the adult stage (Jacobsen et al. 1999b).

Adolf et al. (2012) investigated the response of Danish, Bolivian, and Peruvian varieties to salinity and noted a considerable range in physiological responses and relative salinity tolerance. Despite the complex variation exhibited in the study, Danish varieties of Chilean lowland background were of average or lesser salinity tolerance as measured by relative declines in height and biomass, whereas the "Real"-type quinoa varieties from Southern Bolivia rated more tolerant for these measures.

Investigation of salinity tolerance within Chilean lowland germplasm will prove important in the use of quinoa as a salt tolerant crop. These varieties are most adapted to the conditions found in areas of North America affected by soil salinity. Many salt-affected areas in western North America experience high summer temperatures characteristic of a continental climate. Breeding varieties with greater heat tolerance will be necessary as the susceptibility of quinoa to heat precludes its cultivation in these areas. Chilean lowland varieties have the greatest heat tolerance and characterizing their levels of salinity tolerance will be crucial in developing quinoa as a successful halophytic crop for these areas.

Two germination studies conducted with Chilean germplasm provided evidence for a significant geographical trend in salinity tolerance (Delatorre-Herrera and Pinto 2009; Ruiz-Carrasco et al. 2011). Delatorre-Herrera and Pinto (2009) found higher tolerance among a Chilean highland variety compared to a Chilean lowland variety, measured in terms of seed germination, and related this to adaptation to the saline soils found in the highlands of northern Chile. Ruiz-Carrasco et al. (2011) tested varieties within the range of the Chilean lowland ecotype, comparing varieties from central Chile with a variety from southern Chile. The authors found significant differences among these varieties in growth characteristics, proline accumulation, polyamine response, and expression levels of sodium transporter genes CqSOS1 and CqNHX1. These differences were generally indicative of lower salinity tolerance in the quinoa variety from Southern Chile and were linked by the authors to the gradient of decreasing exposure to salinity and increasing precipitation that runs north to south in Chile.

A recent study measured the salinity tolerance among four quinoa varieties of Chilean lowland origin (Peterson 2013). All four varieties were significantly more tolerant to Na_2SO_4 than they were to NaCl, as measured by yield decreases at 32 dS/m relative to a nonsalt control. Yield declines for quinoa varieties ranged from 43.7% to 65.4% under 32 dS/m NaCl and from 10.8% to 51.9% under 32 dS/m Na_2SO_4. Of particular significance is that these differences generally appeared to match the latitude of origin of the varieties in Chile, in accordance to the findings of previous studies. In terms of yield, the northernmost variety, UDEC-1, was generally most tolerant and the southernmost variety, QQ065, was most susceptible to salinity (Peterson 2013).

Further investigation to confirm a geographical gradient of saline tolerance among Chilean lowland accessions could prove invaluable in identifying the most saline-tolerant Chilean lowland germplasm. In addition, given the increasing temperatures that are also found at more northern and lower latitudes in Chile (Climatologia; Ruiz-Carrasco et al. 2011), greater heat tolerance may be found in conjunction with greater salinity tolerance.

A field study performed in Southern Italy found that the Danish quinoa variety "Titicaca" had no significant difference in yield when irrigated with saline water of 22 dS/m, mixed to approximate a 1 : 1 ratio of seawater to freshwater, compared to a fresh water control (Pulvento et al. 2012). In comparison, barley, considered a saline tolerant grain, has a salinity threshold of 8 dS/m (Maas 1986). When quinoa was grown on a saline-sodic soil in Greece with an EC of 6.5 dS/m, the seedlings had poor establishment, which the authors linked to high pH, high Na^+, and poor soil physical characteristics due to soil sodicity. Varieties responded differently when grown on a nonsaline soil, indicating varietal differences in tolerance. Overall, yield was substantially decreased, with the best performing variety under saline-sodic conditions yielding 1.27 t/ha (1,130 lbs/ac), whereas the best performing variety under nonsaline conditions yielded 2.30 t/ha (2,050 lbs/ac) (Karyotis et al. 2003). The added effects of sodicity to salinity likely explain the large differences in yield response in that experiment compared to the findings by Pulvento et al. (2012). As soil crusting reduces quinoa emergence, the tendency of sodic soils to form a crust may be an impediment to the potential of quinoa to be grown on these soils. The larger effect of sodicity compared to salinity is supported by an early experiment by Torres (1955), which compared the response of quinoa to saline, saline-sodic, and sodic soils. Sodicity was found to have a more detrimental impact on quinoa biomass yield compared to salinity, though differences in sodicity tolerance were seen between the two varieties examined.

PRODUCTION ASPECTS

Variety selection

Quinoa is traditionally divided into five main ecotypes: Valley, which is grown between 2,000 and 4,000 m elevation from Central Peru northward; Altiplano, grown at 4,000 m elevation around Lake Titicaca; Salar, grown at 4,000 m around salt flats in southern Bolivia; Subtropical, from the Yungas region of Bolivia; and Sea level, from temperate latitudes in southern Chile (Tapia et al. 1980 cited in Risi and Galwey 1984). The Sea level ecotype, also referred to as Chilean lowland ecotype, is recognized as the best adapted to temperate latitudes and high summer temperatures. After initial trials, Chilean varieties were chosen to form the basis of breeding programs in Europe and in Colorado (Johnson 1990; Jacobsen 1999). In initial trials, Chilean lowland varieties were the best adapted to cultivation in Washington state (unpublished data), and only Chilean and southern Bolivian varieties set seed in Colorado (Johnson 1990). Bertero (2003) concluded that Chilean lowland and Altiplano varieties, characterized by low photoperiod sensitivity and a short basic vegetative period, were most suited for cool, temperate latitudes.

However, there are reports of non-Chilean varieties producing good yields at high latitude locations in Europe. A wide range of Peruvian and Bolivian varieties produced seed in a field trial in Finland, at two locations above 60°N (Carmen 1984). Likewise, Valley and Altiplano varieties produced significant yields in England, as did Bolivian and Peruvian varieties in Denmark. However, non-Chilean varieties generally needed longer times to reach maturity (Risi and Galwey 1991b; Jacobsen and Stølen 1993; Jacobsen 1997). Christiansen et al. (2010) found that an 18-h-long photoperiod induced a "stay green" reaction in "Real," a Bolivian variety.

Overall, non-Chilean varieties failed to set seed in field trials in Washington state (unpublished data). Bertero (2003) also noted failure of these varieties to reach maturity in Argentina.

In an experiment using "Kanckolla," an Altiplano variety, long daylength and high temperatures were found to decrease the amount of seed filling. The two factors, taken separately, had small but significant effects. However, when 16-h daylength was combined with 28°C (82°F), seed size decreased by 73% (Bertero et al. 1999). This relatively low threshold for high temperatures, combined with the negative influence of long daylength, suggests that most non-Chilean varieties likely have negligible potential for cultivation in temperate regions of North America.

Fertilization

Fertilizer recommendations for quinoa in Colorado were 135 kg N/ha (120 lbs N/acre), based on results from early trials (Johnson and Croissant 1985). However, after more extensive field research, the recommendation was increased to 170–200 kg N/ha (150–180 lbs N/acre). Higher levels of nitrogen were found to have adverse effects, leading to lodging and delayed maturity (Oelke et al. 1992). Under Danish conditions, yield was generally highest at 160 kg N/ha (143 lbs N/ac) compared to the lower levels of 120 kg N/ha (107 lbs N/ac), 80 kg N/ha (71 lbs N/ac), and 40 kg N/ha (36 lbs N/ac). However, increase in yield response under higher fertilization levels was minimal. At 40 kg N/ha, yield was only 24.1% lower than at 160 kg N/ha (Jacobsen et al. 1994). Trials in Denmark and the Netherlands showed differences in nitrogen response across locations and years, although general recommendations of 100–150 kg N/ha (89–134 lbs N/ha) were given. A fertilization level of 50 kg N/ha resulted in poor yields, whereas nitrogen in excess of 150 kg N/ha was found to have meager benefits (Darwinkel and Stølen 1997).

A study performed by Schulte auf'm Erley et al. (2005) in southern Germany found quinoa quite responsive to increased nitrogen fertilization. Yields at no fertilization produced 1,790 kg/ha (1,597 lbs/ac), whereas 120 kg N/ha (107 lbs N/ac) caused yields to almost double to 3,495 kg/ha (3,118 lbs/ac). Nitrogen utilization efficiency (g grain/g N_{plant}) did not change under increasing fertilization, though a significant difference was seen between the two varieties tested (Schulte auf'm Erley et al. 2005).

In another experiment, conducted under greenhouse conditions, nitrogen utilization efficiency showed significant decreases under increasing N, in contrast to the results observed by Schulte auf'm Erley et al. (2005). This difference is notable in that one of the varieties tested, "Faro," was common to both experiments, indicating different nitrogen dynamics due to environment. Again, there was a significant difference between varieties, but this time in nitrogen uptake efficiency (g N_{plant}/g $N_{mineral}$) (Thanapornpoonpong 2004).

The large variation in nitrogen dynamics observed at different locations, as well as significant varietal differences, pose several challenges and opportunities for breeding quinoa for increased nitrogen use efficiency. Quinoa has historically thrived in marginal areas, and varieties that are adapted to low soil nitrogen might prove valuable for low-input agriculture. Traditional cultivation of quinoa on the Peruvian Altiplano generally involves no fertilization. In crop rotations, quinoa is grown after potatoes and scavenges for remaining nutrients in the soil (Aguilar and Jacobsen 2003).

As organic quinoa makes up the bulk of quinoa demand, with organic quinoa constituting 69% of the total US quinoa imports in 2013 (Arco de Nuñez 2014), investigating the response of quinoa varieties to fertilization and management under organic conditions is crucial. Scant data exists in this area of study; however, Bilalis et al. (2012) reported yield increases of 6% with compost application and of 10% with cow manure application compared to a nonfertilized control.

Seed composition is influenced by fertilization. In the aforementioned study by Bilalis et al. (2012), saponin content increased under organic fertilization as compared to a nonfertilized control. Johnson and Ward (1993) found that protein content was responsive to nitrogen applications, increasing by 0.1% per kg of ammonium nitrate.

Soil texture is reported to have a major impact on nitrogen uptake and nitrogen use efficiency. In a lysimeter experiment where 120 kg N/ha (107 lbs N/ac) was applied, nitrogen uptake was 134 kg N/ha (120 lbs N/ac) under a sandy clay loam but only 77 kg N/ha (69 lbs N/ac) under a sandy soil, leading to differing yields of 3,300 kg/ha (2,900 lbs N/ac) and 2,300 kg/ha (2,100 lbs N/ac), respectively (Razzaghi et al. 2012).

In most studies, nitrogen was applied in a single application, although several studies report splitting applications between planting and a later stage (Aguilar and Jacobsen 2003; Schulte auf'm Erley et al. 2005; Pulvento et al. 2010). Darwinkel and Stølen (1997) suggest splitting fertilizer applications exceeding 150 kg N/ha (134 lbs N/ac) to prevent excessive salt exposure. However, there is no published research on the effects of split fertilizer applications on nitrogen uptake in quinoa.

There are contradictory reports on the response of quinoa to phosphorus fertilization. In Chile, yield increased with high levels of phosphorus fertilization (100 and 200 kg N/ha, 89 and 178 lbs N/ac) when accompanied by nitrogen fertilization (Delatorre-Herrera 2003). Recommendations from South America indicate that quinoa responds well to 80 kg N/ha (71 lbs N/ac) and 80 kg P/ha (71 lbs P/ac) (Mujica 1977 cited in Aguilar and Jacobsen 2003). However, Gandarillas (1982 cited in Johnson and Ward 1993) found no yield response to phosphorus or potassium application. In Colorado, phosphorus applications of 34 kg P_2O_5/ha (30 lbs P_2O_5/acre) did not affect yield (Oelke et al. 1992). Darwinkel and Stølen (1997) reported requirements of 70 kg P_2O_5/ha (62 lbs P_2O_5/ac) for quinoa before seed filling and noted that existing levels of phosphorus in many agricultural soils are likely sufficient. They also note a fairly large requirement for potassium, with uptake of 500 kg K_2O/ha (446 lbs K_2O/ac) and recommend application of 100–200 kg K/ha (89–178 lbs K/ac). Data on the micronutrient requirements of quinoa is almost nonexistent. However, Darwinkel and Stølen (1997) report that manganese is a critical micronutrient for quinoa.

Planting/spacing

Row spacing for quinoa varies widely. In South America, a variety of planting techniques are used. When quinoa is planted in rows, spacing ranges from 0.4 m (16 in) and 0.8 m (31 in). Wide ranges of optimal sowing density have also been reported (Tapia et al. 1980 cited in Risi and Galwey 1984).

Recommendations for Colorado were based on a plant density of 320,000 plants/ha (130,000 plants/acre) and a sowing density of 0.6–0.8 kg/ha (0.5 to 0.75 lbs/acre). There was no set row spacing recommended, although spacing greater than 36 cm (14 in) was beneficial (Johnson and Croissant 1985). In experiments in

England, highest yields were seen for the Chilean variety 'Baer' under the closest row spacing tested, at 20 cm (8 in) spacing seeded at 20 kg seed/ha (18 lbs seed/ac). Increasing plant density decreased the amount of branching on plants and also resulted in earlier maturity in the field (Risi and Galwey 1991a).

Jacobsen et al. (1994) investigated row spacings of 50, 25, and 12.5 cm and found no significant differences in yield between them. The widest spacing of 50 cm (20 in) was combined with mechanical hoeing to control weeds, and plants under this spacing appeared healthiest, whereas plants at the 25 cm (10 in) and 12.5 cm (5 in) row spacings were best able to crowd out weeds. Moreover, optimal plant density was calculated to be 327 plants/m^2 (1,320,000 plants/acre) with a standard deviation of 220 plants/m^2 (890,000 plants/acre), leading the authors to conclude that yield could remain stable over a wide range of plant densities. On the basis of later experiments in Denmark and the Netherlands, Darwinkel and Stølen (1997) confirmed the stability of yield across a wide range of row spacings, ranging from 12.5 cm (5 in) to 75 cm (30 in), and plant densities ranging from 30 to 250 plants/m^2 (25 to 210 plants/yd). On the basis of harvesting considerations and weed competition, a seeding rate of 6 kg/ha (5.4 lbs/ac) with target plant densities of 100 to 150 plants/m (85 to 125 plants/yd) was recommended for quinoa cultivation in northern Europe.

The large range of plant densities, row spacing, and seeding rates found in these studies indicate the large influence that location and weed pressure, as well as variety characteristics, have on yield. Given the wide range of conditions found across temperate North America, further research will be necessary to determine the proper spacing and seeding rates for new locations.

Planting depth for quinoa is also variable depending on soil type and conditions. Oelke et al. (1992) recommended planting at a depth of 1.3 to 2.5 cm (0.5 to 1 in) in Colorado. Excessively deep sowings were noted to lead to poor emergence due to waterlogging, whereas planting depths that were too shallow left seeds vulnerable to soil drying. In Europe, high germination required adequate contact of seeds with soil with the proper moisture level (Darwinkel and Stølen 1997). Aufhammer et al. (1994) found that soil crusting had a highly damaging effect on seedling emergence. They stressed the importance of a light-textured seedbed for successful germination. Quinoa germination rates were reported to drop steeply at depths greater than 2 cm (0.8 in). At shallower depths, greater emergence was observed in a sandy soil than a loamy clay soil.

Reports from Europe confirm the need for a fine-textured and even seedbed and likewise indicate that heavy clay soils can cause problems with early establishment of the root system. Planting depths of 1 to 2 cm (0.4 to 0.8 in) were recommended (Darwinkel and Stølen 1997).

Temperature has a significant effect on the percentage and rate of germination but there are varying reports on the optimum temperature needed for successful germination. In Denmark, seed is usually planted when the soil temperature is 7–8°C (45–46°F). An experiment with a Danish variety of Chilean lowland origin showed maximum germination between 15 and 20°C (59–68°F). However, high rates of germination were seen for a wide range of temperatures, including 8°C (46°F), the lowest temperature tested (Jacobsen and Bach 1998). Recommended soil temperatures for planting in Colorado were 7–10°C (45–50°F) (Johnson and Croissant 1985). For England, reports indicated that soil temperatures of 5–8°C (41–46°F) were suitable for planting quinoa in the spring (Galwey 1989). Darwinkel and Stølen (1997) recommended soil temperatures exceeding 10°C (50°F) for planting and note that lower temperatures inhibit proper germination and establishment.

In a germination experiment with the Danish variety "Olav," temperatures of 6°C (43°F) inhibited germination, resulting in only 25% of the germination found at 20°C (68°F) (Jacobsen et al. 1999a). The study also showed that the date of harvest was a significant factor for germination at low temperatures. For seed that had been harvested early, there was 0% germination. The germination percentage increased to a maximum of 45% with a later time of harvest.

Christiansen et al. (1999) investigated five Peruvian quinoa lines for their ability to germinate at low temperatures. In contrast to the results reported by Jacobsen, high germination was seen at temperatures as low as 2°C (36°F), indicating that variation exists in quinoa for minimum germination temperature.

For some areas in North America, sufficiently high soil temperatures may be the determining factor for planting date. The temperature germination study by Jacobsen et al. (1999a) investigated only one variety of Chilean lowland origin. Further screening of a wide range of Chilean lowland varieties for their ability to germinate at lower temperatures could identify varieties suited for earlier planting. If this trait is not identified in the germplasm screened, it could potentially be brought in from non-Chilean varieties such as those identified by Christiansen et al. (1999). Earlier planting could mean earlier establishment, enhanced competitiveness with weeds, and earlier harvest time. Germination under low temperatures would also need to be combined with seedling vigor and proper establishment under such conditions. For instance, early seeding under northwestern European conditions showed not only poor germination but also poor emergence and, subsequently, poor growth (Darwinkel and Stølen 1997).

Maturity and harvesting

Days-to-maturity is a critical yet highly variable factor for the successful cultivation of quinoa. In Colorado, days to maturity range from 90 to 125 days (Johnson and Croissant 1985), whereas days to maturity have ranged from 100 to 130 days for different varieties grown in Eastern Washington. Maturity has been observed to vary significantly depending on environmental variation in the field (unpublished data). In Greece, maturation time was shown to vary from 110 to 116 days compared to the 110 to 180 days reported for Denmark (Jacobsen 2003). Within the same location, maturity time has been shown to vary widely from year to year (Gesinksi 2000; Jacobsen 1998). The wide range of maturity time by location and environmental conditions emphasizes the importance of testing quinoa in new locations over multiple years.

Quinoa can be mechanically harvested through the use of a combine. However, adjustments are needed for the small seed size and large stems. In Colorado, quinoa harvest recommendations are similar to sorghum. Combining does not result in a totally clean product and further processing is necessary after combining. Recommendations include the use of a fanning mill and gravity separator (Oelke et al. 1992). Darwinkel and Stølen (1997) also confirm the necessity of cleaning after combining, although they note that the percentage of chaff can be kept under 5% if done under optimum field conditions with a properly adjusted combine.

Jacobsen et al. (1994) compared the efficacy of swathing compared to combining and found no significant differences in yield between the two methods when performed at their respective optimal harvest time. Swathing did allow for harvest earlier in the season than combining, when the plants were still somewhat green. This could be advantageous in situations where early harvest is important, particularly during the stage when mature seed has formed but plants have not fully senesced and cannot be combined. This has been an issue for quinoa grown in western Oregon and western Washington state, where early onset of seasonal rains in late summer and early autumn can cause preharvest sprouting.

In Colorado, steady yields of 1,340 kg/ha (1,200 lbs/acre) were reported over a 3-year period (Johnson and Croissant 1990). With proper management, however, this can increase to over 2,000 kg/ha (1,800 lbs/acre) (Oelke et al. 1992). In Saskatchewan, yield is reported to range from 840 to 1,400 kg/ha (750 to 1,250 lbs/acre), although yields exceeding 2,200 kg/ha (2,000 lbs/acre) have been reported. On the other hand, complete losses can also occur (AAFRD 2005).

Quinoa yields reported in Europe exceed those of Colorado. Yields in Denmark typically range from 2,000 to 3,000 kg/ha (1,800 to 2,700 lbs/ac) (Jacobsen et al. 2010). Under two field experiments in England, the variety "Baer" yielded 5,140 kg/ha (4,590 lbs/ac) (Risi and Galwey 1991b). In Southern Italy, the variety "Regalona Baer" yielded 3,420 kg/ha (3,050 lbs/ac) and 3,000 kg/ha (2,680 lbs/ac) over 2 years (Pulvento et al. 2010). In Chile, "Baer" is reported to produce yields of 3,000 kg/ha (2,680 lbs/ac) under field conditions and 6,500 kg/ha (5,800 lbs/ac) under experimental conditions (Delatorre-Herrera 2003).

This wide variation in yield indicates the large effect of location and variety on the performance of quinoa. At present, yield data in temperate North America are available from only two locations. Further variety testing will help determine yield potential for quinoa in new areas.

CHALLENGES TO QUINOA PRODUCTION

As a relatively new crop in North America, quinoa faces a diverse range of challenges as it continues to be tested across a range of growing environments across North America. The most limiting of these challenges is high summer temperatures, which greatly restricts the range of quinoa cultivation across the continent to areas with mild summer temperatures.

In areas where quinoa seed set is not threatened by high temperatures, other major challenges, both abiotic and biotic in nature, exist and pose potential threats to successful quinoa cultivation.

Waterlogging and preharvest sprouting

The native range of quinoa cultivation includes areas characterized by high rainfall. For instance, a quinoa variety was collected from a location on Chiloé Island in Chile (Wilson 1978), an island characterized by high levels of annual precipitation (2,500–3,000 mm or 100–120 in) (Vera 2006). However, when rainfall coincides with seed maturity, preharvest sprouting can occur and cause harvest loss among susceptible varieties. Preharvest sprouting was observed in quinoa trials held in 2010 in Olympia, Washington. Here, a rare heavy rainfall event in late summer caused substantial sprouting in many of the 44 varieties tested. However, a few accessions proved more resistant to sprouting, including PI 614880 (unpublished data 2010), which had been collected from Chiloé Island. This accession has been shown to exhibit seed dormancy that confers preharvest sprouting resistance (Ceccato et al. 2011). Variety trials in 2013 at two locations in western Washington state were also affected by rains in early September, causing sprouting in many varieties (Fig. 10.1) and also revealing a spectrum of preharvest sprouting resistance among the range of varieties tested.

Preharvest sprouting was also a challenge for quinoa breeding efforts in the Netherlands (Mastebroek and Limburg 1997). Quinoa varieties in the quinoa program at CPRO-DLO in Wageningen were successfully screened for preharvest sprouting resistance by testing the dormancy of seeds relative to ripening. Variability was found within the accessions and selection for tolerance to preharvest sprouting was successfully achieved in the CPRO-DLO breeding program.

In a greenhouse experiment, González et al. (2009) noted a negative impact from waterlogging, which reduced total plant dry weight and leaf area more severely than that of drought stress. Unnecessary irrigation of seedlings can cause stunting and damping off, and irrigation of quinoa during later growth stages is reported to increase vegetative growth without a corresponding boost in

Fig. 10.1 Preharvest sprouting.

seed production (Oelke et al. 1992). Waterlogging in the germination phase had a deleterious impact on quinoa in a trial in the United Kingdom; varietal differences in tolerance to this stress were observed (Risi and Galwey 1989).

Disease

Owing to its geographical isolation, quinoa in North America has escaped many of the disease pressures common to quinoa in its native range. However, at least one major pathogen of quinoa, *Peronospora variabilis*, which causes downy mildew, has been reported in quinoa grown in Canada (Tewari and Boyetchko 1990). Downy mildew has also been observed in field trials in Washington state (unpublished data, 2012) and in Pennsylvania (Testen et al. 2012). Testen et al. (2014) identified widespread infection of North American quinoa. Molecular screening confirmed *P. variabilis* infection in samples of commercially produced quinoa from Canada, Oregon, and Colorado. The strains of *P. variabilis* infecting North American quinoa were also found to be genetically distinct from *P. variabilis* strains isolated from South America quinoa.

The downy mildew pathogen infecting quinoa has been shown to be closely related to strains that infect *Chenopodium album*, a common weed in North America (Choi et al. 2010). The authors suggest the possibility that *C. album* may be a potential inoculum reservoir for downy mildew in quinoa. Using PCR-based methods, Testen et al. (2014) detected *P. variabilis* from a weedy *C. album* and an unidentified *Chenopodium* sp. collected in Ecuador. More investigation is needed on the potential of *C. album* and other closely related *Chenopodium* spp. to host strains of *P. variabilis* to which *C. quinoa* is susceptible. As Choi et al. (2010) demonstrated, downy mildew exhibits a high level of host specificity. No threat is posed from strains of downy mildew that infect related crops such as spinach and beets, as these strains belong to *Peronospora farinosa*, a separate species from *P. variabilis*.

Variation for resistance to downy mildew exists within the quinoa germplasm. Chilean lowland varieties have been reported to carry more resistance to the pathogen than other ecotypes (Fuentes et al. 2008). However, reports from Denmark of *P. variabilis* infecting Dutch and Danish varieties indicate that problematic levels of susceptibility to *P. variabilis* do exist among varieties with a Chilean lowland background (Danielsen et al. 2002).

The close relative *C. berlandieri* has been noted to have high resistance to downy mildew (Jellen et al. 2011). The domesticate *C. berlandieri* subsp. *nuttalliae* has been crossed with quinoa with the goal of introgressing downy mildew resistance (Bonifacio 2004). In a study screening downy mildew resistance among quinoa varieties and *Chenopodium* spp. under field conditions in India, accessions of *C. berlandieri* subsp. *nuttalliae* and the North American species *C. bushianum* were reported to be immune to downy mildew (Kumar et al. 2006). The authors also observed diverse responses, ranging from lack of disease to high susceptibility, from a wide range of quinoa varieties, leading the authors to postulate the existence of multiple pathotypes. Reports from breeding efforts in the CPRO-DLO program indicated that downy mildew resistance was dominantly inherited (Mastebroek and van Loo 2000).

The reported diversity for downy mildew resistance should allow for the breeding of resistant varieties. As *C. berlandieri* subsp. *nuttalliae* and *C. bushianum* have both been successfully crossed with quinoa, these species represent another source for downy mildew resistance (Wilson 1980).

Danielsen and Munk (2004) developed a standardized method, which they called the "three-leaf technique," to determine the severity of downy mildew infection in terms of yield reduction. They also noted the preference of downy mildew fungi for cool, moist conditions. The Salares-type variety "Utusaya" was found to be highly susceptible to downy mildew. However, in its usual range of cultivation, "Utusaya" escapes infection because of dry climate (Danielsen and Munk 2004). Escape from downy mildew infection should be possible in western regions of the United States characterized by dry summers. This is supported by reports of reduced infection in Europe under dry summer conditions as compared to more humid summers (Jacobsen 1999). In semi-arid conditions of eastern Washington state, downy mildew has been seen on young plants after the early spring rains, but infection stopped with seasonal summer drought and appears to be a minor issue. Areas characterized by wetter conditions and high humidity throughout the growing season may find downy mildew to be a more significant problem. As downy mildew in quinoa has been shown to be transmissible by seed due to oospores that form in the seed pericarp, care must be taken when introducing quinoa in new areas to prevent introducing the pathogen as well (Danielsen et al. 2004).

Additional *Peronospora* spp. may threaten quinoa. Testen et al. (2014) report infection of Ecuadorian quinoa from an undescribed *Peronospora* sp., which produce unique symptoms in plants characteristic of a systemic infection.

Other fungal pathogens have also been reported to infect quinoa in North America. *Sclerotium rolfsii* causes damping-off of quinoa seedlings, as well as seed rot in a fall quinoa variety trial in Southern California (Beckman 1980). More recently, two new pathogens, *Passalora dubia* and an *Ascochyta* sp., have been found on quinoa in Pennsylvania (Testen et al. 2013a, 2013b). The potential impact of these pathogens for quinoa cultivation is currently unknown, and no reports exist of these pathogens outside of research contexts.

Insect pests

Many insect pests have been reported for quinoa in its native range in South America, the most damaging of these being the quinoa moth (*Eurysacca melanocampta* and *Eurysacca quinoae*) (Rasmussen et al. 2003). The potential for transmission of pests from South America to North America is currently unknown. This deserves further investigation, particularly as some farmers trying to grow quinoa in North America plant quinoa seed that have been sold as food. This presents a possible route for the introduction of seed-borne pests of quinoa, and also the potential spread of these pests to closely related *Chenopodium* spp.

Existing reports of insect pest pressures in quinoa from Colorado and South America indicate little overlap in species (Cranshaw et al. 1990; Rasmussen et al. 2003). Although two insect species (*Macrosiphum euphorbiae* and *Helicoverpa zea*) reported in Peru and Bolivia were also found in fields in Colorado, these species originate from North America. Subsequent pest identifications from Washington and Maine (I. Milosavljevic, personal communication, 2013; Conant 2002) have not revealed pests in common with those reported by Rasmussen et al. (2003). Thus, existing evidence suggests that transmission of insect pests from South America to temperate locations in North America has not occurred.

As quinoa continues to be grown in North America, it is likely that native pest species will extend their host range to include quinoa. This has occurred in Europe, where two pests that normally feed on *C. album*, *Scrobipalpa atriplicella* and *Cassida nebulosa*, have begun to attack quinoa (Sigsgaard et al. 2008).

Oelke et al. (1992) reports that insect pressure was not determined to be a significant factor for quinoa grown in Colorado. However, a study of pest pressures in Colorado found a wide range of insect pests on quinoa several years after its introduction. Many of the reported pests were also pests of *C. album* and *Beta vulgaris*. The major insect pests attacking seedlings were *Melanotrichus coagulatus* (Uhler) and the false cinch bug, *Nysius raphanus* Howard. Beet armyworm, *Spodoptera exigua* (Hübner), caused large-scale defoliation at one location near Crestone, Colorado. Another damaging foliar feeder is the boat gall aphid *Hayhurstia atriplicis* (L.). *Lygus* spp. were noted as injurious seed feeding pests, whereas the sugarbeet root aphid (*Pemphigus populivenae* Fitch) caused yield declines (Cranshaw et al. 1990). Oelke et al. (1992) recorded sugarbeet root aphid as a significant quinoa pest whose points of entry are cracks in the soil. In addition, flea beetles and aphids were reported on quinoa grown in Minnesota (Robinson 1986). During a quinoa trial in Maine, a plutellid moth (species unidentified) was found to attack plants (Conant 2002).

Darwinkel and Stølen (1997) noted some overlap in the pests of beets and quinoa in Europe. Beet flea beetles (*Chaetocnema concinna* and *Chaetocnema tibialis*) and beet carrion beetles (*Aclypea opaca*) were found to feed on quinoa grown in sandy soil, though no data is given on the relative seriousness of these pests. Black bean beetle (*Aphis fabae*) was also reported as a pest likely causing decreases in yield.

In Washington state, aphids and *Lygus* sp. have been the major pests at test plots throughout the state since trials were initiated in 2010. Harvesting aphid-infested plants results in quinoa seed mixed with honeydew particles, which are considerably difficult to remove from harvested quinoa seed. *Lygus* sp. has been observed to take shelter in more compact inflorescences (unpublished data, 2010). In 2013, buckthorn aphid (*Aphis nasturtii*), yellow striped armyworm (*Spodoptera ornithogalli*), and *Crambus* sp. were identified as pests in the field (I. Milosavljevic, personal communication 2013).

Various control methods for quinoa pests in South America have been recommended, although much research remains to be done to develop appropriate IPM strategies. Current pest control strategies that are most transferable to North American conditions are crop rotation, intercropping, and the use of biocontrol agents. Varietal resistance for pests such as quinoa moth has been observed (Rasmussen et al. 2003).

As pest pressures change and develop with expanding quinoa production in North America, strategies for pest control will likewise change. Given the current pest pressures reported, some recommendations can be made. In Colorado, Cranshaw et al. (1990) noted the syrphid larvae (*Diaeretiella rapae*) and the convergent lady bird beetle (*Hippodamia convergens*) as predators of the Chenopod aphid, *Hayhurstia atriplicis*. Variations in aphid pressures between sites were also observed. The aforementioned species and other aphid parasitoids should be investigated as potential biocontrol agents. Furthermore, the factors that determine the severity of aphid infestation require further study.

As quinoa and other *Chenopodium* species share common pests, studying the potential of weedy *Chenopodium* sp. as alternate hosts is warranted. Close rotations of quinoa with related crops such as beets should be carefully monitored to ensure that pests common to both crops do not spread.

Weed control

The slow growth of quinoa after germination makes weed management particularly challenging. Related *Chenopodium* spp. often pose a major weed pressure, due to their similarities to quinoa in early growth habit and appearance. Common lambsquarter was also noted as a major weed pressure in southern Colorado, along with pigweed, kochia, and sunflower (Oelke et al. 1992). In Colorado, weed management recommendations for growers included early sowing, planting in areas of low weed pressure, and cultivation (Johnson and Croissant 1985). Tilling the soil for planting after irrigating was found to be effective, particularly with controlling *Chenopodium* spp. (Oelke et al. 1992). Late planting of quinoa in England led to complete losses due to heavy weed pressure (Risi and Galwey 1991a), confirming the importance of early establishment of quinoa.

Grasses were found to cause major yield reductions in quinoa. Johnson and Ward (1993) saw reductions in yield from 1,822 kg/ha (1,626 lbs/ac) to 640 kg/ha (571 lbs/ac) due to competition from grassy weeds.

Jacobsen et al. (2010) compared the efficacy of inter-row hoeing and harrowing for weed control under Danish conditions. They found that inter-row hoeing, when combined with a row spacing of 50 cm, was most effective at reducing weed pressure. Harrowing had a significant but less pronounced impact. Although both methods had a significant decrease on quinoa stands during one of the 2 years the study was conducted, yield significantly increased with both methods. *Chenopodium album* was the predominant weed species, indicating that these methods were effective in controlling weedy *Chenopodium* spp.

A study by Risi and Galwey (1991b) in which 10 quinoa varieties were grown in England showed distinct differences in the negative impact of weed pressure on plant height. Their observations indicated that some quinoa varieties might have greater ability to compete with weeds.

Cross-pollination of *C. quinoa* with weedy *C. berlandieri* has been confirmed at a farm in northern Washington state (Wilson and Manhart 1993). Oelke et al. (1992) reported putative hybrids between quinoa and common lambsquarter in Colorado (species unspecified). Such pollination could have consequences in terms of maintaining purity of quinoa varieties. Control of *C. berlandieri* will prove critical in preventing unwanted cross-pollination.

Saponins

Many varieties of quinoa contain saponins in the seed pericarp. Saponins give quinoa a bitter taste and make it unpalatable. In addition, saponins have potential negative health effects and have been shown to be disruptive to intestinal membranes in rats (Gee et al. 1993). For quinoa to be marketable, saponins must be removed after harvest. This is generally done through abrasion or by washing the saponin from the seed (Johnson and Ward 1993). Alternatively, brushing has been used as a saponin removal method (Darwinkel and Stølen 1997). Removal of the saponin through abrasion can result in loss of minerals such as calcium (Konishi et al. 2004).

While industrial uses for quinoa saponin have been proposed (Jacobsen 2003), the presence of saponins currently poses postharvest difficulties rather than an economic opportunity. The infrastructure required for removal of saponins remains a challenge for small-scale quinoa growers. Commercial production of quinoa in Colorado only began once machinery to remove saponin was obtained (Johnson 1990).

Saponin-free varieties of quinoa do exist, though they are not found in the original Chilean lowland germplasm. The CPRO-DLO quinoa breeding program successfully introgressed the saponin-free trait into improved varieties (Mastebroek and Marvin 1999). The presence of saponin is both qualitatively and quantitatively controlled. Its presence is governed by a single gene with two alleles. Saponin production is dominant, with homozygous recessive plants producing seeds lacking saponin (Ward 2001). When the dominant phenotype is expressed, saponin content is then governed quantitatively. Efforts to breed low saponin varieties quantitatively have not been successful, due to lack of sufficient response to selection (Ward 2000). Future efforts to develop saponin-free varieties will likely rely on generating plants that are homozygous recessive for the qualitative trait. However, this comes with the potential drawback of saponin production being restored through cross-pollination from other varieties that contain saponin. Proper separation of varieties lacking saponin from those with saponin will be necessary to prevent cross-pollination.

While the removal of saponin remains an obstacle for quinoa production, saponin-free varieties have drawbacks. Reports from South America and Europe indicate that saponin-free varieties can suffer yield losses, some quite severe, due to feeding by birds (Risi and Galwey 1991b; Rasmussen et al. 2003; Darwinkel and Stølen 1997). However, saponin-containing varieties may also suffer yield losses due to birds, as rain has been reported to wash away saponins (Oelke et al. 1992). This was observed at a trial site in Chimacum, Washington, where feeding damage was observed following rains in late summer (unpublished data, 2013)

Drought and salinity can have significant effects on saponin content. Irrigation with saline water resulted in a 30% increase in saponin content compared to freshwater irrigation. At the lowest irrigation level, saponin content decreased 42% in comparison to the full irrigation control (Gómez-Caravaca et al. 2012).

ALTERNATIVE USES OF QUINOA

Forage

In addition to the worldwide interest in quinoa for its seed, quinoa has also attracted attention as a forage crop. Carlsson et al. (1984) investigated quinoa in southern Sweden and concluded that it was a promising crop for the production of green liquid protein concentrate. Dry matter production was found to increase with high levels of fertilization, up to 470 kg N/ha (420 lbs N/ac).

Quinoa biomass had high crude protein and low fiber when harvested near the flowering period and was competitive with alfalfa and grass when compared on a cost basis. Quinoa silage was found to be an acceptable method of storing quinoa fodder long term. Increased nitrogen applications were recommended for forage quinoa, with 100 kg N/ha (90 lbs N/ac) applied at sowing followed by an additional 200 kg N/ha (180 lbs N/ha) applied 5 weeks later (Darwinkel and Stølen 1997). In contrast, tests in the Netherlands showed that quinoa performed poorly as a forage crop in comparison with grass and clover (van Schooten and Pinxterhuis 2003).

Feed

The balance of amino acids in quinoa seed has generated interest in its utility as an animal feed. However, the value of quinoa for human consumption currently exceeds its value as animal feed. This would likely restrict any role of quinoa as a feed to low-quality quinoa and to organic farmers looking for a high-quality, protein-rich alternative to maize–soy mixes.

Carlson et al. (2012) tested quinoa hull meal, which contains the saponin-rich pericarp, as a feed additive for pigs. South American quinoa hull meal was supplied at 100, 300, and 500 mg/kg and Danish-grown quinoa at 300 mg/kg; feed intake and utilization, as well as growth rate, were all unaffected compared to those of pigs fed a control diet lacking quinoa hull meal, despite the effects of South American quinoa hull meal on intestinal epithelial physiology detected *ex vivo*.

Several studies have been conducted on the use of quinoa as poultry feed but with mixed results. Improta and Kellems (2001) tested diets based on raw, polished, and washed quinoa for broiler feed. The survival and growth rates of chicks fed a diet of raw quinoa were severely reduced, and this effect was attributed to antinutritional factors (Improta and Kellems 2001). With the polished quinoa diet, this effect was much reduced, whereas the effect of the washed quinoa diet was comparable to that of the maize/soybean meal control diet. Improved performance was also seen in chicks fed diets with higher crude protein percentage, although this may also be due to the lower percentage of quinoa provided with increased supplementation of soybean meal.

Jacobsen et al. (1997) found that increasing levels of both unprocessed and dehulled quinoa in a mixed broiler diet caused depression in broiler growth. In contrast to the previous study, the chicks had poor growth when dehulled quinoa was included in the diet. The authors suggested that aside from the bitter saponins present in the hulls, other antinutritional factors were also responsible for this effect on the chicks. They recommended that quinoa comprise no more than 150 g/kg (15% by weight) of broiler diets. In a study investigating the combination of two-layer diets with a supplemental forage rotation, a 34-day-long quinoa forage period consisting of two quinoa varieties, one of which was saponin-free, was included with no negative effects on layers reported (Horsted and Hermansen 2007).

CONCLUSION

Expanded production of quinoa in temperate areas of North America poses many challenges and opportunities. Despite the main obstacle of heat susceptibility, breeding efforts utilizing within-species diversity may result in varieties with improved heat tolerance. Should that prove insufficient, the North American *Chenopodium* spp. within quinoa's secondary gene pool may prove to be a promising source for introgression of heat tolerance genes. Existing genetic diversity has been identified that should enable farmers and scientists to overcome challenges posed by downy mildew, saponins, and preharvest sprouting. Many aspects of quinoa cultivation in North America remain unknown, particularly disease and pest pressures, because of quinoa's relative novelty to North America and its geographical isolation from pest pressures found in its native range of cultivation. The high level of abiotic stress tolerance in quinoa may give it a distinct advantage in areas with marginal agricultural lands, specifically those affected by soil salinity. Interest and demand for domestically produced quinoa remains high, and with further research and plant breeding efforts, there is considerable potential for expanded quinoa production in North America.

ACKNOWLEDGMENTS

Special thanks go to Ivan Milosavljevic for insect identification and to Hannah Walters for providing preliminary findings. We would also like to acknowledge the dedication of the diverse group of individuals who have helped lay the foundational work for quinoa as a crop in North America. Last but not least, we would like to express our sincere gratitude and respect for the many indigenous peoples of South America who, living in close relationship to quinoa, have grown and shaped this dynamic crop over millennia.

REFERENCES

Adolf VI, Shabala S, Andersen MN, Razzaghi F, Jacobsen SE. Varietal differences of quinoa's tolerance to saline conditions. Plant Soil 2012;357(1–2):1–13.

Aguilar TN. 1968. Growth responses of *Chenopodium quinoa* Willd. to changes in temperature and oxygen tension. Master's Thesis, University of Illinois, Chicago, Illinois, pp. 1–51.

Aguilar PC, Jacobsen SE. Cultivation of quinoa on the Peruvian Altiplano. Food Rev Int 2003;19(1–2):31–41.

[AAFRD] Alberta Agriculture, Food and Rural Development. Quinoa: the next cinderella crop for Alberta? Edmonton, AB: Alberta Agriculture, Food and Rural Development; 2005.

[ARD] Alberta - Agriculture and Rural Development. [Internet]. 2004. Salinity classification, mapping and management in Alberta. Edmonton, AB: Alberta Agriculture and Rural Development (cited 2013 Jan 15). Available from http://www1.agric.gov.ab.ca/$department/deptdocs.nsf/all/sag3267#prairies

Arco de Nuñez S. 2014. Note: I don't know the details of the current book to fill in the rest of the citation. Could you please complete this citation for me? Thanks!

Aufhammer W, Kaul H, Kruse M, Lee J, Schwesig D. Effects of sowing depth and soil conditions on seedling emergence of amaranth and quinoa. Eur J Agron 1994;3(3):205–210.

Ballón E. Low-input agriculture in the Southwestern United States. Fort Collins, CO: Rocky Mountain Forest and Range Experiment Station. General Technical Report RM-GTR-198. pp.; 1990. p 116–119.

Beckman PM. Seed rot and damping-off of *Chenopodium quinoa* caused by *Sclerotium rolfsii*. Plant Dis 1980;64(5):497.

Benlhabib O, Atifi M, Jellen EN, Jacobsen S. The introduction of a new Peruvian crop "quinoa" to a rural community in Morocco. In: Jacobsen SE, Jensen CR, Porter JR, editors. VIII ESA congress: European agriculture in a global context; 2004 July 11–15. Copenhagen, Denmark: Samfundslitteratur; 2004. p 881–884.

Bertero HD. Response of developmental processes to temperature and photoperiod in quinoa (*Chenopodium quinoa* Willd.). Food Rev Int 2003;19(1–2):87–97.

Bertero HD, King RW, Hall AJ. Photoperiod-sensitive development phases in quinoa (*Chenopodium quinoa* Willd.). Field Crop Res 1999;60(3):231–243.

Bhardwaj HL, Hankins A, Mebrahtu T, Mullins J, Rangappa M, Abaye O, Welbaum GE. Alternative crops research in Virginia. In: Janick J, editor. Progress in new crops. Alexandria, VA: ASHS Press; 1996. p 87–96.

Bilalis D, Kakabouki I, Karkanis A, Travlos I, Triantafyllidis V, Hela D. Seed and saponin production of organic quinoa (*Chenopodium quinoa* Willd.) for different tillage and fertilization. Not Bot Horti Agrobo 2012;40(1):42–46.

Bonifacio A. 1995. Interspecific and intergeneric hybridization in Chenopod species. Master's Thesis, Brigham Young University, Provo, Utah.

Bonifacio A. 2004. Genetic variation in cultivated and wild Chenopodium species for quinoa breeding. Ph.D. Dissertation, Brigham Young University, Provo, Utah.

Carlson D, Fernandez JA, Poulsen HD, Nielsen B, Jacobsen SE. Effects of quinoa hull meal on piglet performance and intestinal epithelial physiology. J Anim Physiol An N 2012;96(2):198–205.

Carlsson R, Hanczakowski P, Kaptur T. The quality of the green fraction of leaf protein concentrate from *Chenopodium quinoa* Willd. grown at different levels of fertilizer nitrogen. Anim Feed Sci Tech 1984;11(4):239–245.

Carmen ML. Acclimatization of quinoa (*Chenopodium quinoa* Willd.) and canihua (*Chenopodium pallidicaule* Allen) to Finland. Ann Agr Fenn 1984;23:135–144.

Ceccato DV, Bertero DH, Batlla D. Environmental control of dormancy in quinoa (*Chenopodium quinoa*) seeds: two potential genetic resources for pre-harvest sprouting tolerance. Seed Sci Res 2011;21(2):133.

Choi YJ, Danielsen S, Lübeck M, Hong SB, Delhey R, Shin HD. Morphological and molecular characterization of the causal agent of downy mildew on quinoa (*Chenopodium quinoa*). Mycopathologia 2010;169(5):403–412.

Christiansen JL, Jacobsen SE, Jørgensen ST. Photoperiodic effect on flowering and seed development in quinoa (*Chenopodium quinoa* Willd.). Acta Agr Scand B-S P 2010;60(6):539–544.

Christiansen JL, Ruiz-Tapia EN, Jornsgard B, Jacobsen SE. 1999. Fast seed germination of quinoa (*Chenopodium quinoa*) at low temperature. In: Mela T, Christiansen J, Kontturi M,

Pahkala K, Partala A, Sahramaa M, Sankari H, Topi-Hulmi M, Pithan K. Proceedings of COST 814-workshop: alternative crops for sustainable agriculture; 1999 June 13–15; Turku, Finland. Luxembourg: Office of Official Publications of the European Communities. 220–225.

Climatología [Internet]. date unknown. Santiago, Chile: Dirección Meteorológica de Chile (cited 2013 Mar 23). Available from http://164.77.222.61/climatologia/php/menuAnuarios.php

Conant N. Quinoa introduction in the River Valley, farmer/grower grant final report. College Park, MD: SARE; 2002. ; Project FNE02-406.

Cranshaw WS, Kondratieff BC, Qian T. Insects associated with quinoa, *Chenopodium quinoa*, in Colorado. J Kans Entomol Soc 1990;63(1):195–199.

Danielsen S, Jacobsen SE, Hockenhull J. First report of downy mildew of quinoa caused by *Peronospora farinosa* f. sp. *chenopodii* in Denmark. Plant Dis 2002;86:1175.

Danielsen S, Mercado VH, Ames T, Munk L. Seed transmission of downy mildew (*Peronospora farinosa* f.sp. *chenopodii*) in quinoa and effect of relative humidity on seedling infection. Seed Sci Technol 2004;32(1):91–98.

Danielsen S, Munk L. Evaluation of disease assessment methods in quinoa for their ability to predict yield loss caused by downy mildew. Crop Prot 2004;23(3):219–228.

Darwinkel A, Stølen O. Understanding the quinoa crop: guidelines for growing in temperate regions of N.W. Europe. Brussels: European Commission; 1997.

Delatorre-Herrera J. Current use of quinoa in Chile. Food Rev Int 2003;19(1–2):155–165.

Delatorre-Herrera J, Pinto M. Importance of ionic and osmotic components of salt stress on the germination of four quinua (*Chenopodium quinoa* Willd.) selections. Chil J Agric Res 2009;69(4):477–485.

Dyck E. 2012. Quinoa trial for Northeast upland farms [Internet]. Report No.: FNE12-760. SARE, College Park, MD. Available from: http://mysare.sare.org/mySARE/ProjectReport.aspx?do=viewRept&pn=FNE12-760&y=2012&t=0

Flynn RP. 1990. Growth characteristics of quinoa (*Chenopodium quinoa* Willd.) and yield response to increasing soil water deficit. Master's Thesis, Colorado State University, Fort Collins, Colorado, pp. 69p.

Fuentes F, Bhargava A. Morphological analysis of quinoa germplasm grown under lowland desert conditions. J Agron Crop Sci 2011;197(2):124–134.

Fuentes FF, Martinez EA, Hinrichsen PV, Jellen EN, Maughan PJ. Assessment of genetic diversity patterns in Chilean quinoa (*Chenopodium quinoa* Willd.) germplasm using multiplex fluorescent microsatellite markers. Conserv Genet 2008;10(2):369–377.

Galwey NW. Exploited plants - quinoa. Biologist 1989;36(5): 267–274.

Gandarillas H. 1982. Quinoa production. IBTA-CIID (Trans Sierra-Blanca Assoc, Denver, CO. 1985).

Gee JM, Price KR, Ridout CL, Wortley GM, Hurrell RF, Johnson IT. Saponins of quinoa (*Chenopodium quinoa*): effects of processing on their abundance in quinoa products and their biological effects on intestinal mucosal tissue. J Sci Food Agric 1993;63(2):201–209.

Geerts S, Raes D, Garcia M, Vacher J, Mamani R, Mendoza J, Huanca R, Morales B, Miranda R, Cusicanqui J, Taboada C. Introducing deficit irrigation to stabilize yields of quinoa (*Chenopodium quinoa* Willd.). Eur J Agron 2008;28(3):427–436.

Gesinksi K. Potential for *Chenopodium quinoa* acclimatisation in Poland. In: Parente G, Frame J, editors. Abstracts/Proceedings of COST 814 conference, crop development for cool and wet regions of Europe; 2000 May 10–13; Pordenone, Italy. Luxembourg: Office of Official Publications of the European Communities; 2000. p 547–552.

Gómez-Caravaca AM, Iafelice G, Lavini A, Pulvento C, Caboni MF, Marconi E. Phenolic compounds and saponins in quinoa samples (*Chenopodium quinoa* Willd.) grown under different saline and nonsaline irrigation regimens. J Agric Food Chem 2012;60(18):4620–4627.

Gómez-Pando LR, Álvarez-Castro R, Eguiluz-de la Barra A. SHORT COMMUNICATION: effect of salt stress on Peruvian germplasm of *Chenopodium quinoa* Willd.: a promising crop. J Agron Crop Sci 2010;196(5):391–396.

González JA, Gallardo M, Hilal M, Rosa M, Prado FE. Physiological responses of quinoa (*Chenopodium quinoa* Willd.) to drought and waterlogging stresses: dry matter partitioning. Bot Stud 2009;50(1):35–42.

Hariadi Y, Marandon K, Tian Y, Jacobsen SE, Shabala S. Ionic and osmotic relations in quinoa (*Chenopodium quinoa* Willd.) plants grown at various salinity levels. J Exp Bot 2011;62(1):185–193.

Horsted K, Hermansen JE. Whole wheat versus mixed layer diet as supplementary feed to layers foraging a sequence of different forage crops. Animal 2007;1:575–585.

Iliadis C, Karyotis T, Jacobsen SE. Adaptation of quinoa under xerothermic conditions and cultivation for biomass and fibre production. In: Jacobsen SE, Portillo Z, editors. International Potato Center (CIP) memorias, primer taller internacional sobre quinua – recursos geneticos y sistemas de producción; 1999 May 10–14. Lima: UNALM; 2001. p 371–378.

Improta F, Kellems RO. Comparison of raw, washed and polished quinoa (*Chenopodium quinoa* Willd.) to wheat, sorghum or maize based diets on growth and survival of broiler chicks. Livest Res Rural Dev 2001;13:1–10.

Jacobsen SE. Adaptation of quinoa (*Chenopodium quinoa*) to Northern European agriculture: studies on developmental pattern. Euphytica 1997;96(1):41–48.

Jacobsen SE. Developmental stability of quinoa under European conditions. Ind Crops Prod 1998;7:169–174.

Jacobsen SE. Potential for quinoa (*Chenopodium quinoa* Willd.) for cool and wet regions of Europe. In: Mela T, Christiansen J, Kontturi M, Pahkala K, Partala A, Sahramaa M, Sankari H, Topi-Hulmi M, Pithan K, editors. Proceedings of COST

814-workshop: alternative crops for sustainable agriculture; 1999 June 13–15; Turku, Finland. Luxembourg: Office of Official Publications of the European Communities; 1999. p 87–99.

Jacobsen SE. The worldwide potential for quinoa (*Chenopodium quinoa* Willd.). Food Rev Int 2003;19(1–2):167–177.

Jacobsen SE. Quinoa's world potential. In: Ochatt S, Jain SM, editors. Breeding of neglected and under-utilized crops, spices and herbs. Enfield, New Hampshire: Science Publishers; 2007. p 109–122.

Jacobsen SE. The situation for quinoa and its production in southern Bolivia: from economic success to environmental disaster. J Agron Crop Sci 2011;197(5):390–399.

Jacobsen SE, Bach AP. The influence of temperature on seed germination rate in quinoa (*Chenopodium quinoa* Willd.). Seed Sci Technol 1998;26(2):515–524.

Jacobsen SE, Christiansen JL, Rasmussen J. Weed harrowing and inter-row hoeing in organic grown quinoa (*Chenopodium quinoa* Willd.). Outlook Agr 2010;39(3):223–227.

Jacobsen SE, Jørgensen I, Stølen O. Cultivation of quinoa (*Chenopodium quinoa*) under temperate climatic conditions in Denmark. J Agri Sci 1994;122(1):47.

Jacobsen SE, Jørnsgård B, Christiansen JL, Stølen O. Effect of harvest time, drying technique, temperature and light on the germination of quinoa (*Chenopodium quinoa*). Seed Sci Technol 1999a;27:937–944.

Jacobsen SE, Monteros C, Christiansen JL, Bravo LA, Corcuera LJ, Mujica A. Plant responses of quinoa (*Chenopodium quinoa* Willd.) to frost at various phenological stages. Eur J Agron 2005;22(2):131–139.

Jacobsen SE, Monteros C, Corcuera LJ, Bravo LA, Christiansen JL, Mujica A. Frost resistance mechanisms in quinoa (*Chenopodium quinoa* Willd.). Eur J Agron 2007;26(4):471–475.

Jacobsen SE, Mujica A, Jensen CR. The resistance of quinoa (*Chenopodium quinoa* Willd.) to adverse abiotic factors. Food Rev Int 2003;19(1–2):99–109.

Jacobsen EE, Skadhauge B, Jacobsen SE. Effect of dietary inclusion of quinoa on broiler growth performance. Anim Feed Sci Tech 1997;65:5–14.

Jacobsen SE, Stølen O. Quinoa – morphology, phenology and prospects for its production as a new crop in Europe. Eur J Agron 1993;2:19–29.

Jacobsen SE, Quispe H, Mujica A. 1999b. Quinoa: an alternative crop for saline soils in the Andes. In: Scientists and farmer–partners in research for the 21st century. CIP Program Report 1999–2000, pp. 403–408.

Jellen EN, Kolano BA, Sederberg MC, Bonifacio A, Maughan PJ. Wild crop relatives: genomic and breeding resources. In: Kole C, editor. Chenopodium. Berlin: Springer; 2011. p 35–61.

Jensen CR, Jacobsen SE, Andersen MN, Núñez N, Andersen SD, Rasmussen L, Mogensen VO. Leaf gas exchange and water relation characteristics of field quinoa *Chenopodium quinoa* Willd.) during soil drying. Eur J Agron 2000;13(1):11–25.

Johnson DL. New grains and pseudograins. In: Janick J, Simon JE, editors. Advances in new crops. Portland, Oregon: Timber Press; 1990. p 122–127.

Johnson DL, Croissant RL. Quinoa production in Colorado. Fort Collins, Colorado: Colorado State University Cooperative Extension; Service-In-Action Sheet: No.112; 1985.

Johnson DL, Croissant RL. Alternate crop production and marketing in Colorado. Fort Collins, Colorado: Colorado State University Cooperative Extension; 1990. ; Technical Bulletin LTB90-3.

Johnson DL, Ward S. Quinoa. In: Janick J, Simon JE, editors. New crops. New York, NY: Wiley; 1993. p 219–221.

Karyotis TH, Iliadis C, Noulas CH, Mitsibonas TH. Preliminary research on seed production and nutrient content for certain quinoa varieties in a saline–sodic soil. J Agron Crop Sci 2003;189(6):402–408.

Konishi Y, Hirano S, Tsuboi H, Wada M. Distribution of minerals in quinoa (*Chenopodium quinoa* Willd.) seeds. Biosci Biotechnol Biochem 2004;68(1):231–234.

Koyro HW, Eisa SS. Effect of salinity on composition, viability and germination of seeds of *Chenopodium quinoa* Willd. Plant Soil 2007;302:79–90.

Kumar A, Bhargava A, Shukla S, Singh HB, Ohri D. Screening of exotic *Chenopodium quinoa* accessions for downy mildew resistance under mid-eastern conditions of India. Crop Prot 2006;25(8):879–889.

Maas EV. Salt tolerance of plants. Appl Agric Res 1986;1(1):12–26.

Martínez EA, Veas E, Jorquera C, San Martín R, Jara P. Re-introduction of quínoa into arid Chile: cultivation of two lowland races under extremely low irrigation. J Agron Crop Sci 2009;195(1):1–10.

Mastebroek HD, Limburg H. Breeding harvest security in *Chenopodium quinoa*. In: Stølen O, Bruhn K, Pithan K, Hill J, editors. Proceedings of the COST 814 workshop on small grain cereals and pseudo-cereals; 1996 Feb 22–24; Copenhagen, Denmark. Luxembourg: Office of Official Publications of the European Communities; 1997. p 79–86.

Mastebroek HD, Marvin H. 1999. Progress in breeding of sweet quinoa. Mela T, Christiansen J, Kontturi M, Pahkala K, Partala A, Sahramaa M, Sankari H, Topi-Hulmi M, Pithan K. Proceedings of COST 814-workshop: alternative crops for sustainable agriculture; 1999 June 13–15; Turku, Finland. Luxembourg: Office of Official Publications of the European Communities. 100–107.

Mastebroek HD, van Loo R. 2000. Breeding of quinoa – state of the art. Parente G, Frame J. Abstracts/Proceedings of COST 814 conference, crop development for cool and wet regions of Europe; 2000 May 10–13; Pordenone, Italy. Luxembourg: Office of Official Publications of the European Communities. 491–496.

Mujica A. Tecnología del cultivo de la quinua. Fondo Simón Bolivar. Ministerio de Alimentación. Zona Agraria XII. Puno: IICA, UNTA; 1977.

National Research Council. Lost crops of the Incas: little-known plants of the Andes with promise for worldwide cultivation. Washington, DC: National Academy Press; 1989. p 1–415.

NCDC. 2011. Annual mean extreme maximum temperature [Internet]. Asheville, NC: NCDC (cited 2013 March 23). Available from http://hurricane.ncdc.noaa.gov/climaps/tmp13a13.pdf

Oelke EA, Putnam DH, Teynor TM, Oplinger ES. 1992. Quinoa. Alternative field crops manual. University of Wisconsin Cooperative Extension Service, University of Minnesota Extension Service, Center for Alternative Plant & Animal Products.

Ohlsson I. During late development stages the quinoa crop is not always successful. In: Parente G, Frame J, editors. Abstracts/Proceedings of COST 814 conference, crop development for cool and wet regions of Europe; 2000 May 10–13; Pordenone, Italy. Luxembourg: Office of Official Publications of the European Communities; 2000. p 497–499.

Ohlsson I, Dahlstedt L. Quinoa - potential in Sweden. In: Mela T, Christiansen J, Kontturi M, Pahkala K, Partala A, Sahramaa M, Sankari H, Topi-Hulmi M, Pithan K, editors. Proceedings of COST 814-workshop: alternative crops for sustainable agriculture; 1999 June 13–15; Turku, Finland. Luxembourg: Office of Official Publications of the European Communities; 1999. p 139–145.

Peterson A. 2013. Salinity tolerance and nitrogen use efficiency of quinoa for expanded production in temperate North America. Master's Thesis, Washington State University, Pullman, Washington, DC, pp. 1–189.

Prego I, Maldonado S, Otegui M. Seed structure and localization of reserves in *Chenopodium quinoa*. Ann Bot 1998;82:481–488.

Pulvento C, Riccardi M, Lavini A, d' Andria R, Iafelice G, Marconi E. Field trial evaluation of two *Chenopodium quinoa* genotypes grown under rain-fed conditions in a typical Mediterranean environment in South Italy. J Agron Crop Sci 2010;196(6):407–411.

Pulvento C, Riccardi M, Lavini A, Iafelice G, Marconi E, d' Andria R. Yield and quality characteristics of quinoa grown in open field under different saline and non-saline irrigation regimes. J Agron Crop Sci 2012;198(4):254–263.

Rasmussen C, Lagnaoui A, Esbjerg P. Advances in the knowledge of quinoa pests. Food Rev Int 2003;19(1–2):61–75.

Razzaghi F, Plauborg F, Jacobsen SE, Jensen CR, Andersen MN. Effect of nitrogen and water availability of three soil types on yield, radiation use efficiency and evapotranspiration in field-grown quinoa. Agr Water Manage 2012;109:20–29.

Risi JC, Galwey NW. The Chenopodium grains of the Andes – Inca crops for modern agriculture. Adv Appl Biol 1984;10:145–216.

Risi JC, Galwey NW. The pattern of genetic diversity in the Andean grain crop quinoa (*Chenopodium quinoa* Willd). I. Associations between characteristics. Euphytica 1989;41(1):147–162.

Risi JC, Galwey NW. Effects of sowing date and sowing rate on plant development and grain yield of quinoa (*Chenopodium quinoa*) in a temperate environment. J Agr Sci 1991a;117(03):325.

Risi JC, Galwey NW. Genotype x environment interaction in the Andean grain crop quinoa (*Chenopodium quinoa*) in temperate environments. Plant Breed 1991b;107(2):141–147.

Robinson RG. Amaranth, quinoa, ragi, tef, and niger: tiny seeds of ancient and modern interest. St. Paul, MN: Agricultural Experiment Station, University of Minnesota; 1986. ; Station Bulletin AD-SB-2949.

Romero S, Shahriari S. A food's global success creates a quandary at home. The New York Times 2011, March 20, Sect A:6.

Ruiz-Carrasco K, Antognoni F, Coulibaly AK, Lizardi S, Covarrubias A, Martínez EA, Molina-Montenegro MA, Biondi S, Zurita-Silva A. Variation in salinity tolerance of four lowland genotypes of quinoa (*Chenopodium quinoa* Willd.) as assessed by growth, physiological traits, and sodium transporter gene expression. Plant Physiol Biochem 2011;49(11): 1333–1341.

van Schooten HA, Pinxterhuis JB. 2003. Quinoa as an alternative forage crop in organic dairy farming. Kirilov A, Todorov N, Katerov I. Optimal forage systems for animal production and the environment: proceedings of the 12th symposium of the European Grassland Federation; 2003 May 26–28; Pleven, Bulgaria pp. 445–448.

Schulte auf'm Erley G, Kaul HP, Kruse M, Aufhammer W. Yield and nitrogen utilization efficiency of the pseudocereals amaranth, quinoa, and buckwheat under differing nitrogen fertilization. Eur J Agron 2005;22(1):95–100.

Sigsgaard L, Jacobsen SE, Christiansen JL. Quinoa, *Chenopodium quinoa*, provides a new host for native herbivores in Northern Europe: case studies of the moth, *Scrobipalpa atriplicella*, and the Tortoise Beetle, *Cassida nebulosa*. J Insect Sci 2008;8(50):1–4.

Small E. New crops for Canadian agriculture. In: Janick J, editor. Perspectives on new crops and new uses. Alexandria, VA: ASHS Press; 1999. p 15–52.

Smith BD. *Chenopodium berlandieri* ssp. *jonesianum*: evidence for a Hopewellian domesticate from Ash Cave, Ohio. SE Archaeology 1985;4(2):107–133.

Swenson EM. 2006. Genetic diversity of Bolivian *Peronospora farinosa* f. sp. *chenopodii* (downy mildew) and quinoa's resistance response. Master's Thesis, Brigham Young University, Provo, Utah.

Tapia ME, Mujica S, Canahua A. Origen distribución geográfica y sistemas de producción en quinua. Primera Reunion sobre Genética y Fitomejoramiento de la Quinua. Puno: Proyecto PISCA/UNTA/IBTA/IICA/CIID; 1980. p A1–A8.

Testen AL, del Mar Jiménez-Gasco M, Ochoa JB, Backman PA. Molecular detection of *Peronospora variabilis* in quinoa seed and phylogeny of the quinoa downy mildew pathogen in South America and in the United States. Phytopathology 2014;104(4):379–386.

Testen AL, McKemy JM, Backman PA. First report of quinoa downy mildew caused by *Peronospora* variabilis in the United States. Plant Dis 2012;96(1):146.

Testen AL, McKemy JM, Backman PA. First report of Passalora leaf spot of quinoa caused by *Passalora dubia* in the United States. Plant Dis 2013a;97(1):139.

Testen AL, McKemy JM, Backman PA. First report of Ascochyta leaf spot of quinoa caused by *Ascochyta* sp. in the United States. Plant Dis 2013b;97(6):844.

Tewari JP, Boyetchko SM. Occurrence of *Peronospora farinosa* f. sp. *chenopodii* on quinoa in Canada. Can Plant Dis Surv 1990;70(2):127–128.

Thanapornpoonpong S. 2004. Effect of nitrogen fertilizer on nitrogen assimilation and seed quality of amaranth (*Amaranthus* spp.) and quinoa (*Chenopodium quinoa* Willd.). Ph.D. Dissertation, Georg-August University of Göttingen, Göttingen, Germany.

Torres JD. 1955. Comparative study of resistance of alfalfa, brome grass and quinoa seedlings to synthetic saline soils, saline alkali and alkali soils. Master's Thesis. Cornell University, Ithaca, NY.

USDA. The second RCA appraisal: soil, water, and related resources on nonfederal land in the United States. Washington, DC: U.S. Government Printing Office; 1989.

USDA. 2011 RCA appraisal. Washington, DC: USDA; 2011.

Vera RR. [Internet]. 2006. Chile. Rome: FAO (cited 2013 Mar 11). Available from: http://www.fao.org/ag/AGP/AGPC/doc/Counprof/Chile/cile.htm

Ward SM. 1994. Developing improved quinoa varieties for Colorado. Ph.D. Dissertation, Colorado State University, Fort Collins, CO.

Ward SM. Response to selection for reduced grain saponin content in quinoa (*Chenopodium quinoa* Willd.). Field Crop Res 2000;68(2):157–163.

Ward SM. A recessive allele inhibiting saponin synthesis in two lines of Bolivian quinoa (*Chenopodium quinoa* Willd.). J Hered 2001;92(1):83–86.

Wilson HD. *Chenopodium quinoa* Willd.: variation and relationships in southern South America. Natl Geogr Res Rep 1978;19:711–721.

Wilson HD. Artificial hybridization among species of *Chenopodium* sect. *Chenopodium*. Syst Bot 1980;5(3):253–263.

Wilson HD, Manhart J. Crop/weed gene flow: *Chenopodium quinoa* Willd. and *C. berlandieri* Moq. Theor Appl Genet 1993;86(5):642–648.

Chapter 11
Nutritional Properties of Quinoa

Geyang Wu

School of Food Science, Washington State University, Pullman, WA, USA

INTRODUCTION

Quinoa is known as a "complete food" and has been called "mother grain" in the Andes (Abugoch 2009; Vega-Gálvez et al. 2010) due to its high nutritional value (Table 11.1) (Jancurová et al. 2009). The quinoa seed is as small as a millet seed, flat and oval shaped, with color ranging from dark red to pale yellow. The protein content in quinoa is higher than that in other cereals and accounts for about 16.5% of its dry weight. Lipids are about 6.3% and higher than the lipid content in other cereals. Carbohydrates are the major composition in quinoa and comprise around 69%. The fiber and ash content of quinoa are 3.8%, comparable to those in other cereals.

In this chapter, the chemical composition and nutritional properties of quinoa is reviewed. The content, composition and quality of protein, carbohydrates, lipids, vitamins, minerals, antinutritional factors, and bioactive compounds are also summarized.

PROTEIN

Quinoa seed is considered a complete food (Abugoch 2009), primarily because its protein is high in content and quality, with a balanced amino acid profile. The net protein utilization (NPU) of quinoa is 68, digestibility (TD) is 95, biological value (BV) is 71, and all of these are indicators of the high quality of quinoa protein (Ruales et al. 2002).

The protein content of quinoa ranges from 8% to 22% among different varieties (Valencia-Chamorro 2003), higher than that of other cereals such as wheat (9–14%) (Khan and Shewry 2009; Jancurová et al. 2009), barley (8–14%) (Cai et al. 2013), and rice (6–15%) (Juliano 1985) and lower than that of soy beans (28–36%) (Table 11.2). The reason for the high variability in protein content among quinoa varieties is not clear yet, though both the genotype and the growing environment can affect protein content.

About 57% protein exists in the embryo of quinoa seed, 39% in the perisperm, and only 4% in the bran (including seed coat and pericarp) (Ando et al. 2002) (Table 11.2). Prego et al. (1998) studied the structural form of protein in quinoa seed. Fig. 11.1 shows protein bodies (PBs) (black particles) in spongy tissue of a quinoa cotyledon (a) and in endosperm cells (b). The diameter of the PBs in the cotyledon ranges from 0.3 to 3 μm and 1 to 3 μm in the endosperm. These PBs contain one or more globoid crystals that store minerals P, K, and Mg.

The amino acids of quinoa are well balanced and comparable to those of soy protein, casein, and wheat (Table 11.3) (Wang et al. 1999; USDA 2011, 2005; Tang et al. 2006; Abugoch et al. 2008). Quinoa seed has high levels of arginine, glycine, lysine, methionine, threonine, and tryphtophan

Quinoa: Improvement and Sustainable Production, First Edition. Edited by Kevin Murphy and Janet Matanguihan.

Table 11.1 Chemical Composition of Quinoa and Cereals (Jancurová et al. 2009).

	Quinoa	Barley	Maize	Rice	Wheat	Oat	Rye
Protein	16.5	10.8	10.22	7.6	14.3	11.6	13.4
Lipid	6.3	1.9	4.7	2.2	2.3	5.2	1.8
Fiber	3.8	4.4	2.3	6.4	2.8	10.4	2.6
Ash	3.8	2.2	11.7	3.4	2.2	2.9	2.1
Carbohydrates	69	80.7	81.1	80.4	78.4	69.8	80.1
kcal/100 g	399	383	408	372	392	372	390

Table 11.2 Protein Content in Quinoa Seed Fraction (Ando et al. 2002).

	Whole Grain	Milled Grain	Bran	Perisperm	Embryo
Protein content	12.9	13.3	6.1	7.2	23.5
Proportion %	100	96	4	39	57

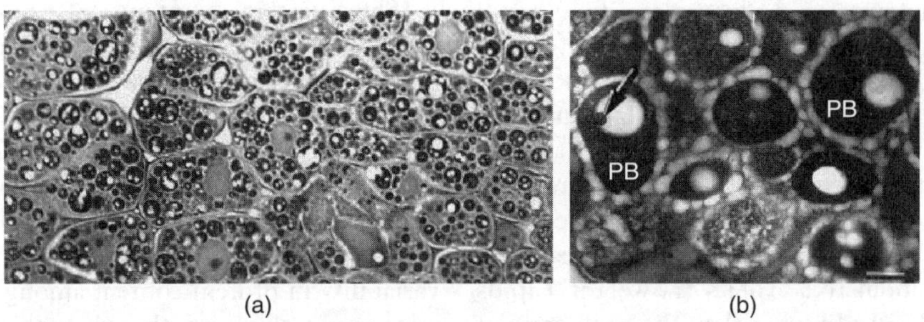

Fig. 11.1 (a) Spongy tissue of quinoa cotyledon and (b) endosperm cell showing protein bodies (PBs) with globoid crystal (black arrow) (Prego et al. 1998).

compared to wheat. Glutamic acid and cysteine are lower in quinoa than they are in wheat, whereas histidine, isoleucine, phenylalanine, serine, and valine contents are comparable. Soy bean protein and casein are considered a nutritionally balanced protein with high quality (Tang et al. 2006). The amino acids in quinoa are comparable to those in casein, another indication that quinoa has excellent amino acid and protein quality.

Quinoa seed is a great source of amino acids as it contains all the essential amino acids (Table 11.4) (Friedman and Brandon 2001; Abugoch et al. 2008). Moreover, all of the essential amino acid content in quinoa are higher than the requirements suggested by FAO/WHO.

The protein in cereal is divided into four groups, namely albumin, globulin, prolamins, and glutelins. Albumin and globulins are the major proteins in quinoa (76.6%), whereas the content of glutelins and prolamins are much lower, at 12.7% and 7.2%, respectively (Table 11.5). Gluten is the protein complex formed by linkages between proteins such as gliadins and glutenins in wheat or glutelins and prolamins in other cereals (Koziol 1992). Gluten is critical to dough elasticity, loaf shape, and in imparting a chewy texture to the end product. However, a significant proportion of the population in Western countries suffer from gluten intolerance and coeliac disease. Coeliac disease is a chronic small intestine inflammatory disease caused by genetic intolerance to gluten, such as gliadin of wheat and prolamins of rye and barley (Gallagher et al. 2004). The prevalence of coeliac disease used to be significantly underestimated from epidemiological studies (Hovdenak et al. 1999; Fasano and Catassi 2001). Sabatino and Corazza (2009) discussed a more accurate method of estimating the true prevalence of coeliac disease using serological tests for IgA antigliadin antibodies. With this method, studies showed that coeliac

Table 11.3 Amino Acids of Quinoa, Soy, Casein, and Wheat (mg/g).

Amino Acid	Quinoa[a]	Soy Protein[b]	Casein[c]	Wheat[d]
Arginine	99.7	41.0	37.0	43
Aspartic acid	80.1	118.1	63.0	51
Glycine	53.8	38.6	16.0	37
Glutamic acid	163.6	212.9	190.0	322
Histidine	25.8	29.0	27.0	22
Isoleucine	43.3	44.8	49.0	40
Leucine	73.6	70.0	84.0	68
Lysine	52.5	53.9	71.0	26
Methionine	21.8	9.3	26.0	13
Cystine	5.5	0.6	0.4	23
Phenylalanine	44.9	53.0	45.0	48
Tyrosine	35.4	37.1	55.0	–
Serine	52.1	54.8	46.0	45
Threonine	43.9	41.0	37.0	28
Tryphtophan	38.5	–	14.0	18
Valine	50.6	44.1	60.0	43
Alanine	38.2	38.3	27.0	36

[a]Abugoch et al. (2008).
[b]FAO/WHO suggested requirement (Friedman and Brandon 2001).
[c]USDA (2011, 2005).
[d]Tang et al. (2006).

Table 11.4 Essential Amino Acid of Quinoa Protein and FAO/WHO Suggested Requirement (mg/g protein).

Essential Amino Acid	Quinoa Protein[a]	FAO/WHO Suggested Requirement[b]
Histidine	25.8	18
Isoleucine	43.3	25
Leucine	73.6	55
Lysine	52.5	51
Methionine and cysteine	27.3	25
Phenylalanine and tyrosine	80.3	47
Threonine	43.9	27
Tryptophan	38.5	7
Valine	50.6	32

[a]Abugoch et al. (2008).
[b]FAO/WHO suggested requirement (Friedman and Brandon 2001).

disease in adult populations of United Kingdom and United States are 1:87 (West et al. 2003) and 1 : 105 (Fasano et al. 2003), respectively. The disease occurs in Caucasians in much higher frequency compared to Africans and Asians (Hoffenberg et al. 2003; Di Sabatino and Corazza 2009). The coeliac disease is life-long and the only effective treatment is to consume gluten-free foods (Gallagher et al. 2004). Gluten-free food products are fast emerging in the western market and the demand keeps increasing. Only small amounts of prolamins in quinoa participate in gluten formation; thus, quinoa is recognized as gluten-free (Alvarez-Jubete et al. 2010a) and considered a potential ingredient in gluten-free products and diet.

Table 11.5 Subgroups of Protein from Quinoa, Maize, Rice, and Wheat (% Total Protein) (Koziol 1992).

	Albumin + Globulins	Glutenins/ Glutelins	Gliadins/ Prolamins
Quinoa	76.6	12.7	7.2
Maize	38.3	37.2	24.5
Rice	19.2	71.9	8.9
Wheat	17.1	54.4	28.5

The quality and property of protein can be affected by pH and temperature during extraction or processing. Abugoch et al. (2008) isolated quinoa protein by alkaline of pH9 and 11 and then studied and compared the physicochemical and functional properties of both isolations. Proteins from the two pH levels showed similar patterns in native/sodium dodecyl sulfate polyacrylamide gel electrophoresis (SDS-PAGE) and scanning electron microscope (SEM). Protein from pH11 showed lower fluorescence intensity of tryptophan. Protein from pH11 showed no endotherm, whereas protein from pH9 showed an endotherm with denaturation temperature of 98.1°C and denaturation enthalpy of 12.7 J/g (Abugoch et al. 2008). Abugoch et al. (2009) also studied the stability of quinoa flour protein during storage. Protein solubility and water absorption were studied on quinoa flour stored at different conditions. The results indicated that quinoa flour can be stored at 20 to 30°C in double kraft paper bags for 2 months with no significant change in functional properties.

Quinoa protein exhibits bioactive properties. Takao et al. (2005) provided evidence to indicate that dietary supplement of quinoa protein (5%) could control total cholesterol levels in both plasma and liver of mice. The quinoa diet suppressed enzymes that catalyze cholesterol synthesis and stimulated cholesterol hydroxylase. In addition, lower molecular weight peptide in quinoa protein hydrolysate exhibited increased radical scavenging activity and inhibited activity to angiotensin-converting enzyme, an important factor leading to hypertension (Kim et al. 2001; Wu and Ding 2001; Aluko and Monu 2003).

Protein plays a significant role in the functional properties of quinoa. The functional properties of food ingredients refer to the technological characteristics during processing, including solubility, emulsifying activity index, and foaming property. These are critical to food manufacturing as these affect the taste and mouth feel. Enzymatic hydrolysis is a common method to improve functional properties of food. Aluko and Monu (2003) hydrolyzed quinoa protein with alcalase and fractionated the hydrolysate by ultrafiltration of 10,000 and 5,000 molecular weight cutoff membranes. The functional properties of quinoa protein concentration, protein hydrolysate, and membrane permeates were studied. Protein hydrolysate and membrane permeates exhibited higher solubility and lower emulsifying ability. Hydrolyzed peptides had low foaming property.

CARBOHYDRATES

Starch

Starch is the major component and predominant energy reserve in cereal seed, providing 70–80% of human diet calories (Damodaran et al. 2008). The main source of starch in the human diet is cereal grain seed such as corn, rice, and wheat and roots such as potato and cassava. Starch comprises approximately 55% of quinoa seed (Lindeboom et al. 2005).

Starch exists in nature as granules, small particles with diameters ranging from less than 1 μm to more than 100 μm (Lindeboom et al. 2004). Starch granules are categorized into three size ranges: large (>15 μm), medium (5–15 μm), and small (<5 μm) (Wilson et al. 2006). In quinoa seed, starch granules are polygonal in shape, with a diameter of 0.08–2.0 μm (Ando et al. 2002), which is significantly smaller than the diameter of starch granules in other cereals such as wheat, rice, and amaranth, a pseudo-cereal (Table 11.6) (Tang et al. 1998). Chen et al. (2003) indicated that starch granule size significantly affects the processing ability and quality of noodle. Noodles made from starch with smaller granule size offer better quality, which may be due to higher granule surface area.

Most quinoa starch granules exist in perisperm cells (Figs. 11.2 and 11.3) (Prego et al. 1998; Ando et al. 2002). The pericarp and seed coat cover the seed (Fig. 11.2). The endosperm consists of two cotyledons and one hypocotyl–radicle axis. The

Table 11.6 Granule Size of Starch from Quinoa and Other Cereals (Lindeboom et al. 2004).

Source	Diameter, μm
Quinoa	0.6–2
Amaranth	1–2
Rice	2–10
Oat	2–14
Buckwheat	2–14
Wheat	<10 and 10–35
Rye	2–3 and 22–36
Barley	2–3 and 12–32

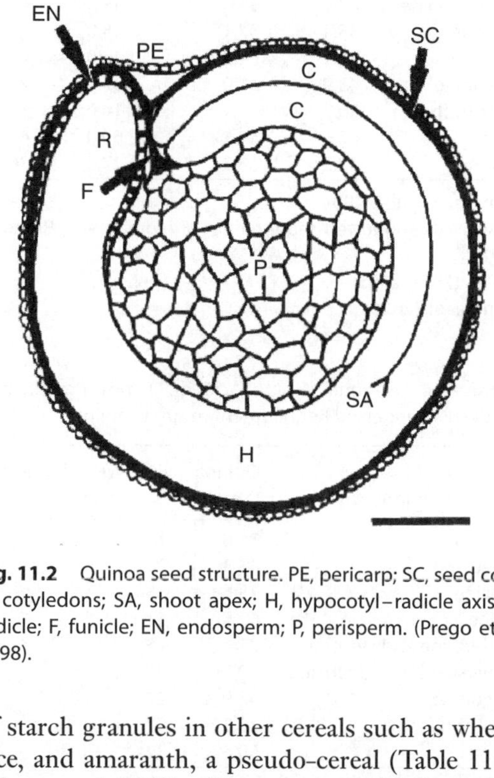

Fig. 11.2 Quinoa seed structure. PE, pericarp; SC, seed coat; C, cotyledons; SA, shoot apex; H, hypocotyl–radicle axis; R, radicle; F, funicle; EN, endosperm; P, perisperm. (Prego et al. 1998).

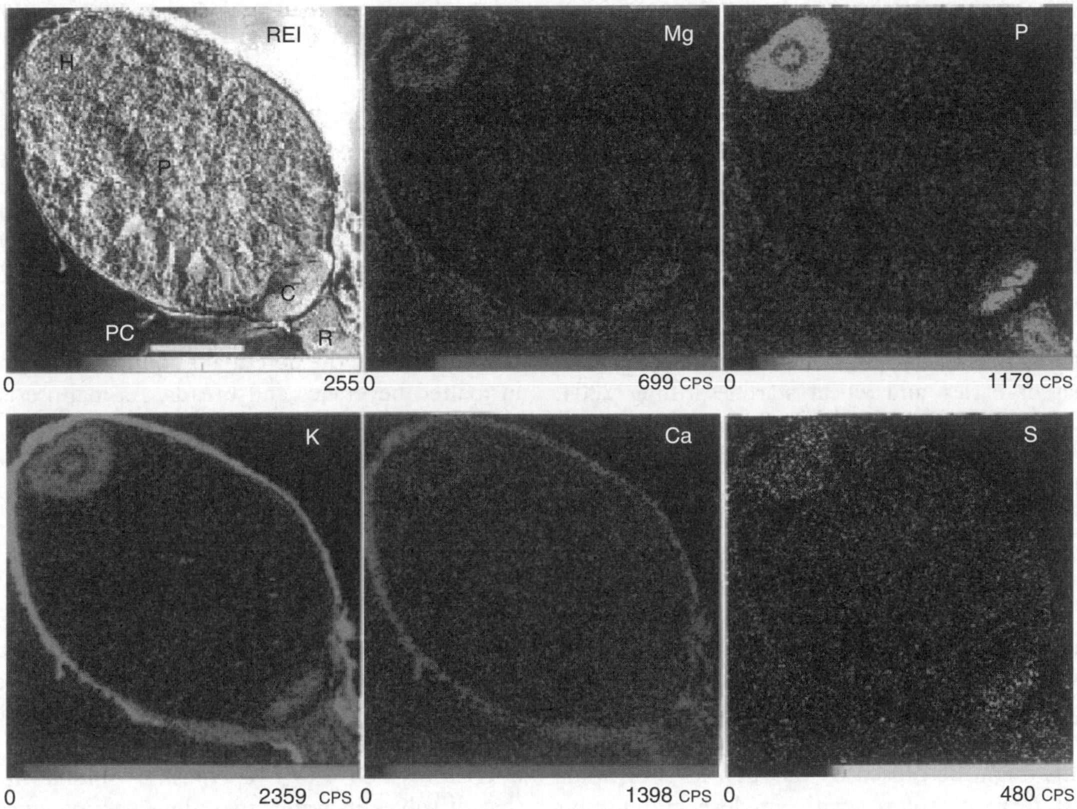

Fig. 11.3 Refractive electron image (REI) and element mapping of a quinoa seed (P, perisperm; PC, pericarp; C, cotyledon; R, radicle; H, hypocotyl–radicle axis) (Konishi et al. 2004).

perisperm is in the middle part of the seed and consists of 58.8% of grain weight (Ando et al. 2002). Starch aggregates and granules are packed densely in the cell of the perisperm (Ando et al. 2002).

Starch granules consist of two types of molecules, namely amylose, which is linear, and amylopectin, which is highly branched. Amylose is composed of linear polysaccharides of (1 → 4)-linked α-D-glucopyranosyl units with average molecule weight of 106 (Damodaran et al. 2008). Amylopectin is one the largest molecules in nature, with molecule weight of 107. It is composed of branched polysaccharides with 4–5% α-D-(1 → 6) branch point linkage. Most grain starches contain about 25% amylose and about 75% amylopectin. Amylose content in quinoa ranges from 3 to 20% (Lindeboom et al. 2005), lower than that of wheat, corn, and potato (Damodaran et al. 2008). Amylose and amylopectin percentage affects starch properties significantly, including pasting, gelatinization, and retrogradation. Low amylose starch is more likely to yield sticky food texture.

X-ray diffraction graphic divided starches to three types: type A is present in cereal starch; type B exists in potatoes, roots, and resistant starch; and type C is found in legume starch and mixture of corn and potato starch. The X-ray diffraction pattern of quinoa starch is type A, the type common to most cereals (Table 11.7) (Ando et al. 2002).

Gelatinization temperature is the temperature at which starches absorb water and start to swell. It reflects the amount of time it takes for cereals to cook. Gelatinization temperature is determined using differential scanning calorimetry (DSC). The onset gelatinization temperature (To) of

Table 11.7 Quinoa Starch Characteristics (Ando et al. 2002).

Analysis	Range
X-ray diffraction pattern	A-type
Gelatinization, °C	
To	54.0
Tp	62.2
Tc	71.0
Enthalpy, ΔH, J/g	11.0

Table 11.8 Sugar Content of Quinoa Flour (mg/100 g sample) (Ogungbenle 2003).

Sugar	Content
Glucose	19.00
Fructose	19.60
D-Ribose	72.00
D-Galactose	61.00
Maltose	101.00
D-Xylose	120.00

quinoa starches is 54.0°C and the peak temperature (Tp) is 62.2°C, both of which are lower than that of barley and wheat starches (Tang 2004; Tang et al. 2005). The conclusion of gelatinization temperature (Tc) is 71.0°C and enthalpy ΔH is 11.0 J/g.

The swelling power of starch is an important reflection of end-use properties such as noodle quality (Khan and Shewry 2009). It is calculated as percentage of the sedimental paste weight to the sample weight (Ahamed et al. 1996). The swelling power of quinoa starch at 95°C is 8.54 g H_2O/g sample, lower than that of corn starch at the same temperature (21.0 g H_2O/g sample). This might be caused by the high level of lipids that form complexes with amylose and inhibit mobility of amylose. Freeze thaw stability of quinoa starch is higher than that of wheat and barley starch (Watanabe et al. 2007). Tolerance of freeze thaw suggests that quinoa can be used in novel food products.

On the basis of its properties, starch from quinoa has been extracted and utilized in novel ways. Araujo-Farro et al. (2010) utilized quinoa starch to make film, which exhibited excellent mechanical properties, low solubility, and good barrier properties. The quinoa starch film is edible and recyclable and can be utilized widely in food packaging.

Sugar

Sugar content of quinoa flour was determined by Ogungbenle (2003) (Table 11.8). Glucose and fructose can significantly increase serum glucose in human body. Their contents in quinoa are low, at 19 mg/100 g and 19.6 mg/100 g, respectively. Maltose content is as high as 101.00 mg/100 g, which indicates the potential of quinoa to be used in malted beverages and breads. As maltose is a sweet disaccharide with sweetness that is 46% of the sucrose standard, quinoa with high maltose can also be used in weaning foods (Ogungbenle 2003).

Dietary fiber

Dietary fiber refers to the family of carbohydrates in plants that cannot be digested in the human small intestine but can be partially metabolized by microbiota in the large intestine. Dietary fiber is divided into two types: (i) soluble fiber that dissolves in water, including pectins, gums, mucilages, and some hemicellulose and (ii) insoluble fiber that does not dissolve in water, including cellulose, lignin, lignin, and most hemicellulose.

Soluble fiber exhibits high water-holding capacity and creates viscous solutions. Fiber delays gastric emptying, slows down digestion, creates postprandial satiety, and changes small intestine transit time. Thus, a high fiber diet has been shown to help with weight control.

Fiber can bond with lipids and cholesterols, resulting in decreased lipid absorption and serum cholesterol, thus helping prevent hypoglycemia and hypolipidemia. High fiber intake increases fecal volume to control constipation and other gastrointestinal disorders such as irritable bowel syndrome and gallstone formation. Fiber can be degraded and fermented in the large intestine. This property allows fiber to improve mucosal cell profile, maintain a balance of gastrointestinal microbiota, and protect against colon cancer.

Cereals, vegetables, and fruits are major sources of fiber. Fiber content in quinoa is

8.8–10.3% (Wright et al. 2002), comparable to that of wheat (12.2%). The daily value of fiber intake recommended by the USDA is 25 g. Quinoa is a high fiber food, as its fiber content is 35.2–41.2% of the daily recommended value.

LIPIDS

Lipids in quinoa seed is about 6.9%, which is higher compared to other cereals (Table 11.9) (Ogungbenle 2003). The iodine value is 54%, which indicates a proximate amount of unsaturated fatty acid fraction. Quinoa oil has a high proportion of unsaturated fatty acids that provide more health benefits compared to saturated fatty acids. The acid value is 0.5% and peroxide value is 2.44%, which indicate that the lipids are relatively stable and not easily oxidized when exposed to heat and oxygen. Molecular weight or chain length of fatty acids can be represented by saponification value, which is the number of milligrams of KOH required to saponify 1 g of fat under specific conditions. The saponification value of quinoa lipids is 192%, lower than this is that of butter fat (241%) and comparable to soybean oil (194%) (Ogungbenle 2003).

The composition of lipids from quinoa is shown in Table 11.10. Palmitic acid is the main saturated fatty acid, comprising approximately 10% of the total fatty acid content. Unsaturated fatty acids occupy about 87% of total lipids, including oleic acid (C18 : 1), linoleic acid (C18 : 2), and linolenic acid (C18 : 3). Linoleic acid (C18 : 2) is the major group, making up 52.0% of the total lipids. Quinoa seed embryo is higher in fatty acid content (10.2%) compared to the perisperm, pericarp, and seed coat.

The total fatty acid of quinoa is 6.5%, higher than that of corn oil (4%) and lower than that of wheat germ oil (11%) (Table 11.11) (Ando et al. 2002). Total fatty acid of other commercialized oil origins are much higher, ranging from 20% in olive oil to 60% in walnut oil. Saturated fatty acid in quinoa oil is 11% of total lipids, including 10.2% palmitic acid and 0.8% stearic acid. High

Table 11.9 Properties of Quinoa Lipids (%) (Ogungbenle 2003).

Component Value	Content
Acid	0.50
Iodine	54.0
Peroxide	2.44
Saponification	192.0

Table 11.10 Fatty Acid Composition of Quinoa (100% Dry Basis) (Ando et al. 2002).

Fatty Acid	Milled Grain	Perisperm	Embryo	Whole Grain
Total fatty acid	6.7	5.0	10.2	6.5
Myristic (C14:0)	0.2	0.1	0.2	0.2
Palmitic (C16:0)	10.3	10.8	9.5	10.2
Stearic (C18:0)	0.8	0.7	0.9	0.8
Oleic (C18:1)	25.6	29.5	19.7	24.9
Linoleic (C18:2)	52.0	49.0	56.4	52.5
Linolenic (C18:3)	9.8	8.7	11.7	10.1
Others	1.3	1.2	1.6	1.4

Table 11.11 Comparison of Percent Fatty Acid from Quinoa and Oil from Other Crops.

Fatty Acid	Quinoa	Corn	Wheat Germ	Olive Oil	Canola	Flax	Coconut Oil	Walnut
Total fatty acid	6.5	4	11	20	30	35	35	60
Saturated FA	11.0	17	18	16	7	9	91	16
Palmitic (C16:0)	10.2	–	0	–	–	5	91	11
Stearic (C18:0)	0.8	17	18	16	7	4	0	5
Unsaturated FA	87.5	83	80	83	91	91	9	84
MUFA								
Oleic (C18:1)	24.9	24	25	75	54	19	6	28
PUFA	62.6	59	55	8	37	72	3	56
Linoleic (C18:2)	52.5	59	50	8	30	14	3	51
Linolenic (C18:3)	10.1	–	5	–	7	58	0	5

intake of saturated fatty acid causes human health issues such as obesity (van Dijk et al. 2009), cardiovascular disease (CVD) (Sacks and Katan 2002), and type II diabetes (Hunnicutt et al. 1994). Stearic acid was reported to decrease levels of high-density lipoprotein (HDL), the good lipoprotein that protects against CVD (Hu et al. 2001). Saturated fatty acid proportion in quinoa oil is lower than that of other oils except for canola and flax seed oil. Steric acid is only 0.8%, much lower than that found in other oils.

Unsaturated fatty acid is divided into two categories, monounsaturated fatty acid (MUFA) and polyunsaturated fatty acid (PUFA). The MUFA has only one double carbon bond in lipid chain, for example, oleic acid (C18 : 1). On the other hand, PUFA has two or more double carbon bonds, such as linoleic (C18 : 2) and linoleic (C18 : 2). MUFA has been reported to exhibit a negative correlation to total mortality and CVD death (Jacobs et al. 1992). Olive oil is rich in MUFA (75%) and is the major source of fat in the Mediterranean diet. The death rate due to heart disease in Mediterranean populations is very low, which can be explained by their MUFA-rich diet (Chait et al. 1993). MUFA in quinoa oil is approximately 25%, lower than that of olive oil and canola but comparable to other oils.

Healthy PUFAs normally refer to n-6 polyunsaturated fatty acid linoleic (C18:2) and n-3 polyunsaturated fatty acid linoleic (C18:2). On the basis of the metabolic studies, increasing intake of linoleic acid can improve lipid profile, lower cholesterol (Chait et al. 1993), and enhance insulin sensitivity (Lovejoy and DiGirolamo 1992; Lovejoy 1999). Taking linoleic fatty acid not only decreases the risk of coronary heart disease (Chait et al. 1993) and significantly reduces incidence of type II diabetes (Salmerón et al. 2001) but also decreases the risk of heart arrhythmia (Abeywardena et al. 1991). Linoleic acid content in quinoa oil is 52.5%, comparable to that in corn oil (59%), wheat germ oil (55%), and walnut oil (51%) and much higher than that in olive oil, canola, flax oil, and coconut oil.

n-3 fatty acid α-linoleic (C18:2) (shortened as ALA) is another functional food component that has been drawing attention from consumers and the food industry. It is even more essential to infants and children, because of its critical function on brain and nerve development (Cunnane and Thompson 1995). ALA can generate long-chain n-3 fatty acid such as docosahexaenoic (DHA) and eicosapentaenoic (EPA) through desaturation-chain elongation pathway (Cunnane and Thompson 1995). This effect of ALA is important because DHA and EPA enhances brain function and boosts the immune system. ALA also inhibits the transformation of linoleic acid to arachidonic acid (C20:4, n-6) and controls inflammation in the human body (Renaud and Nordøy 1983). Excessive inflammation can lead to obesity (Wymann and Solinas 2013), type II diabetes (Oliver et al. 2010), CVD (Crumpacker 2010), and gastrointestinal diseases (Ohman and Simrén 2010). Epidemiologic studies have provided convincing evidence for the association of ALA's effect against CVD and arrhythmia (Hu et al. 1999). Although ALA in quinoa oil (10.1%) is lower than that in flaxseed oil (58%), it is relatively rich in ALA when compared to other oil products, such as wheat germ oil (5%), canola oil (7%), coconut oil (0%), and walnut oil (5%) (Table 11.11).

Linoleic acid and ALA compete for the desaturation and chain elongation pathway (Garg et al. 1989; Cunnane and Thompson 1995); thus, the balance between linoleic acid and ALA is critical for human health (Chan et al. 1993). An ALA:linoleic ratio of between 1 : 10 to 1 : 5 has been shown to help control coronary heart disease (Hu et al. 1999). The ratio of ALA and linoleic acid in quinoa oil is close to 1 : 5.

As a whole, the components of quinoa oil are well balanced and beneficial to human health. Quinoa oil is low in saturated fatty acids and high in MUFA and PUFAs. Quinoa can be considered a potential oil seed with quality comparable to other commercialized functional oils.

VITAMINS

Cereals are one of the major sources of vitamin B complex. Cereal-based food products, such as bread and ready-to-eat cereal and rice, contribute

Table 11.12 Vitamin Content in Quinoa, Wheat, Rice, and Barley (mg/100 g).

	Quinoa[a–d]	RDI	Wheat[e]	Rice[f]	Barley[f]
Thiamin (B_1)	0.29–0.38	1.5	0.48	0.47	0.49
Riboflavin (B_2)	0.30–0.39	1.7	0.12	0.10	0.20
Niacin (B_3)	1.06–1.52	20	3.6	5.98	5.44
Pyridoxine (B_6)	0.487	2.0	0.43	NR	NR
Folate (B_9)	0.781	0.4	0.054	NR	NR
Ascorbic acid (C)	4.0	60	0	0	0
α-Tocopherol (VE) (IU)	5.37	30	1.0	0.18	0.35
β-Carotene	0.39	NR	0.02	NR	0.01

[a]Koziol (1992).
[b]Ruales et al. (2002).
[c]Ranhotra et al. (1993).
[d]USDA (2011, 2005).
[e]Khan and Shewry (2009).
[f]Jancurová et al. (2009).

approximately 45%, 30%, 28%, 14%, and 19% of the total daily intake of thiamin, riboflavin, niacin, vitamin B_6, and folate, respectively (Khan and Shewry 2009). Vitamin B provides an essential function for the human body, whereas folate is known to affect the health of the neural system. When folate-enriched cereal products were promoted in the United States, the rate of neural tube defects decreased (Yang et al. 2010). Studies showed that vitamin B_6 and folate supplement can decrease the risk of coronary heart disease (Rimm et al. 1998), hyperhomocysteinemia in men (Ubbink et al. 1993), Alzheimer's disease (Rieder and Fricke 2001), and depression (Skarupski et al. 2010).

Riboflavin (B_2) and folate (B_9) in quinoa is 0.30–0.39 and 0.781, respectively, significantly higher than that in wheat, rice, and barley (Table 11.12). Pyridoxine (B_6) in quinoa is 0.487 mg/100 g, comparable to that in wheat, almost 25% of RDI, which shows that quinoa is a good resource of pyridoxine (B_6). Folate, known as folic acid and vitamin B_9, plays an essential role in embryo development and the production of red blood cells. As folate cannot be synthesized by the human body, it has to be supplied from the diet. Folate in quinoa is 0.781 mg/100 g, higher than in wheat, and even RDI. Therefore, quinoa is a great resource of folate and should be recommended to pregnant women.

Vitamin E refers to "tocols," including α-, β-, γ-, and δ-tocopherols and α-, β-, γ-, and δ-totrienols, which are nonacyl lipids (Khan and Shewry 2009). The relative activity of vitamin E is calculated by tocol contents as α-tocopherol equivalents (Eitenmiller and Landen 1999). Vitamin E is the chain-breaking antioxidant that protects the human body from free radical oxidation (Maxwell and Lip, 1997), thus reducing the risk of CVD (Pryor 2000; Trumbo 2005) and cancer (Papas and Vos 2001; Kline et al. 2007). Vitamin E in quinoa also protects the lipid from oxidation. Quinoa seed oil is more stable during processing and storage, even with high percentage of unsaturated fatty acids, due to the protection of vitamin E (Ng et al. 2007).

MINERALS

In a typical western diet, approximately 50% of iron and manganese, 30% of copper and magnesium, and 20% of zinc and phosphorus come from cereals and cereal products (Cotton et al. 2004). Quinoa is a rich source of minerals (Table 11.13) (Ando et al. 2002). The potassium, magnesium, phosphorus, iron, magnesium, and copper contents in quinoa whole grains are relatively high, whereas zinc is deficient as determined from the USDA daily intake reference. Most minerals in quinoa have higher content

Table 11.13 Mineral Content of Quinoa Grain Fraction (mg/mg%) (Ando et al. 2002).

	Whole Grain	RDI[a]	Milled Grain	Bran	Perisperm	Embryo
K	825.7	NR	639.3	2908.5(29%)	387.9(28%)	1125.4(43%)
Mg	452.6	400	415.2	958.3(17%)	215.2(29%)	750.2(54%)
Ca	121.3	1000	91.8	481.3(30%)	71.8(34%)	139.7(36%)
P	359.5	1000	360.2	350.8(8%)	286.6(50%)	482.6(42%)
Fe	9.5	18	9.2	14.3(13%)	7.2(48%)	11.3(39%)
Mn	3.7	NR	3.4	10.8(19%)	2.4(38%)	5.1(43%)
Cu	0.7	2	0.6	1.4(19%)	0.5(44%)	0.8(37%)
Zn	0.8	15	0.8	0.7(9%)	0.6(48%)	1.1(43%)
Na	1.3	NR	1.2	3.2(21%)	0.5(31%)	1.5(48%)

[a]Reference daily intake (USDA, 2011).

compared to that in wheat, which is deficient in iron, copper, manganese, and zinc (Khan and Shewry 2009).

Konishi et al. (2004) studied mineral distribution in quinoa seed by energy-dispersive X-ray microanalysis (EDX) and scanning electron microscopy (Fig. 11.3). Phosphorus, potassium, and magnesium are located in the embryo and contribute to formation of phytic acid and phytate in phytin globoids. Calcium is mainly in the pericarp and associated with pectin to develop the thick cell wall. Consequently, dehulling of quinoa would result in decreased calcium content.

ANTI-NUTRITIONAL FACTORS OF QUINOA

There are several anti-nutritional factors in quinoa, including saponins (0.1–5%), phytic acid (1.05–1.35%), and protease inhibitors (<50 ppm) (Jancurová et al. 2009; Vega-Gálvez et al. 2010). These factors could cause negative effects to the nutritional, sensory, and quality aspects of quinoa products. Phytic acid could bind minerals and affect their metabolisms and functions. Quinoa also contains small amounts of trypsin inhibitors, though at a much lower content than that in other grains (Vega-Gálvez et al. 2010).

Saponins, found in the pericarp of the quinoa seed, are the most significant anti-nutritional factor in quinoa (Vega-Gálvez et al. 2010). Saponins are plant glycosides that impart a bitter taste and form insoluble complexes with minerals that affect mineral absorption. As saponins are contained in the pericarp of seed, they are easily removed through mechanical abrasion. Rinsing the seed in cold alkaline water would also help remove the saponins (Jancurová et al. 2009).

Saponins are glycosylated secondary metabolites that consist of a hydrophilic carbohydrate chain and a lipophilic triterpene aglycone (Wink 2004). Saponins in quinoa consist of carbohydrate chains attached to a hydrophobic aglycone by glycosidic links. The carbohydrate chains of quinoa saponins normally include arabinose, glucose, galactose, glucuronic acid, xylose, and rhamnose (Kuljanabhagavad and Wink 2009). Soponins are categorized into two groups (i) monodesmosidic saponins with single carbohydrate chain and (ii) bidesmosidic saponins with two carbohydrate chains (Kuljanabhagavad and Wink 2009). Monodesmosidic saponins are emulsifying agents and form stable foam in water. They also exhibit higher toxicity compared to the bidesmosidic saponins.

Although saponins are considered antinutritional factors, they also have beneficial properties, including analgesic, anti-inflammatory, antimicrobial, antioxidant, antiviral, cytotoxic and hemolytic activity, and immunostimulatory effects, plus neuroprotective action. Saponins also increase the permeability of the intestinal mucosa and reduce fat absorption (Abugoch 2009).

Table 11.14 Chemical Structure of the Aglycones in Quinoa (Kuljanabhagavad and Wink 2009).

Aglycone	R_1	R_2	R_3	Formula	MW
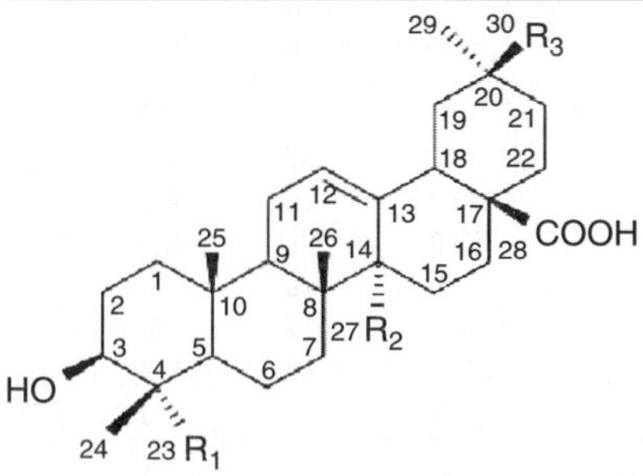					
Oleanolic acid (I)	CH_3	CH_3	CH_3	$C_{30}H_{48}O_3$	456
Hederagenin (II)	CH_2OH	CH_3	CH_3	$C_{30}H_{48}O_4$	472
Phytolaccagenic acid (III)	CH_2OH	CH_3	$COOCH_3$	$C_{31}H_{48}O_6$	516
Serjanic acid (IV)	CH_3	CH_3	$COOCH_3$	$C_{31}H_{48}O_5$	500
3β-Hydroxy-23-oxo-olean-12-en-28-oic acid (V)	CHO	CH_3	CH_3	$C_{30}H_{46}O_4$	470
3β-Hydroxy-27-oxo-olean-12-en-28-oic acid (VI)	CH_3	CHO	CH_3	$C_{30}H_{46}O_4$	470
3β,23α,30β-Trihydroxy-olean-12-en-28-oic acid (VII)	CH_2OH	CH_3	CH_2OH	$C_{30}H_{48}O_5$	488

Recent studies show that there is a growing interest in the bioactivity and health benefits of saponins. The beneficial effects of saponins are influenced by various functional groups at different position of aglycones skeletons (Table 11.14) (Kuljanabhagavad and Wink 2009). Monodesmosidic saponins can affect the fluidity and permeability of cell membrane (Kuljanabhagavad and Wink 2009). High concentrations of monodesmosidic saponins are toxic because they can lyse, or break down, animal cells. However, moderate amounts of saponins can have antifungal and antibacterial activity by lysing bacterial and fungal cells, such as those of *Candida albicans* (Woldemichael and Wink 2001). The antifungal activity of saponins is dependent on the functional group of aglycone skeletons. Stuardo and San Martín (2008) pointed out that carbohydrate chain attached at C3 is critical for antifungal properties. Use of a saponin-based molluscicide to control *Pomacea canaliculata* snails was tried in Southern Brazil and the results showed that it could be effective in rice fields that are not irrigated with heavily polluted water (San Martin et al. 2008).

Inflammation is the nonimmune response to body injury due to pathogens, damaged cells, or irritants, with accompanying signs of increased blood flow, vasodilation, and release of soluble mediators (Ferrero-Miliani et al. 2007). Ordinary inflammation is an important defense for bodily health. However, inflammation can also be one of the basic reasons for many potentially fatal chronic diseases such as CVD (Myasoedova et al. 2011), type II diabetes (Donath and Shoelson 2011), eye disorders (Donoso et al. 2006), gastrointestinal disease (Camilleri et al. 2011), obesity (Mathieu et al. 2010; Sun and Karin 2012), cancer (Kozlov 2009), and neurodegenerative diseases, such as Alzheimer's and Parkinson's diseases (Jones 2001; Cameron and Landreth, 2010; Przedborski 2010). Successive studies of Ghosh et al. (1983) and Mujica (1994) indicated that saponins had anti-inflammatory activity and may explain why plant saponins were used as anti-inflammatory remedies in traditional medicine (Song and Hu 2009; Keller et al. 2011; Raju and Rao 2012). Monodesmosidic saponins in quinoa such as 3-*O*-β-D-glucopyranosyl oleanolic acid has the potential to inhibit inflammation (Ma et al. 1989) and 3-*O*-β-D-glucopyranosyl hederagenin has been reported to exhibit antioxidant activity (Ilhami et al. 2006).

BIOACTIVE COMPOUNDS

Phenolic compounds

Phenolic compounds containing at least one phenol ring are a major group of phytochemicals in cereals, located primarily in the outer layer of grain (Gross 1980). Phenolics are divided into simple phenols and polyphenols based on the number of phenol rings. Simple phenols include phenolic acids containing one aromatic ring with one or more hydroxyl groups. Polyphenols include phenolic acid dehydrodimers, flavonoids, lignans, and tannins that contain three or more phenol rings (Khan and Shewry 2009). Recent studies showed that phenolics exhibited high antioxidant activity (Djordjevic et al. 2011; Min et al. 2012; Guo and Beta 2013), which was determined by a combination of two methods (i) the DPPH (2,2-diphenyl-1-picylhydrazyl) scavenging capacity assay and (ii) ferric reducing antioxidant power (FRAP) assay (Karadag et al. 2009).

Total phenolics content can be expressed as gallic acid equivalent (GAE). Total phenolic compounds in quinoa seed have been reported to be 71.7 mg GAE/100 g, which is higher than that in wheat (53.1 mg GAE/100 g) and amaranth (21.2 mg GAE/100 g) (Alvarez-Jubete et al. 2010b). Red seeded quinoa was shown to contain 50% more total phenolic content, 90% more total flavonoid content, and 150% more FRAP than yellow seeded quinoa (Brend 2012).

Phenolic acid

Phenolic acids are a group of phenolic compounds with single phenol ring. Phenolic acids are mostly bound to cell structural components such as cellulose, protein, lignin, or flavonoids and sugars by the ester, ether, and acetyl linkages (Dervilly-Pinel et al. 2001; Yuan et al. 2005). Only a small fraction of phenolic acids are free acids. Phenolic acids provide diverse functions to plant, such as nutrient uptake, enzyme activity, microbial habitation, and protection against pathogens (Ikegawa et al. 1996; Kroon and Williamson, 1999). Gallic acid and rosmarinic acid are present with highest potential of antioxidant activity (Soobrattee et al. 2005). Ferulic acid, which is one of the main phenolic acids in cereal, not only exhibits high antioxidant activity but also cross links with arabinoxylans to form the cell wall and insoluble dietary fiber (Faulds et al. 2004; Zhou et al. 2005; Yadav et al. 2007).

Table 11.15 Phenolic Acid Content in Quinoa and Other Cereals (μg/g wb).

	Quinoa[a]	Whole Wheat[b]	Corn[b]	Rice[b]
Ferulic acid	440	890	380	240
Caffeic acid	40	37	26	ND
Vanillic acid	43.4	15	4.6	7.8
Gallic acid	320	ND	ND	ND
Cinnamic acid	10	ND	ND	ND
p-Coumaric acid	ND	37	31	76
p-Hydroxybenzoic acid	76.8	7.4	5.7	15
Syringic acid	ND	13	7.8	ND

[a] Paśko et al. (2008).
[b] Mattila et al. (2005).

Phenolic acids in quinoa were analyzed qualitatively and quantitatively by HPLC method (Mattila et al. 2005). Table 11.15 shows the type and content of phenolic acids in quinoa and other cereals. Ferulic acid in quinoa is lower than it is in whole wheat, comparable to corn, and higher than it is in rice. Gallic acid is the main phenolic acid in quinoa. *p*-Hydroxybenzoic acid, vanillic acid, *p*-coumaric acid, and cinnamic acid are also found in quinoa with significantly higher content compared to those in whole wheat, corn, and rice.

Flavonoids

Flavonoids are another group of phenolic compounds with 2-phenyl-1,4-benzopyrone backbone and divided into subgroups as flavones, isoflavones, flavan, proanthocyanidins, and anthocyanidins. Flavonoids such as quercetin and epicatechin exhibit antioxidant activity (Leopoldini et al. 2006; Yin et al. 2012; Giménez et al. 2013) and has a negative correlation with the risk of developing coronary heart disease (Kris-Etherton and Keen 2002; McCullough et al. 2012) and type II diabetes (Wedick et al. 2012). Flavonoids have the potential to benefit cognitive performance both acutely and chronically (Lamport et al. 2012). Evidence indicates that flavonoids have potential to reduce the risk of cervical cancer (Ju et al. 2012), lung cancer (Khan et al. 2012), leukemia (Spagnuolo et al. 2012), breast cancer (Zamora-Ros et al. 2013a), colorectal cancer (Zamora-Ros et al. 2013b), and prostate cancer (Adhami et al. 2012).

Flavonoids in quinoa include quercetin glycosides (43.4 μmol/100 g dw) and kaempferol glycosides (36.7 43.4 μmol/100 g dw) and are higher than that in buckwheat, which has 30.1 μmol/100 g dw quercetin glycosides and nondetectable kaempferol glycosides (Alvarez-Jubete et al. 2010b). The temperature and length of process can cause the loss of flavonoids in cereals (Sensoy et al. 2006). However, bread made from 100% quinoa flour maintained most of the flavonoids content (Sensoy et al. 2006).

Isoflavones are a group of flavonoids that act as phytoestrogens in animals and human beings. They also stimulate osteoprogerin secretion, improve bone health, reduce arterial resistance, and exhibit antioxidant activities (Hsu et al. 2001; Teede et al. 2003; Li et al. 2013). Quinoa seed has been shown to contain isoflavones, particularly daidzein and genistein (Vega-Gálvez et al. 2010).

Anthocyanins are a group of flavonoid constituents that provide the bright red-orange to blue-violet colors in the plants. Epidemiological studies showed that intake of anthocyanins decreases the development of CVD (Mink et al. 2007). The role of anthocyanins in CVD relate strongly with protection against oxidative stress (Wallace 2011). They provide protection from DNA cleavage, estrogenic activity, enzyme inhibition, increased cytokine production, decreased capillary permeability and fagility, and membrane strengthening (Ramirez-Tortosa 2001; Acquaviva et al. 2003; Lazze et al. 2003; Rossi, et al. 2003; Lefevre et al. 2008). A low dose of anthocyanins intake can reduce ischemia, lower blood pressure and lipid levels, and reduce inflammation in patients with CVD. Anthocyanin content in quinoa has been reported at 102.4 mg CGE/100 g DW (Paśko et al. 2009), higher than that of jasmine rice, amaranth, soy bean (Gorinstein et al. 2007), and sorghum (Awika et al. 2005).

Carotenoids

Carotenoids are antioxidants, presenting provitamin A activity, colorants, and essential components of plant (Dini et al. 2010). They can quench singlet oxygen, which is an important reactive oxygen species in light-induced oxygen. Carotenoids are also reported to prevent age-related macular degeneration (AMD) and benefit eye health (Alves-Rodrigues and Shao 2004). They can reduce the risk of coronary heart disease, protect from ischemic stroke, and protect skin from UV-induced damage (Alves-Rodrigues and Shao 2004). The carotenoids content in sweet quinoa is 0.4 mg/10 g, greater than the content in einkorn (0.1 mg/10 g) (Hidalgo and Brandolini 2008) and corn (0.07 mg/10 g) (Scot and Eldridge 2005). The carotenoids in bitter quinoa are about 0.08 mg/10 g.

SUMMARY

Quinoa is called "a complete food" for several reasons. First, quinoa is high in protein, both quantitatively and qualitatively. It contains all the essential amino acids, whereas many other crops are deficient in some of them. For this reason, quinoa is a perfect protein supplement, especially for vegetarians. Second, the lipids of quinoa are composed of a large proportion of unsaturated fatty acids, which bring more health benefits than saturated fatty acids. Third, quinoa is rich in vitamin B complex, vitamin E, and minerals, which are essential to human health. Last but not least, bioactive compounds such as phenolic compounds, flavonoids, and carotenoids in quinoa are the critical antioxidants in diet and provide vital health benefits.

REFERENCES

Abeywardena MY, McLennan PL, Charnock JS. 1991. Differential effects of dietary fish oil on myocardial prostaglandin I2 and thromboxane A2 production. Am J Physiol - Heart Circ Physiol 260(2):379–385.

Abugoch LEJ. 2009. Chapter 1: Quinoa (*Chenopodium quinoa* Willd.). Adv Food Nutr Res 58:1–31.

Abugoch LE, Castro E, Tapia C, Añón MC, Gajardo P, Villarroel A. 2009. Stability of quinoa flour proteins (*Chenopodium quinoa* Willd.) during storage. Int J Food Sci Tech 44(10):2013–2020.

Abugoch LE, Romero N, Tapia CA, Silva J, Rivera M. 2008. Study of some physicochemical and functional properties of quinoa (*Chenopodium quinoa* Willd) protein isolates. J Agric Food Chem 56(12):4745–4750.

Acquaviva R, Russo A, Galvano F, Galvano G, Barcellona ML, Li VG, Vanella A. 2003. Cyanidin and cyanidin 3-O-beta-D-glucoside as DNA cleavage protectors and antioxidants. Cell Biol Toxicol 19(4):243–252.

Adhami VM, Syed DN, Khan N, Mukhtar H. 2012. Dietary flavonoid fisetin: a novel dual inhibitor of Pi3k/Akt and Mtor for prostate cancer management. Biochem Pharmacol 84(10):1277–1281.

Ahamed NT, Singhal RS, Kulkami PR, Palb M. 1996. Physicochemical and functional properties of *Chenopodium orcinoa* starch. Carbohyd Polym 31:99–103.

Aluko RE, Monu E. 2003. Functional and bioactive properties of quinoa seed protein hydrolysates. J Food Sci 68(4):1254–1258.

Alvarez-Jubete L, Arendt EK, Gallagher E. 2010a. Nutritive value of pseudocereals and their increasing use as functional gluten-free ingredients. Trends Food Sci Tech 21(2):106–113.

Alvarez-Jubete L, Wijngaard H, Arendt EK, Gallagher E. 2010b. Polyphenol composition and *in vitro* antioxidant activity of amaranth, quinoa, buckwheat and wheat as affected by sprouting and baking. Food Chem 119(2):770–778.

Alves-Rodrigues A, Shao A. 2004. The science behind lutein. Toxicol Lett 150(1):57–83.

Ando H, Chen YC, Tang H, Shimizu M, Watanabe K, Mitsunaga T. 2002. Food components in fractions of quinoa seed. Food Sci Technol 8(1):80–84.

Araujo-Farro PC, Podadera G, Sobral GAP, Menegalli FC. 2010. Development of films based on quinoa (*Chenopodium quinoa* Willd.) starch. Carbohyd Polym 81(4):839–848.

Awika JM, Rooney LW, Waniska RD. 2005. Anthocyanins from black sorghum and their antioxidant properties. Food Chem 90(2):293–301.

Brend Y. 2012. Total phenolic content and antioxidant activity of red and yellow quinoa (*Chenopodium quinoa* Willd.) seeds as affected by baking and cooking conditions. Food Nutr Sci 3(08):1150–1155.

Cai S, Yu G, Chen X, Huang Y, Jiang X, Zhang G, Jin X. 2013. Grain protein content variation and its association analysis in barley. BMC Plant Biol 13:35.

Cameron B, Landreth GE. 2010. Inflammation, microglia, and Alzheimer's Disease. Neurobiol Dis 37(3):503–509.

Camilleri M, Carlson P, McKinzie S, Zucchelli M, D'Amato M, Busciglio I, Burton D, Zinsmeister AR. 2011. Genetic susceptibility to inflammation and colonic transit in lower functional gastrointestinal disorders: preliminary analysis. Neurohastroenterol Motil: Off J Eur Gastrointest Motil Soc 23(10):e935–e938.

Chait A, Brunzell JD, Denke MA, Eisenberg D, Ernst ND, Franklin FA, Ginsberg H. 1993. Rationale of the diet-heart statement of the American Heart Association. Report of Nutrition Committee. Circulation 88(6):3008–3029.

Chan JK, McDonald BE, Gerrard JM, Bruce VM, Weaver BJ, Holub BJ. 1993. Effect of dietary alpha-linolenic acid and its ratio to linoleic acid on platelet and plasma fatty acids and thrombogenesis. Lipids 28(9):811–817.

Chen Z, Schols HA, Voragen AGJ. 2003. Starch granule size strongly determines starch noodle processing and noodle quality. Food Chem Toxicol 68(5):1584–1589.

Cotton PA, Subar AF, Friday JE, Cook A. 2004. Dietary sources of nutrients among us adults, 1994 to 1996. J Am Diet Assoc 104(6):921–930.

Crumpacker CS. 2010. Invited commentary: human cytomegalovirus, inflammation, cardiovascular disease, and mortality. Am J Epidem 172(4):372–374.

Cunnane SC, Thompson LU. 1995. Flaxseed in human nutrition. In: Chamberlain JG, Nelson GJ, editors. The effect of dietary alpha-linolenic acid on blood lipids and lipoproteins in humans. Champaign, IL: AOAC Press. 1–458.

Damodaran S, Parkin KL, Fennema OR. 2008. Fennema's food chemistry. Boca Raton, FL: CRC Press. 1–1160.

Dervilly-Pinel G, Rimsten L, Saulnier L, Andersson R, Åman P. 2001. Water-extractable arabinoxylan from pearled flours of wheat, barley, rye and triticale. Evidence for the presence of ferulic acid dimers and their involvement in gel formation. J Cereal Sci 34(2):207–214.

van Dijk SJ, Feskens EJ, Bos MB, Hoelen DW, Heijligenberg R, Bromhaar MG, de Groot LC, de Vries JH, Müller M, Afman LA. 2009. A saturated fatty acid-rich diet induces an obesity-linked proinflammatory gene expression profile in adipose tissue of subjects at risk of metabolic syndrome. Am J Clin Nutr 90(6):1656–1664.

Dini I, Tenore GC, Dini A. 2010. Antioxidant compound contents and antioxidant activity before and after cooking in sweet and bitter *Chenopodium quinoa* seeds. LWT - Food Sci Technol 43(3):447–451.

Di Sabatino A, Corazza GR. 2009. Coeliac disease. Lancet 373: 1480–1493.

Djordjevic TM, Šiler-Marinkovic SS, Dimitrijevic-Brankovic SI. 2011. Antioxidant activity and total phenolic content in some cereals and legumes. Int J Food Prop 14(1):175–184.

Donath MY, Shoelson SE. 2011. Type 2 diabetes as an inflammatory disease. Nature Rev Immunol 11(2):98–107.

Donoso LA, Kim D, Frost A, Callahan A, Hageman G. 2006. The role of inflammation in the pathogenesis of age-related macular degeneration. Surv Ophthalmol 51(2):137–152.

Eitenmiller RR, Landen WO. 1999. Vitamin analysis for the health and food Sciences. Boca Raton, FL: CRC Press. 1–518.

Fasano A, Berti I, Gerarduzzi T, Tarcisio Not, Colletti RB, Drago S, Elitsur Y, Green PHR, Guandalini S, Hill ID, Pietzak M, Ventura A, Thorpe M, Kryszak D, Fornaroli F, Wasserman SS, Murray JA, Horvath K. 2003. Prevalence of celiac disease in at-risk and not-at-risk groups in the United States: a large multicenter study. Arch Intern Med 163(3):286–292.

Fasano A, Catassi C. 2001. Current approaches to diagnosis and treatment of celiac disease: an evolving spectrum. Gastroenterology 120(3):636–651.

Faulds CB, Mandalari G, LoCurto R, Bisignano G, Waldron KW. 2004. Arabinoxylan and mono-and dimeric ferulic acid release from brewer's grain and wheat bran by feruloyl esterases and glycosyl hydrolases from *Humicola insolens*. App Microbiol Biotechnol 64(5):644–650.

Ferrero-Miliani L, Nielsen OH, Andersen PS, Girardin SE. 2007. Chronic inflammation: importance of Nod2 and Nalp3 in interleukin-1& beta generation. Clini Exp Immunol 147(2):227–235.

Friedman M, Brandon DL. 2001. Nutritional and health benefits of soy proteins. J Agric Food Chem 49(3):1069–1086.

Gallagher E, Gormley TR, Arendt EK. 2004. Recent advances in the formulation of gluten-free cereal-based products. Trends Food Sci Technol 15(3–4):143–152.

Garg ML, Wierzbicki AA, Thomson AB, Clandinin MT. 1989. Dietary saturated fat level alters the competition between alpha-linolenic and linoleic acid. Lipids 24(4):334–339.

Ghosh D, Thejomoorthy P, Veluchamy P 1983. Anti-inflammatory and analgesic activities of oleanolic acid 3-/3-glucoside (Rdg-1) from *Randia dumetorum* (Rubiaceae). Indian J Pharmacol 15(4):331–342.

Giménez B, Moreno S, López-Caballero ME, Montero P, Gómez-Guillén MC. 2013. Antioxidant properties of green tea extract incorporated to fish gelatin films after simulated gastrointestinal enzymatic digestion. LWT - Food Sci Technol 53(2):445–451.

Gorinstein S, Vargas OJM, Jaramillo NO, Salas IA, Ayala ALM, Arancibia-Avila P, Toledo F, Katrich E, Trakhtenberg S. 2007. The total polyphenols and the antioxidant potentials of some selected cereals and pseudocereals. Eur Food Res Technol 225(3–4):321–328.

Gross GG. 1980. Phenolic acids. In: Stumpf PK, Conn Eric E, editors. The biochemistry of plants: a comprehensive treatise. New York, NY: Academic Press. 707 p.

Guo WW, Beta T. 2013. Phenolic acid composition and antioxidant potential of insoluble and soluble dietary fibre extracts derived from select whole-grain cereals. Food Res Int 51(2):518–525.

Hidalgo A, Brandolini A. 2008. Protein, ash, lutein and tocols distribution in einkorn (*Triticum monococcum* L. subsp. *monococcum*) seed fractions. Food Chem 107(1):444–448.

Hoffenberg EJ, MacKenzie T, Barriga KJ, Eisenbarth GS, Bao F, Haas JE, Erlich H, Bugawan TL, Sokol RJ, Taki I, Norris JM, Rewers M. 2003. A prospective study of the incidence of childhood celiac disease. J Pediatr 143(3):308–314.

Hovdenak N, Hovlid E, Aksnes L, Fluge G, Erichsen MM, Eide J. 1999. High prevalence of asymptomatic coeliac disease in Norway. Eur J Gastroenterol Hepatol 11(2):185–188.

Hsu CS, Shen WW, Hsueh YM, Yeh SL. 2001. Soy isoflavone supplementation in postmenopausal women. Effects on plasma lipids, antioxidant enzyme activities and bone density. J Reprod Med 46(3):221–226.

Hu FB, Manson JE, Willett WC. 2001. Types of dietary fat and risk of coronary heart disease: a critical review. J Am Coll Nutr 20(1):5–19.

Hu FB, Stampfer MJ, Manson JE, Rimm EB, Wolk A, Colditz GA, Hennekens CH, Willett WC. 1999. Dietary intake of a-linolenic acid and risk of fatal ischemic heart disease among women. Am J Clin Nutr 69(5):890–897.

Hunnicutt JW, Hardy RW, Williford J, McDonald JM. 1994. Saturated fatty acid-induced insulin resistance in rat adipocytes. Diabetes 43(4):540–545.

Ikegawa T, Mayama S, Nakayashiki H, Kato H. 1996. Accumulation of diferulic acid during the hypersensitive response of oat leaves *Topuccinia coronata* f.sp. *avenae* and its role in the resistance of oat tissues to cell wall degrading enzymes. Physiol Mol Plant Pathol 48(4):245–256.

Ilhami AU, Mshvildadze V, Gepdiremen A, Elias R. 2006. The antioxidant activity of a triterpenoid glycoside isolated from the berries of *Hedera colchica*: 3-O-(B-D-Glucopyranosyl)-Hederagenin. Phytother Res 20(2):130–134.

Jacobs D, Blackburn H, Higgins M, Reed D, Iso H, McMillan G, Neaton J, Potter J, Rifkind B. 1992. Report of the conference on low blood cholesterol: mortality associations. Circulation 86(3):1046–1060.

Jancurová M, Minarovicova L, Dandar A. 2009. Quinoa – a review. Czech J Food Sci 27(2):71–79.

Jones RW. 2001. Inflammation and Alzheimer's disease. Lancet 358(9280):436–437.

Ju HK, Lee HW, Chung KS, Choi JH, Cho JG, Baek NI, Chung HG, Lee TK. 2012. Standardized flavonoid-rich fraction of *Artemisia princeps* Pampanini cv. Sajabal induces apoptosis via mitochondrial pathway in human cervical cancer Hela cells. J Ethnopharmacol 141(1):460–468.

Juliano BO. 1985. Rice: chemistry and technology. Saint Paul, MN: American Association of Cereal Chemists. 774 p.

Karadag A, Ozcelik B, Saner S. 2009. Review of methods to determine antioxidant capacities. Food Anal Methods 2(1):41–60.

Keller AC, Ma J, Kavalier A, He K, Brillantes AM, Kennelly EJ. 2011. Saponins from the traditional medicinal plant *Momordica charantia* stimulate insulin secretion *in vitro*. Phytomed 19(1):32–37.

Khan N, Afaq F, Khusro FH, Mustafa Adhami V, Suh Y, Mukhtar H. 2012. Dual inhibition of phosphatidylinositol 3-kinase/Akt and mammalian target of rapamycin signaling in human non-small cell lung cancer cells by a dietary flavonoid fisetin. Int J Cancer 130(7):1695–1705.

Khan K, Shewry PR. 2009. Wheat: chemistry and technology. Saint Paul, MN: AACC International. 1–476.

Kim SK, Byun HG, Park PJ, Shahidi F. 2001. Angiotensin I converting enzyme inhibitory peptides purified from bovine skin gelatin hydrolysate. J Agric Food Chem 49(6):2992–2997.

Kline K, Lawson KA, Yu W, Sanders BG. 2007. Vitamin E and cancer. Vitam Horm 76:435–461.

Konishi Y, Hirano S, Tsuboi H, Wada M. 2004. Distribution of minerals in quinoa (*Chenopodium quinoa* Willd.) seeds. Biosci Biotechnol Biochem 68(1):231–234.

Koziol MJ. 1992. Chemical composition and nutritional evaluation of quinoa (*Chenopodium quinoa* Willd.). J Food Comp Anal 5:35–68.

Kozlov SV. 2009. Inflammation and cancer methods and protocols. [Internet] New York, NY: Humana Press. Available from: 10.1007/978-1-60327-530-9

Kris-Etherton PM, Keen CL. 2002. Evidence that the antioxidant flavonoids in tea and cocoa are beneficial for cardiovascular health. Curr Op Lipidol 13(1):41–49.

Kroon PA, Williamson G. 1999. Hydroxycinnamates in plants and food: current and future perspectives. J Sci Food Agric 79(3):355.

Kuljanabhagavad T, Wink M. 2009. Biological activities and chemistry of saponins from *Chenopodium quinoa* Willd. Phytochem Rev 8(2): 473–490.

Lamport DJ, Dye L, Wightman JD, Lawton CL. 2012. The effects of flavonoid and other polyphenol consumption on cognitive performance: a systematic research review of human experimental and epidemiological studies. Nutr Aging 1(1):5–25.

Lazze MC, Pizzala R, Savio M, Stivala LA, Prosperi E, Bianchi L. 2003. Anthocyanins protect against DNA damage induced by tert-butyl-hydroperoxide in rat smooth muscle and hepatoma cells. Mutat Res Genet Toxicol Environ Mutagen 535(1):103–115.

Lefevre M, Wiles JE, Zhang X, Howard LR, Gupta S, Smith AA, Ju ZY, DeLany JP. 2008. Gene expression microarray analysis of the effects of grape anthocyanins in mice: a test of a hypothesis-generating paradigm. Metabolism 57:S52-S57.

Leopoldini M, Russo N, Chiodo S, Toscano M. 2006. Iron chelation by the powerful antioxidant flavonoid quercetin. J Agric Food Chem 54(17):6343–6351.

Li Y, Kong D, Ahmad A, Bao B, Sarkar FH. 2013. Antioxidant function of isoflavone and 3,3-diindolylmethane: are they important for cancer prevention and therapy? Antioxid Redox Signaling 19 (2):139–150.

Lindeboom N, Chang PR, Falk KC, Tyler RT. 2005. Characteristics of starch from eight quinoa lines. Cereal Chem 82(2):216–222.

Lindeboom N, Chang PR, Tyler RT. 2004. Analytical, biochemical and physicochemical aspects of starch granule size, with emphasis on small granule starches: a review. Starch - Stärke 56(34):89–99.

Lovejoy JC. 1999. Dietary fatty acids and insulin resistance. Curr Atheroscler Rep 1(3):215–220.

Lovejoy JC, DiGirolamo M. 1992. Habitual dietary intake and insulin sensitivity in lean and obese adults. Am J Clin Nutr 55(6):1174–1179.

Ma WW, Heinstein PF, McLaughlin JL. 1989. Additional toxic, bitter saponins from the seeds of *Chenopodium quinoa*. J Nat Prod 52(5):1132–1135.

Mathieu P, Lemieux I, Després JP. 2010. Obesity, inflammation, and cardiovascular risk. Clin Pharmacol Ther 87(4):407–416.

Mattila P, Pihlava JM, Hellström J. 2005. Contents of phenolic acids, alkyl- and alkenylresorcinols, and avenanthramides in commercial grain products. J Agric Food Chem 53(21):8290–8295.

Maxwell SRJ, Lip GYH. 1997. Free radicals and antioxidants in cardiovascular disease. Br J Clin Pharmacol 44(4):307.

McCullough ML, Peterson JJ, Patel R, Jacques PF, Shah R, Dwyer JT. 2012. Flavonoid intake and cardiovascular disease mortality in a prospective cohort of us adults. Am J Clin Nutr 95(2):454–464.

Min B, Gu LW, McClung AM, Bergman CJ, Chen MH. 2012. Free and bound total phenolic concentrations, antioxidant capacities, and profiles of proanthocyanidins and anthocyanins in whole grain rice (*Oryza sativa* L.) of different bran colours. Food Chem 133(3):715–722.

Mink PJ, Scrafford CG, Barraj LM, Harnack L, Hong CP, Nettleton JA, Jacobs DR. 2007. Flavonoid intake and cardiovascular disease mortality: a prospective study in postmenopausal women. Am J Clin Nutr 85(3):895–909.

Mujica A. 1994. Andean grains and legumes. Leon J, Bermejo JEH. Neglected crops: 1492 from a different perspective. Rome: FAO, Plant Production and Protection. 1–341.

Myasoedova E, Crowson CS, Kremers HM, Roger VL, Fitz-Gibbon PD, Therneau TM, Gabriel SE. 2011. Lipid paradox in rheumatoid arthritis: the impact of serum lipid measures and systemic inflammation on the risk of cardiovascular disease. Ann Rheum Dis 70(3):482–487.

Ng SC, Anderson A, Coker J, Ondrus M. 2007. Characterization of lipid oxidation products in quinoa (*Chenopodium quinoa*). Food Chem 101(1):185–192.

Ogungbenle HN. 2003. Nutritional evaluation and functional properties of quinoa (*Chenopodium quinoa*) flour. Int J Food Sci Nutr 54(2):153–158.

Ohman L, Simrén M. 2010. Pathogenesis of Ibs: role of inflammation, immunity and neuroimmune interactions. Nature Reviews. Gastroenterol Hepatol 7(3):163–173.

Oliver E, McGillicuddy F, Phillips C, Toomey S, Roche HM. 2010. Postgraduate symposium: the role of inflammation and macrophage accumulation in the development of obesity-induced Type 2 Diabetes mellitus and the possible therapeutic effects of long-chain N-3. Proc Nutr Soc 69(2):232–243.

Papas A, Vos E. 2001. Vitamin E, cancer, and apoptosis. Am J Clin Nutr 73(5):1113–1114.

Paśko P, Bartoń H, Zagrodzki P, Gorinstein S, Fołta M, Zachwieja Z. 2009. Anthocyanins, total polyphenols and antioxidant activity in amaranth and quinoa seeds and sprouts during their growth. Food Chem 115(3):994–998.

Paśko P, Sajewicz M, Gorinstein S, Zachwieja Z. 2008. Analysis of selected phenolic acids and flavonoids in *Amaranthus cruentus* and *Chenopodium quinoa* seeds and sprouts by HPLC. Acta Chromatogr 20(4):661–672.

Prego I, Maldonado S, Otegui M. 1998. Seed structure and localization of reserves in *Chenopodium quinoa*. Ann Bot 82:481–488.

Pryor WA. 2000. Vitamin E and heart disease: basic science to clinical intervention trials. Free Radic Biol Med 28(1):141–164.

Przedborski S. 2010. Inflammation and Parkinson's Disease pathogenesis. Mov Disord 25(1):55–57.

Raju J, Rao CV. 2012. Diosgenin, a steroid saponin constituent of yams and fenugreek: emerging evidence for applications in medicine. [Internet]. INTECH Open Access Publisher. Available from: http://www.intechopen.com/articles/show/title/diosgenin-a-steroid-saponin-constituent-of-yams-and-fenugreek-emerging-evidence-for-applications-in-

Ramirez-Tortosa C. 2001. Anthocyanin-rich extract decreases indices of lipid peroxidation and DNA damage in vitamin E-depleted rats. Free Radic Biol Med 31(9):1033–1037.

Ranhotra GS, Gelroth JA, Glaser BK, Lorenz KJ, Johnson DL. 1993. Composition and protein nutritional quality of quinoa. Cereal Chem 70(3):303–305.

Renaud S, Nordøy A. 1983. Small is beautiful: alpha-linolenic acid and eicosapentaenoic acid in man. Lancet 1(8334):1169.

Rieder CR, Fricke D. 2001. Vitamin B12 and folate in relation to the development of Alzheimer's Disease. Neurology 57(9):1742–1743.

Rimm EB, Willett WC, Hu FB, Sampson L, Colditz GA, Manson JE, Hennekens C, Stampfer MJ. 1998. Folate and vitamin B6 from diet and supplements in relation to risk of coronary heart disease among women. JAMA: J Am Med Assoc 279(5):359–364.

Rossi A, Serraino I, Dugo P, Di Paola R, Mondello L, Genovese T, Morabito D, Dugo G, Sautebin L, Caputi AP, Cuzzocrea S. 2003. Protective effects of anthocyanins from blackberry in a rat model of acute lung inflammation. Free Radic Res 37(8):891–900.

Ruales J, de Grijalva Y, Lopez-Jaramillo P, Nair BM. 2002. The nutritional quality of an infant food from quinoa and its effect on the plasma level of insulin-like growth factor-1 (Igf-1) in undernourished children. Int J Food Sci Nutr 53(2):143–154.

Sacks FM, Katan M. 2002. Randomized clinical trials on the effects of dietary fat and carbohydrate on plasma lipoproteins and cardiovascular disease. Am J Med 113(9B):13–24.

Salmerón J, Hu FB, Manson JE, Stampfer MJ, Colditz GA, Rimm EB, Willett WC. 2001. Dietary fat intake and risk of Type 2 diabetes in women. Am J Clin Nutr 73:1019–1026.

San Martin R, Ndjoko K, Hostettmann K. 2008. Novel molluscicide against *Pomacea canaliculata* based on quinoa (*Chenopodium quinoa*) saponins. Crop Prot 27(3–5):310–319.

Scot CE, Eldridge AL. 2005. Comparison of carotenoid content in fresh, frozen and canned corn. J Food Comp Anal 18(6):551–559.

Sensoy I, Rosen RT, Ho CT, Mukund VK. 2006. Effect of processing on buckwheat phenolics and antioxidant activity. Food Chem 99(2):388–393.

Skarupski KA, Tangney C, Li H, Ouyang B, Evans DA, Morris MC. 2010. Longitudinal association of vitamin B-6, folate, and vitamin B-12 with depressive symptoms among older adults over time. Am J Clin Nutr 92(2):330–335.

Song XM, Hu SH. 2009. Adjuvant activities of saponins from traditional Chinese medicinal herbs. Vaccine 27(36):4883–4890.

Soobrattee MA, Neergheen VS, Luximon-Ramma A, Aruoma OI, Bahorun T. 2005. Phenolics as potential antioxidant therapeutic agents: mechanism and actions. Mutat Res Fund Mol Mech Mut 579(1–2):200–213.

Spagnuolo C, Russo M, Bilotto S, Tedesco I, Laratta B, Russo GL. 2012. Dietary polyphenols in cancer prevention: the example of the flavonoid quercetin in leukemia. Ann NY Acad Sci 1259(1):95–103.

Stuardo M, San Martín R. 2008. Antifungal properties of quinoa (*Chenopodium quinoa* Willd.) alkali treated saponins against *Botrytis cinerea*. Ind Crop Prod 27(3):296–302.

Sun B, Karin M. 2012. Obesity, inflammation and liver cancer. J Hepatol 56(3):704–713.

Takao T, Watanabe N, Yuhara K, Itoh S, Suda S, Tsuruoka Y, Nakatsugawa K, Konishi Y. 2005. Hypocholesterolemic effect of protein isolated from quinoa (*Chenopodium quinoa* Willd.) seeds. Food Sci Technol Res 11(2):161–167.

Tang H. 2004. Relationship between functionality and structure in barley starches. Carbohydr Polym 57(2):145–152.

Tang H, Mitsunaga T, Kawamura Y. 2005. Functionality of starch granules in milling fractions of normal wheat grain. Carbohydr Polym 59(1):11–17.

Tang CH, Ten Z, Wang XS, Yang XQ. 2006. Physicochemical and functional properties of hemp (*Cannabis sativa* L.) protein isolate. J Agric Food Chem 54(23):8945–8950.

Tang H, Yoshida T, Watanabe K, Mitsunaga T. 1998. Some properties of starch granules in various plants. Kinki Daigaku Nougaku Sougou Kenkyusho Houkoku 6:83–89 [in Japanese].

Teede HJ, McGrath BP, DeSilva L, Cehun M, Fassoulakis A, Nestel PJ. 2003. Isoflavones reduce arterial stiffness: a placebo-controlled study in men and postmenopausal women. Arterioscler Thromb Vasc Biol 23(6):1066–1071.

Trumbo PR. 2005. The level of evidence for permitting a qualified health claim: FDA's review of the evidence for selenium and cancer and Vitamin E and heart Disease. J Nutr 135(2):354–356.

Ubbink JB, Vermaak WJ, van der Merwe A, Becker PJ. 1993. Vitamin B-12, Vitamin B-6, and folate nutritional status in men with hyperhomocysteinemia. Am J Clin Nutr 57(1): 47–53.

[USDA] U.S. Department of Agriculture, Agricultural Research Service. (2011) [Internet]. USDA national nutrient database for standard reference, Release 18. Nutrient Data Laboratory Home Page. Available from: http://ndb.nal.usda.gov/.

[USDA] U.S. Department of Agriculture, Agricultural Research Service. (2005) [Internet]. Dietary reference intakes: macronutrients. National Academy of Sciences. Institute of Medicine. Food and Nutrition Board. Available from: http://www.nal.usda.gov/fnic/DRI/DRI_Energy/energy_full_report.pdf.

Valencia-Chamorro SA. 2003. Quinoa. Caballero B. Encyclopedia of food science and nutrition, 8. Amsterdam, The Netherlands: Academic Press. 4895–4902.

Vega-Gálvez A, Miranda M, Vergara J, Uribe E, Puente L, Martínez EA. 2010. Nutrition facts and functional potential of quinoa (*Chenopodium quinoa* Willd.), an ancient Andean grain: a review. J Sci Food Agric 90(15):2541–2547.

Wallace TC. 2011. Anthocyanins in cardiovascular disease. Adv Nutr 2(1):1–7.

Wang M, Hettiarachchy NS, Qi M, Burks W, Siebenmorgen T. 1999. Preparation and functional properties of rice bran protein isolate. J Agric Food Chem 47(2):411–416.

Watanabe K, Peng NL, Tang H, Misunaga T. 2007. Molecular structural characteristics of quinoa starch. Food Sci Technol Res 13(1):73–76.

Wedick NM, Pan A, Cassidy A, Rimm EB, Sampson L, Rosner B, Willett W, Hu FB, Sun Q, van Dam RM. 2012. Dietary flavonoid intakes and risk of Type 2 diabetes in US men and women. Am J Clin Nutr 95(4):925–933.

West J, Logan RFA, Hill PG, Lloyd A, Lewis S, Hubbard R, Reader R, Holmes GKT, Khaw KT. 2003. Seroprevalence, correlates, and characteristics of undetected coeliac disease in England. Gut 52(7):960–965.

Wilson JD, Bechtel DB, Todd TC, Seib PA. 2006. Measurement of wheat starch granule size distribution using image analysis and laser diffraction technology. Cereal Chem J 83(3):259–268.

Wink M. 2004. Phytochemical diversity of secondary metabolites. Encyclopedia of plant & crop science. New York, NY: Marcel Dekker. 1–1329.

Woldemichael GM, Wink M. 2001. Identification and biological activities of triterpenoid saponins from *Chenopodium quinoa*. J Agric Food Chem 49(5):2327–2332.

Woldemichael GM, Wink M. 2009. Biological activities and chemistry of saponins from *Chenopodium quinoa* Willd. Phytochem Rev 8(2):473–490.

Wright KH, Pike OA, Fairbanks DJ, Huber CS. 2002. Composition of *Atriplex hortensis*, sweet and bitter *Chenopodium quinoa* seeds. J Food Sci 67(4):1383–1385.

Wu J, Ding X. 2001. Hypotensive and physiological effect of angiotensin converting enzyme inhibitory peptides derived from soy protein on spontaneously hypertensive rats. J Agric Food Chem 49(1):501–506.

Wymann MP, Solinas G. 2013. Inhibition of phosphoinositide 3-kinase attenuates inflammation, obesity, and cardiovascular risk factors. Ann NY Acad Sci 1280(1):44–47.

Yadav MP, Moreau RA, Hicks KB. 2007. Phenolic acids, lipids, and proteins associated with purified corn fiber arabinoxylans. J Agric Food Chem 55(3):943–947.

Yang Q, Cogswell ME, Hamner HC, Carriquiry A, Bailey LB, Pfeiffer CM, Berry RJ. 2010. Folic acid source, usual intake, and folate and Vitamin B-12 status in US adults: national health and nutrition examination survey (Nhanes) 2003–2006. Am J Clin Nutr 91(1):64–72.

Yin J, Becker EM, Andersen ML, Skibsted LH. 2012. Green tea extract as food antioxidant. Synergism and antagonism with A-tocopherol in vegetable oils and their colloidal systems. Food Chem 135(4):2195–2202.

Yuan XP, Wang J, Yao H. 2005. Antioxidant activity of feruloylated oligosaccharides from wheat bran. Food Chem 90(4):759–764.

Zamora-Ros R, Ferrari P, González CA, Tjønneland A, Olsen A, Bredsdorff L, Overvad K, Touillaud M, Perquier F, Fagherazzi G, Lukanova A, Tikk K, Aleksandrova K, Boeing H, Trichopoulou A, Trichopoulos D, Dilis V, Masala G, Sieri S, Mattiello A, Tumino R, Ricceri F, Bueno-de-Mesquita HB, Peeters PH, Weiderpass E, Skeie G, Engeset D, Menéndez V, Travier N, Molina-Montes E, Amiano P, Chirlaque MD, Barricarte A, Wallström P, Sonestedt E, Sund M, Landberg R, Khaw KT, Wareham NJ, Travis RC, Scalbert A, Ward HA, Riboli E, Romieu I. 2013a. Dietary flavonoid and lignan intake and breast cancer risk according to menopause and hormone receptor status in the European Prospective Investigation into Cancer and Nutrition (EPIC) study. Breast Cancer Res Treat 139(1):163–176.

Zamora-Ros R, Not C, Guinó E, Luján-Barroso L, García RM, Biondo S, Salazar R, Moreno V. 2013b. Association between habitual dietary flavonoid and lignan intake and colorectal cancer in a Spanish case–control study (the Bellvitge Colorectal Cancer study). Cancer Causes Control 24(3): 549–557.

Zhou K, Yin JJ, Yu LL. 2005. Phenolic acid, tocopherol and carotenoid compositions, and antioxidant functions of hard red winter wheat bran. J Agric Food Chem 53(10):3916–3922.

(a) (b) (c)

Plate 3.1 (a–c) Examples of quinoa plants in farmer's field showing the vast diversity of colors and the forms of panicles (Province of Omasuyos, bordered to the south and west by Lake Titicaca, Bolivia) (Del Castillo and Winkel, IRD – CLIFA, 2002–2008).

Quinoa: Improvement and Sustainable Production, First Edition. Edited by Kevin Murphy and Janet Matanguihan.

Plate 5.1 Phytophagous insects associated with quinoa in the Neotropical region. (a) Lepidoptera, larvae feeding on quinoa leaf; (b) *Leptoglossus* sp. in quinoa leaf; (c,d) *Oiketicus kirbyi*, neotenic larvae coming out of bag or basket; (e) *O. kirbyi* bag or basket in quinoa branch; (f) adult *Eurysacca* sp.; (g) *Eurysacca* sp. larvae feeding on quinoa leaf; (h) Pyralidae, larvae; (i) *Nezara viridula*, adult in quinoa panicle; and (j) adult *Epicauta adspersa* in quinoa leaf.

Plate 5.2 Entomophagous insect on quinoa in Amaicha del Valle. (a) Parasitoid *Copidosoma* sp. adult; (b) *Copidosoma* sp. parasitism in *Eurysacca* larvae; (c) *Eurysacca* damaged pupa; (d) parasitoid Ichneumonidae; (e) unidentified parasitoid; (f–l) predator insects; (f) *Eriopis connexa*, immature stage; (g) *Chrysoperla argentina*, adult; (h) *C. externa*, immature stage; (i) *C. externa*, adult; (j) *C. argentina*, preying on *Spodoptera frugiperda* (Lepidoptera) eggs; (k) *C. argentina*, immature stage preying on aphids; and (l) *C. externa*, immature stage preying on aphids.

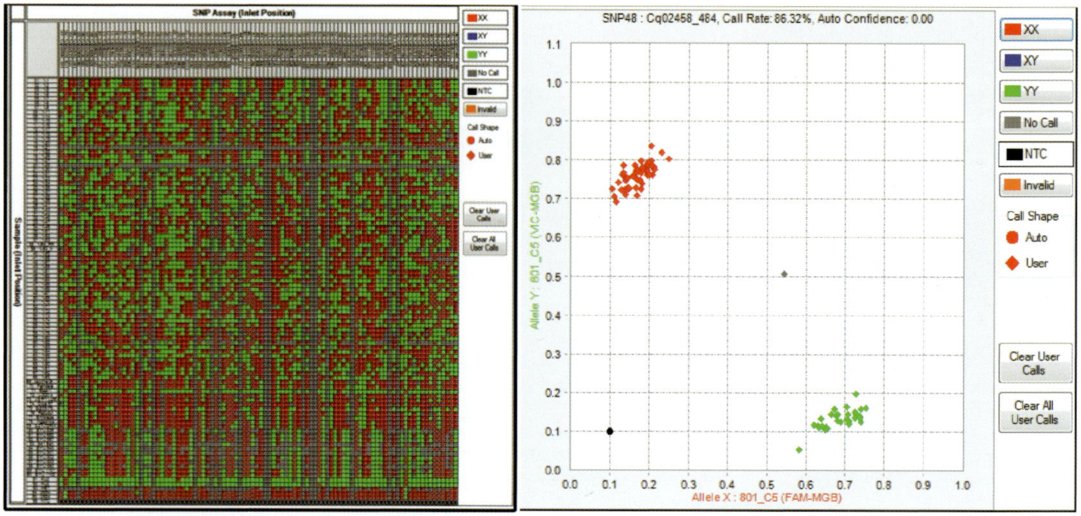

(a) (b)

Plate 7.2 Example of SNP assays using the KASPar^{TM} genotyping chemistry on the Fluidigm access array in the quinoa RIL mapping population. (a) The genotyping across the 96.96 IFC chip (96 DNA samples on the vertical, 96 SNP assays on the horizontal). (b) Individual SNP loci in a Cartesian graph. A no template control (NTC) and a synthetic heterozygote are identified.

Plate 8.2 Geographic distribution of quinoa in the Andean Region.

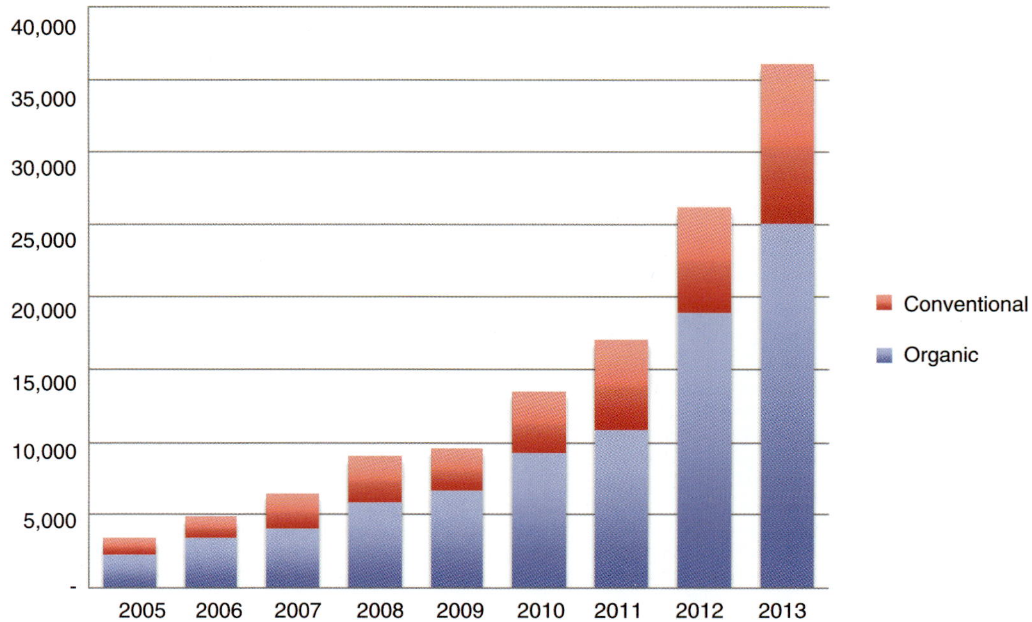

Plate 12.4 US quinoa imports in MT per year, organic versus nonorganic.

Plate 12.6 Multiple quinoa varieties are planted and harvested in the same fields. Photo credit: Vitaliy Prokopets.

Chapter 12
Quinoa's Calling

Sergio Núñez de Arco

Quinoa Specialist and Co-Founder, Andean Naturals, Inc., 393 Catamaran St., Foster City, CA, USA

INTRODUCTION

It is February 20, 2013, at the General Assembly of the United Nations in New York. Quinoa, the rediscovered staple of the Andean civilizations, is thrust into the limelight. It is the launching of the International Year of Quinoa (IYQ). The acting president of the UN opens the 64th Plenary Meeting with these words:

> Too often in the past, we have heard extravagant claims for the benefits of different foods, but it is my opinion that, in quinoa, we truly have a plant that deserves the title of 'super food.' In many ways, this traditional staple of the Andes region represents many of the ideals and goals of the United Nations. It has an important role to play in ensuring food security, boosting nutrition and, ultimately, eradicating poverty. It also brings attention to the importance that the United Nations attaches to indigenous knowledge and practice.

José Graziano da Silva, head of the FAO, further emphasizes the role that quinoa is expected to play: "We are here to recruit a new ally in combating hunger and food insecurity – quinoa."

Fast forward to December 31, 2013, the last day of the Year of Quinoa. A storekeeper in a small open market in La Paz, Bolivia complains that prices for quinoa have gone up so much that sales are slow for this item. "People complain about the price. They can get three bags of rice instead of one of quinoa." Bolivia is the second poorest country in the Western Hemisphere and the world's leading producer of quinoa. Even there, quinoa prices have doubled in 2013 reaching a historical high of \$4/lb.

In the United States, the main consumer of quinoa worldwide, quinoa prices are also set to double. Quinoa packed for retail is going from \$4/lb in September 2013 to \$6–9/lb depending on the presentation and retailer (Fig. 12.1). Quinoa is quickly gaining the reputation of an exclusive fancy food, a food for the wealthy: exactly the opposite of what the United Nations envisions.

As 2013 came to an end, the quinoa market was rife with speculation. Prices were at a peak and no one can predict how this will impact demand. The 2014 quinoa crops in both Peru and Bolivia were expected to be at least 30% larger than those of previous years, and there will be added competition on the market from nontraditional quinoa-producing countries.

This is the quinoa conundrum: The UN desires quinoa to be cheap, widespread, and accessible to the poor; yet 45,000 smallholder farmers in the Andes recently lifted out of poverty are grateful for high prices. Could quinoa both provide fair incomes for thousands of smallholder farmers and become a healthy, inexpensive staple to the poor of the world?

Quinoa: Improvement and Sustainable Production, First Edition. Edited by Kevin Murphy and Janet Matanguihan.

Fig. 12.1 Shelf at a Whole Foods Store in California shows quinoa prices ranging from $5.50 to $9.00/lb. Photo credit: Sergio Nuñez de Arco.

A SNAPSHOT OF THE ECONOMICS OF A SMALLHOLDER FARMER IN BOLIVIA AND THE INTERNATIONAL MARKET

On October 10, 2013 at 6 AM, it is −5°C (23°F) as the sun begins to rise on Miguel Huayllas' 3-ha field, located north of the Uyuni Salt flats, at 13,000 feet in the southern Altiplano of Bolivia. Frost has struck again, but apparently the cold was not strong enough to stop the rabbits from coming into his field and eating the leaves from a few rows. He will have to replant these rows by hand, with a "tempranera" or seed that has a short growth cycle. The rest of the field is fine. "This would have killed potatoes, barley, anything. Only the quinoa can withstand this cold."

Bolivia has an estimated 35,000 smallholder farmers like Miguel who produce quinoa, each with an average of 3 ha. Miguel's production costs per hectare are $720, which includes the llama dung and some organic inputs he is provided on credit (Table 12.1).

Table 12.1 Production costs, based on actual results from production of Jacha Inti suppliers of organic quinoa grown in the southern Altiplano region of Bolivia.

Production costs	USD per ha
Soil preparation	79.14
Organic inputs (seed, fertilizer)	164.53
Labor costs	
Planting	71.94
Weeding	28.78
Fertilizing	14.39
Pest Control 1	14.39
Pest Control 2	14.39
Harvest costs	
Harvest with scythe	103.6
Threshing	89.93
Threshing labor	14.39
Winnowing	14.39
Poly-woven bags	10.07
Transport costs	100.72
Total per hectare	ϖ720.65

From each of his 3 ha, Miguel hopes to harvest 19 "quintales" (100 lbs or 1 CWT apiece). In total, Miguel hopes to harvest 5,700 lbs. The year 2013 has marked a historical high price for quinoa, reaching $2.70/lb farm-gate price by the end of the year. "I could sell all my quinoa right now, the middlemen here are offering us high prices every weekend." Miguel, like most quinoa farmers in Bolivia, only sells a third of his crop at harvest in April and May. He then sells another third over the course of the year to cover living expenses. The remaining third he keeps as insurance, only selling it in late January, once he is sure that his new quinoa crop will be successful. The gradual release of inventories at going market rates by the small farmers keeps the quinoa washing/cleaning (or "processing") plants busy but fails to provide the security and volumes of forward contracting that multinational food companies seek.

If prices stay at the current levels, Miguel's farm could bring in revenues of $15,390 for the 2014 crop year. For a family of five, with three working adults, this is a per capita income of almost $13 per day for the adults, much higher even than the equivalent minimum wage they could earn in the cities, which is $8 per day. Miguel, however, does not have the benefits of an employee; a pension for retirement and health insurance are luxuries that he aspires to.

Things are going well for quinoa farmers, and much has changed in 10 years. Back in 2004, quinoa farmers struggled to survive, producing quinoa for self-consumption and trading. Their only hope for improvement was to move out of the countryside to the cities. Gradually, the average quinoa farm went from producing on less than 1 ha to farming on 3 ha. Prices have gone from 60 cents a pound to $2.70 in the past decade. Quinoa farmers today can send their children to nearby towns for higher education.

"We don't like all the uncertainty. I wish prices would have stayed at the same level they have since the 2008 increase," Miguel tells us as we try to purchase his remaining stocks. "Why should I sell now, if prices are going up almost every week?" Multiply Miguel's position by thousands and the market will be continuously short on quinoa and bidding ever higher prices to secure the stock available. Bolivian exporters have to fill sales contracts after all. Eager to cash in on the quinoa craze, Bolivian exporters (usually processing plants) sold forward contracts to US importers banking on the promises of hundreds if not thousands of small family farmers to deliver at a set price. These contracts were then forward sold to food companies, distributors, and supermarkets. In 2013, this situation came to a crisis when farmers started delivering at higher market prices or not delivering at all. Prices skyrocketed from $3,600/metric ton (MT) in January 2013 to $8,000/MT at the end of the year (for washed, cleaned, export-grade white quinoa). Faced with contract defaults from their vendors at origin, importers started bidding up the price of quinoa to secure the product they had sold forward.

At the beginning of 2014, there were many questions in the minds of stakeholders in the supply chain. With raw material in Bolivia costing almost three times the price that had been stable since the last increase in 2008, farmers were wondering what is going on. They wanted to know who are buying quinoa at these ever-increasing prices? At the other end of the market, food companies wonder how they can include more quinoa in their products when prices are so volatile and they cannot be assured of long-term contracts.

Part of the solution may come from new quinoa-producing countries, including North America, which will follow the more traditional supply chain model that includes annual contracts of large volumes. As the market feeds from this much requested new supply of quinoa, it may seem that quinoa is headed down the path of being a cheaper commodity. It may then fulfill its role as a tool to combat world hunger, as the United Nations calls it to be. But how will Miguel Huayllas and his fellow smallholder quinoa farmers fare in this new environment?

THE QUINOA MARKET: SUPPLY AND DEMAND

Bolivia, Peru, and Ecuador increase quinoa acreage

It is important to note that there is very little market data for quinoa. Most of the numbers here were compiled from data from the FAO, Bolivian National Statistics Bureau and Customs, the Peruvian SUNAT, and US Customs (Figs. 12.2–12.4).

Evolution of quinoa, (Figs. 12.7–12.10 and Fig. 12.3) acreage in Bolivia

There are an estimated 40,000 to 60,000 smallholder family farms in Bolivia that plant quinoa.

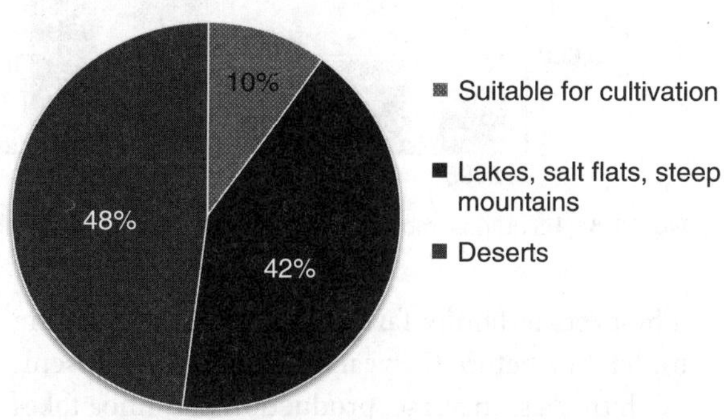

Fig. 12.2 Land considered suitable for cultivation and pasture in the Altiplano.

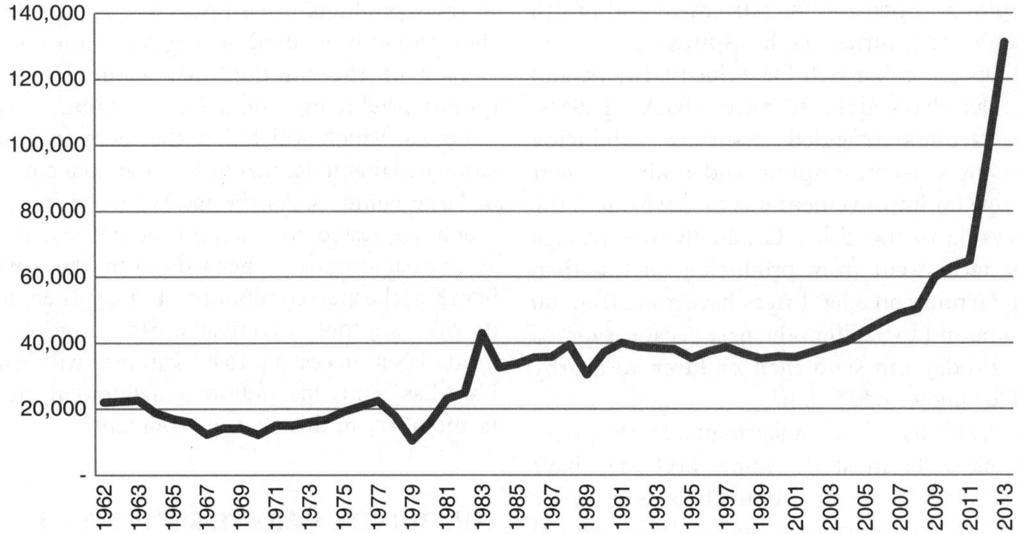

Fig. 12.3 Quinoa production area in hectares.

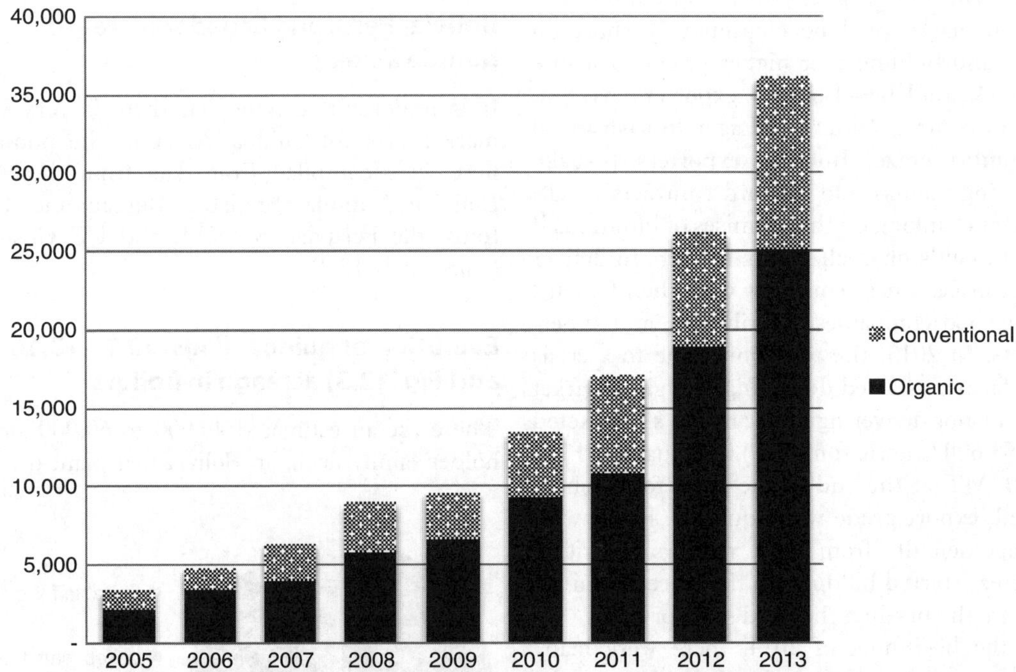

Fig. 12.4 US quinoa imports in MT per year, organic versus nonorganic. (*See color insert for representation of this figure.*)

The average family farm has gone from farming under 1 ha before the year 2000 to 3 ha at present. With this last increase, production of quinoa takes up one-sixth of the marginally productive lands of the Altiplano (Fig. 12.2).

Before the 1980s, Bolivian quinoa producers planted an average of 20,000 ha of quinoa mainly on protected hillsides. These microclimate pockets were planted manually onto rocky hillsides that provided protection against frost. Quinoa was planted there in rotation with corn, potatoes, and beans.

After 1983, spurred by some initial demand from the Asociación Nacional de Productores de Quinua (ANAPQUI) for export to Europe and the United States and by the Peruvian market, planting increased to an average of 35,000 ha (Fig. 12.3). The increase showed that quinoa could be cultivated in a semi-mechanized manner, utilizing tractors to turn over and clear flat lands. These lands, previously not utilized for cultivation, were discovered to be adequate for quinoa production. It is speculated that global warming caused the change in climates, making frosts less frequent in the flatlands and opening these up for quinoa production.

Cesín Curi from the Center for the Promotion of Sustainable Technologies (CPTS) in Bolivia, the nonprofit organization behind the development of the leading quinoa washing and cleaning technology, estimates that there are 250,000 ha that are designated for planting quinoa in the Bolivian highland plateau ("Altiplano"), out of an estimated 1.5 million hectares of marginally productive land.

These marginal lands are in protected areas, such as around the shores of Lake Titicaca. Today, these are used for pasture and the production of potatoes, wheat, barley, tubers, and beans. Although there still seems to be plenty of room to grow quinoa, it will be at the expense of other crops and will encroach on pastureland. As Cesín Curri points out in his organization's quest for funding for experimental fields in arid lands, "If quinoa could be cultivated in the 'non-suitable areas for cultivation,' then the growth potential for quinoa is enormous." Still, the question remains of how cost-competitive cultivating in these desert areas can be. The land there is sandy and low in nutrients and quinoa production in these areas will most likely require synthetic fertilizers.

The US quinoa market and evolution of prices

The United States is the main market for all quinoa exported, consuming an estimated 30% of all quinoa produced in 2013. The amount of quinoa imports and the importance of the United States for quinoa have both increased in the past few years at a rapid pace, as shown in Table 12.2.

Quinoa in the eye of a market storm

Quinoa benefited from the megatrends of whole grains, organic/natural foods, and gluten-free foods.

The US market for quinoa had three stages of growth:

- *Mid 1980s to 2000*. During this initial phase, the quinoa market grew to 1,000 MT. This phase included Hispanic food importers and natural-food pioneers such as Ancient Harvest importing quinoa from Bolivia, Arrowhead Mills importing from Peru, and Inca Organics importing from Ecuador. These companies experienced some growth, but struggled with low market awareness, limited distribution into natural channels, and sourcing difficulties (mainly quality issues and low available volumes due to limited processing plant capacities).
- *2000–2007*. In the early to intermediate phase, the quinoa market grew from 1,000 to 3,300 MT (Table 12.2). This phase was marked by a rapid development of the organic food

Table 12.2 US quinoa imports in metric tons per year and US imports as a percentage of total world production.

	2004	2005	2006	2007	2008	2009	2010	2011	2012	2013
Quinoa imports in MT per year	1,548	2,425	2,479	3,346	**5,824**	8,456	11,901	16,557	26,155	36,038
Growth per year, %		57	2	35	**74**	45	41	39	58	38
Quinoa imported into United States as percentage of world production, %			4	6	10	11	15	21	26	30

industry, which was experiencing annual growth rates of 15–20%. The low-carb trend led by the Atkins diet was fading, and in its place emerged a strong interest in whole grains. In Bolivia, quinoa processing plants developed better washing systems with the support of nonprofit organizations. These in turn were in a better position to offer higher volumes of washed/cleaned quinoa to the market. By 2007, Trader Joe's supermarkets started carrying quinoa nationwide, marking the transition of quinoa from an organic niche product to a specialty food market item.

- *2007–present*. The quinoa market grows from 3,300 to 36,000 MT during this phase (Table 12.2). This growth phase is marked by the entry of quinoa into the mainstream market. Larger facilities in Bolivia and Peru offer a more reliable and larger supply to food companies. With importers and commodity traders willing to take risk on importing large volumes of quinoa to provide long-term contracts to retailers, new quinoa-based lines and brands are launched, particularly at Costco and Whole Foods supermarkets. By December 2013, there were over 200 registered importers of quinoa in the US market purchasing from 169 exporting companies. In 2012, multinational food companies (Kellogg, Pepsico, Mars) launched new quinoa product lines. In 2013, quinoa retail market leader Enray (TruRoots) was acquired by Smuckers.

Today, quinoa is considered a standard bearer of the organic and health food movements. Of all the quinoa sold in the United States, over 65% is certified organic (Fig. 12.4). This is due to the fact that the brand holders and importers who spearheaded the initial growth of the market had a commitment to organics. As one starts looking at new product launches, it is evident that future market growth of quinoa will no longer be in branded bags for retail but as an ingredient in ready-to-eat packaged goods.

Quinoa has three main destinations in stores and markets:

Quinoa in a retail pack for use as a side dish. All of Costco, Trader Joe's and one of seven brands of quinoa on retail at Whole Foods are Bolivian Organic. This is where Bolivian quinoa, with a larger seed and uniform size is preferred (Fig. 12.5). Quinoa in bags available at retailers such as Target, Marshalls, and Home Goods tend to come from Peru.

Quinoa in a grain blend for use as a side dish (e.g., rice–quinoa blend). Of five blended products, only two have Bolivian quinoa. In blended products, size, cooking time, and appearance in general are less important. This includes flash-frozen quinoa that is ready to eat. In this segment, organic certification is less important.

In products made with quinoa (pastas, cookies, bread mixes, cereals, whisky, vodka, and shampoos) where the origin of quinoa is unknown. As the market for quinoa evolves, there is an increased demand for quinoa to be used as an ingredient. In these markets, the quinoa seed size, organic certification, cooking time, and flavor are not as important as the price (Fig. 12.6).

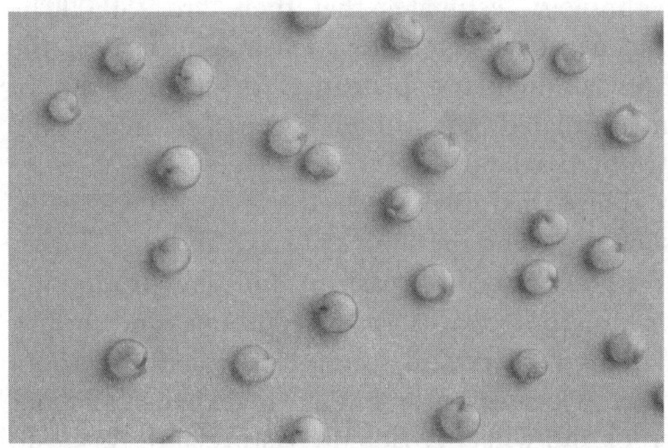

Fig. 12.5 Close-up of white quinoa seeds. Photo credit: Vitaliy Prokopets.

Fig. 12.6 Multiple quinoa varieties are planted and harvested in the same fields. Photo credit: Vitaliy Prokopets. (*See color insert for representation of this figure.*)

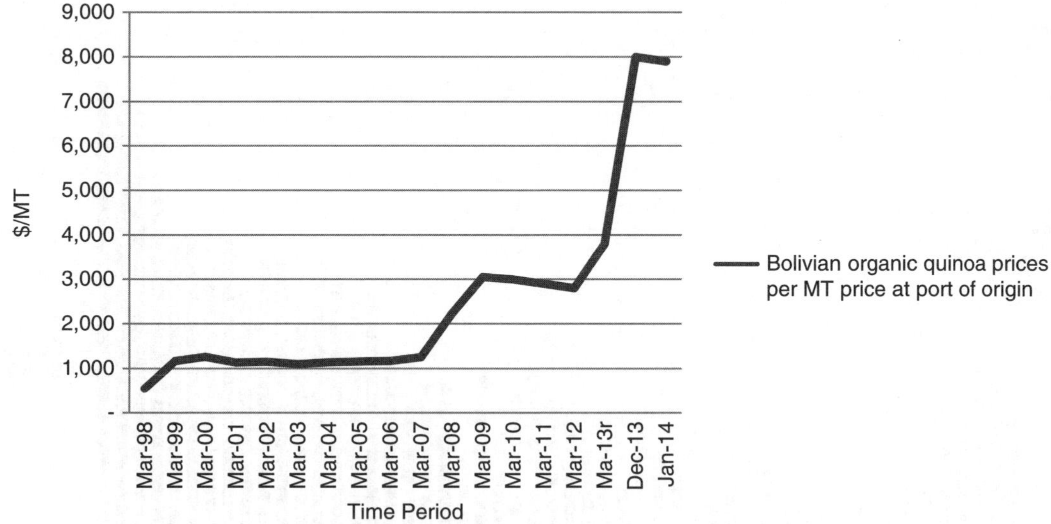

Fig. 12.7 Bolivian organic quinoa prices per MT price at port of origin.

Quinoa prices have been an issue in 2013, after a period of relative stability between 2009 and 2012 (Fig. 12.7). What has been an important departure from recent years, however, is the difference in price between Bolivian and Peruvian quinoa. In December 2013, Bolivian organic quinoa was sold at an average of $8,000/MT, whereas Peruvian organic quinoa went for $7,000/MT. Historically, Peruvian quinoa sold for only $200/MT below Bolivian. This difference may be influenced by multiple factors, one of which is the higher demand and preference for Bolivian quinoa (which, as mentioned earlier, is preferred for its size and uniformity).

Both Bolivian and Peruvian governments are concerned with the high prices of quinoa and are taking action. Rather than taking a hard stance with export controls or nationalization of industries, both countries have opted for subsidies to continue providing quinoa at lower prices in social programs such as school lunches, increased investment in research into quinoa production and processing, and funding made available for increased production (loans for materials, seeds, and working capital to growers).

Bolivia, Peru, and Ecuador, as well as non-traditional quinoa-producing countries, are expanding acreage in 2014 (Fig. 12.8). Bolivia's National Statistics Institute (INE) estimates a 29% increase in quinoa cultivations for 2014. Supply is certainly set to increase for 2014 and the entire supply chain expects a price drop (Figs. 12.9 and 12.10). Farmers will certainly be the first to feel the impact of the supply–demand equation, many fearing a return to subsistence agriculture.

The quinoa grower rises out of poverty

In this section, I will focus on the economics of the family farmers in the Salt Flat Region of Bolivia.

The World Bank has set a poverty line of $1.25/day/adult, which equates to $112/month for an average quinoa-producing family with three working adults. This threshold was passed in 2012 for the average quinoa farm with incomes going from $34/month/farm in 2007 to $240 in 2012. As Bolivian quinoa farmers look forward to 2014 and beyond, their incomes could reach $800/month with increased acreage and improved yields.

Raúl Vera Copa, President of Association of Producers of Quinoa and Camelids (APQC) (Fig. 12.11), an association of 200 producing family farms in Bolivia, shared his views during an interview the *Shared Interest* documentary, which aired October 2013. Below is a translation of an excerpt of Raúl's interview:

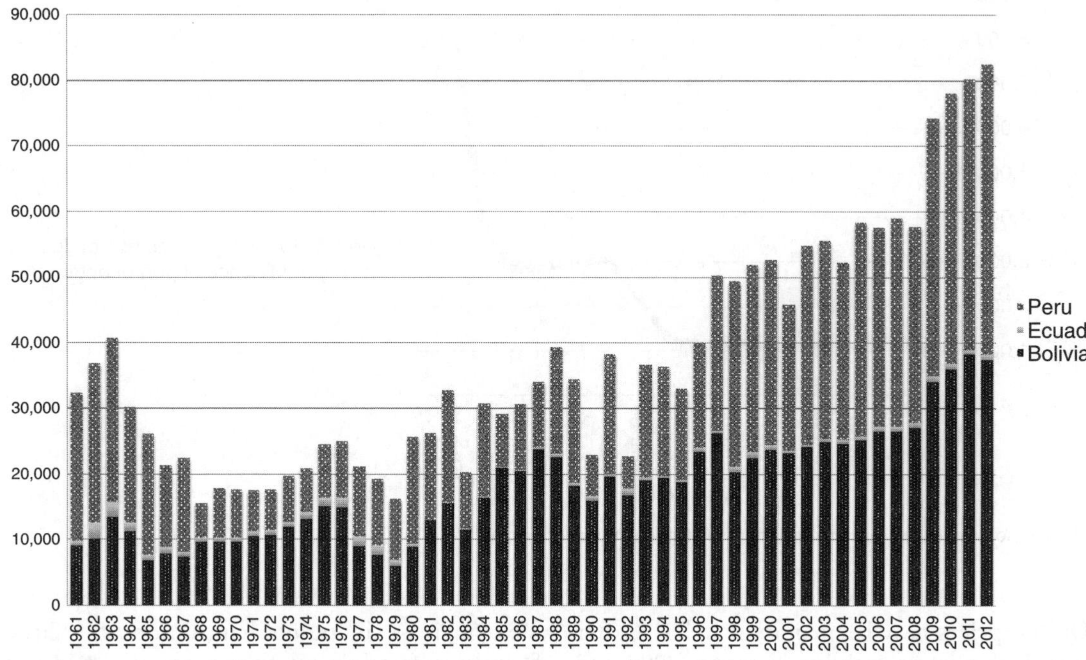

Fig. 12.8a Production per country per year in metric tons (MT).

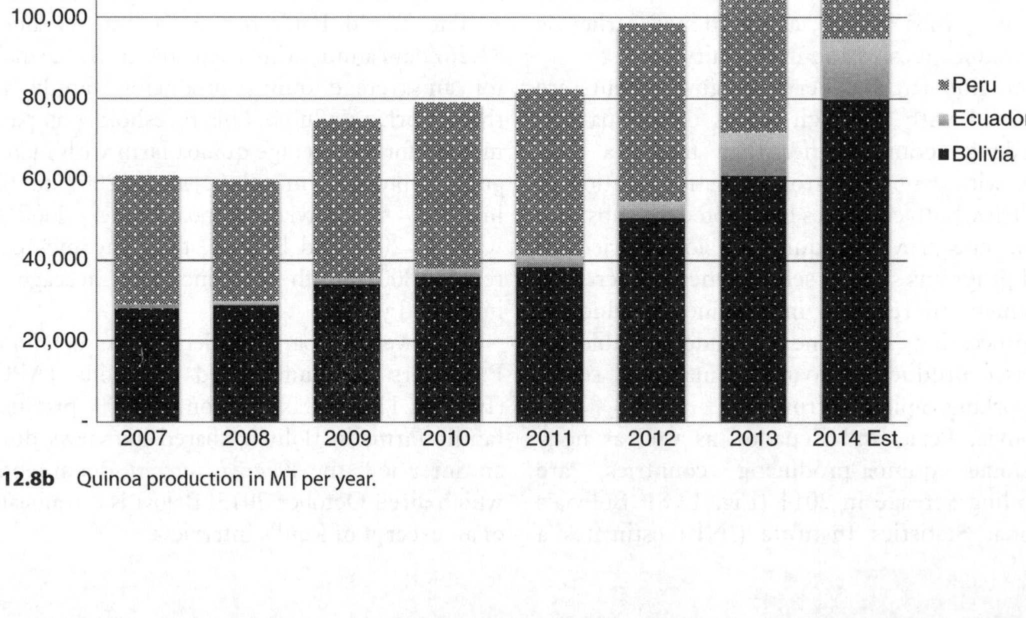

Fig. 12.8b Quinoa production in MT per year.

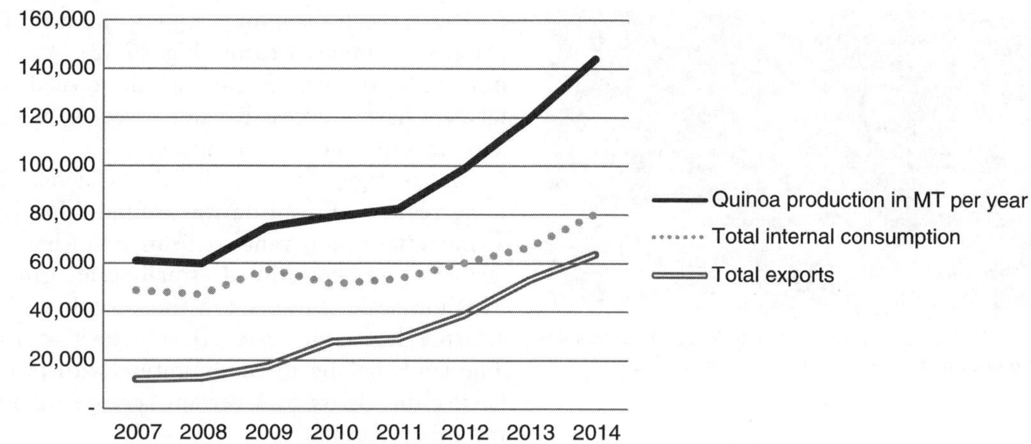

Fig. 12.9 Quinoa production, consumption, and exports in MT per year from main producing countries.

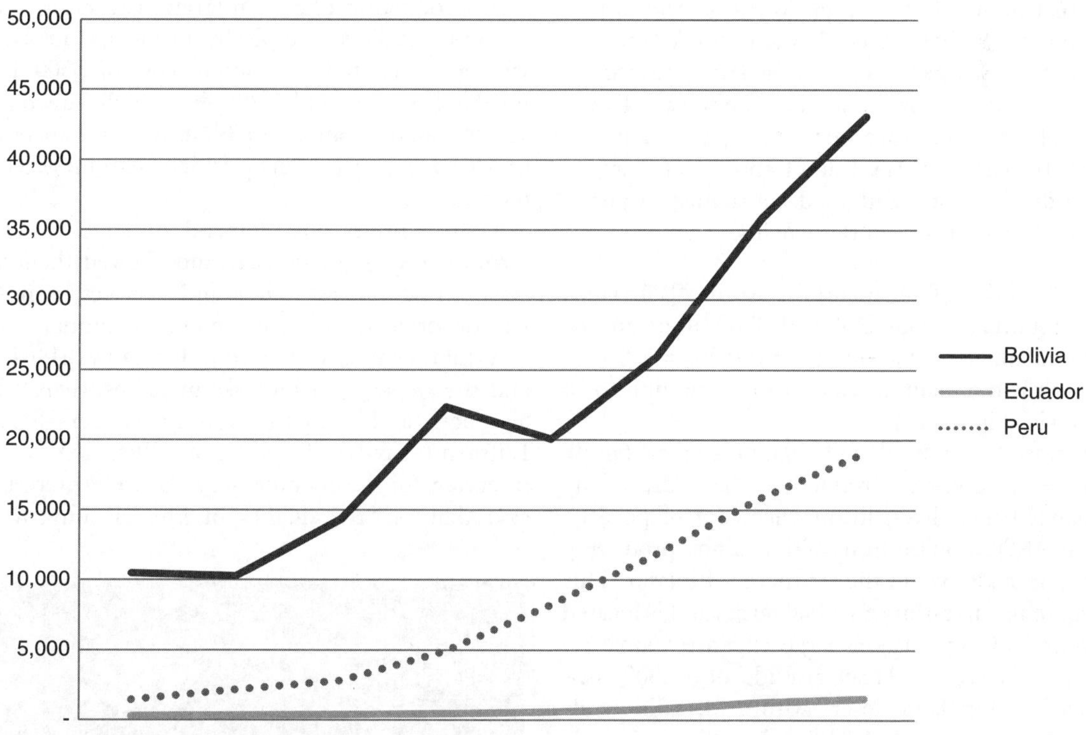

Fig. 12.10 Quinoa exports per country in MT per year.

I remember when there was almost not enough. Sometimes when we were young we didn't even have shoes to go to a party, and to go to Challapata (nearest town). We didn't use shoes before. Little by little things have improved. (…) Before, when there was no Jacha Inti, no cleaning plant, we sold the quinoa to middlemen at 200, 300 or 400, and on top we also invested in clearing lands, in the planting and the care, scarecrows, all

Fig. 12.11 Raúl Vera Copa, President of APQC, an association of 200 producing family farms in Bolivia. Photo credit: Matías Musa.

> of that, and we earned very little. But since there is this demand, now we are satisfied that quinoa has gone up in price. And also these earnings we don't want to take them out to the cities. No, with the resources that we are using now we want … we want to live well. And improve our houses, we want a bathroom and also, I don't know … to have a dining room, and good bedrooms, so our children can stay in these fields.

The prices Raúl mentions are in "Bolivianos per Quintal," about US$0.30–0.60/lb of unprocessed quinoa. One quintal = 100 lbs = 1 CWT, which is the main measure for trading quinoa in Bolivia (Fig. 12.13).

Indeed, for Raúl and 50,000 similar family farmers, quinoa production has drastically changed their lives, lifting them out of poverty. The APQC is in the heart of the quinoa-producing region of Bolivia in the southern Altiplano. This region has an estimated 9,000 farms and is located around the salt flats, a large expanse visible on satellite pictures. At an altitude of 13,000 feet, this is a highland desert with arid, saline soils which receives only 8–12 inches of rain a year. The poor soils, lack of irrigation, and year-round frosts limit the cultivation of alternative crops to some protected microclimates located around the fringes of the mountains. Even today, these lands are distant from the main cities and were not considered suitable for cultivation by the Spanish. The lands were left to the indigenous communities. It takes 10 hours of driving from La Paz to reach the quinoa region, and half of the drive is on unpaved roads (Fig. 12.12). As in the past, today the land is communally owned, with farmers having access to land based on heritage and years of residence in the area.

On average, farmers in the Bolivian Salt Lake region cultivate farms around 6 ha. Most farmers have plots ranging from 1 to 5 ha; these farmers are considered small-scale growers. Medium-scale growers are those with 6–15 ha and the few large-scale growers start at 16 ha. The lands belong to communities, with farmers having the rights to a certain acreage inherited from their parents.

Until 2008, Bolivian quinoa farmers practiced subsistence agriculture, and they barely covered their production costs. Emigration to cities was common. A farmer typically produced on 2 ha, with an estimated production cost of $500/ha. His land would yield 1,800 lbs, which he would sell at a market rate of $0.39/lb for a net income of $404 per year per farm, which is around $34 a month.

Higher prices and demand in 2008 enticed farmers to increase acreage and allowed them to start hiring outside labor and services, mainly tractor services, and invest in organic inputs.

With the increase in quinoa demand worldwide and the subsequent increase in prices, there has also been a shift in the way the farmers lived. Eufraen Huayllas, President of APROCAY, in an interview for the documentary "Shared Interest," said that … "the quality of life has improved

Fig. 12.12 Unpaved roads leading to quinoa fields. Photo credit: Vitaliy Prokopets.

Table 12.3 Evolution of hectares farmed, quinoa production costs, and earnings for Bolivian farmers.

Period	Land area farmed, ha	Cost of production, USD/ha	Yield, lbs/ha	Sales price, USD/lb	Annual profits, USD	Equivalent monthly net income[a], USD
Before 2008	2	500	1,800	0.39	404	34
2008–2012	3	660	1,500	1.07	2,835	236
2013	4	720	1,300	1.70	5,960	497
2014 (projected)	6	800	1,900	1.00	6,600	550

[a]Farmer's equivalent net income: (hectares planted × yield × market price) – (production cost per acre × acreage planted).

because, before, in this community, we only lived and produced for our own consumption. We had our sheep, our llamas and small plots... because when I was small quinoa was not valued." APROCAY is an association with 63 member families in Bolivia.

Table 12.3 shows how the income from a medium-scale quinoa farm would have evolved in the past few years, as demand and prices increased.

The shift in quinoa prices, as seen in Table 12.3, can be divided in four phases: (i) Until 2008, prices were relatively stable, with an initial weak demand driven mainly by the emerging Peruvian market and organic industry pioneers in Europe and the United States (Priméal, Quinoa Corporation). (ii) In 2008, there was a marked increase with prices almost tripling, going from 275 to 800 bolivianos per CWT ($0.40–1.15/lb of unprocessed quinoa) (Fig. 12.13). This increase was caused by strong demand in North America marked by the entry of Costco and Trader Joe's supermarkets to the quinoa market. (iii) After relative stability from 2009 to 2012, prices once again drastically shot up in 2013, this time taking unprocessed quinoa prices from 800 to 1,850 Bolivianos per 100 lbs ($1.15–2.59/lb). This latest price spike was linked to a high demand, a tight supply, and much speculation caused by all the attention given to quinoa during the IYQ. (iv) As Fig. 12.13 shows, the latest price increase is expected to be corrected in 2014, so for the intent of determining the average income in 2014, a value of 700 Bs/CWT ($1/lb) for unprocessed quinoa was taken. This value is completely speculative, as it will depend on the market at any given moment. We estimate, however, that it is unlikely that new sources will compete with Bolivia at less than $1/lb of unprocessed raw materials.

On average, yields per hectare for mechanized cultivation of organic quinoa in Bolivia are 13 CWT (1,300 lbs), yielding seed that costs $0.55/lb. With additional inputs and improved organic management practices, this yield can be increased to 20–25 CWT. This yield is low compared to the Peruvian nonorganic production, which yields 33 CWT/ha (3,300 lbs) and costs $877/ha to produce, yielding seed that costs $0.27/lb to produce.

CURRENT PRODUCTION PRACTICES, INCREASED ACREAGE, AND THOUGHTS ON SUSTAINABILITY

Efigenia Encinas, representative of Sumaj Camaña association of Women Farmers (Fig. 12.14), was interviewed about her farming practices and thoughts on sustainability. Below is a translated excerpt of her interview:

Interviewer: Do you have children, are you married?

Efigenia: No, I am single because I am the eldest daughter and I always have wanted to represent the people and my family. I used to sow quinoa by hand with my grandfather. My grandfather left us this place and I preserve it.

Interviewer: So, since when have you produced quinoa?

Efigenia: Since I was 10 or 12. I used to sow by hand with my grandfather. Because

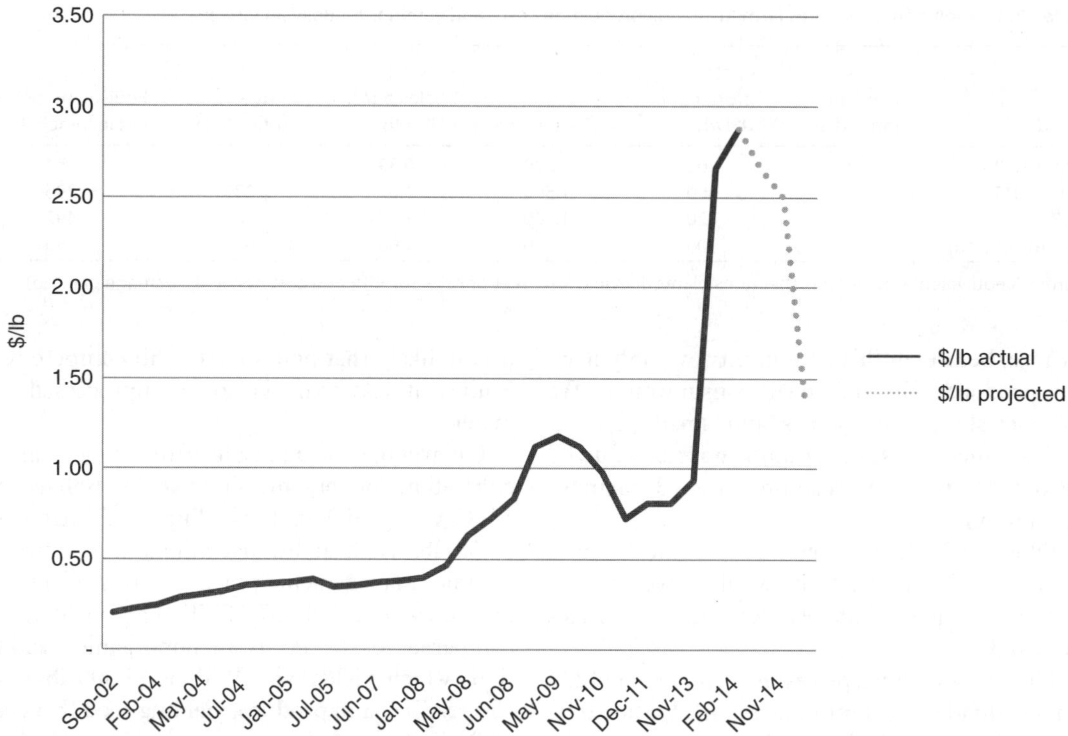

Fig. 12.13 Farm-gate (unprocessed) quinoa prices in Bolivia in $/lb.

this community was very small, now it's bigger.

Interviewer: And now how do you plant?

Efigenia: Now we plant with machinery. And with this "Satiri" tractor as they call them. With that. Before, we planted by hand. In those small fields where we were before, that's where we used to produce.

Interviewer: Do you think the production will improve now with the tractors? How do you see it?

Efigenia: For me there are advantages and disadvantages, because, imagine, there is land erosion that I see and the consequences of winds, the climate factor. Before, there were bushes everywhere, so this land was for grazing. Now sometimes there are no green barriers, so because of that the grooves made by the tractor get swept by the wind. Before we planted by hand, hole by hole, step by step. So these consequences can cause the production to fail.

Interviewer: So, your advice for improving the production …

Efigenia: My advice for improving the production would be to keep the live barriers and also to raise animals in order to **have natural fertilizer,** because before we always sowed with llama and sheep manure, with manure, so I believe everything was natural. Now I know we need a lot of supplies, these organic fertilizers that are already produced. It's like using a shampoo or oil for the production. That's what I see. Because before our sacrifice (monetary investment) was small, but the plant was healthier. There was not this erosion of land. Now, I think that to **maintain the sustainability we should maintain the raising of livestock, also make live barriers**. Plant them where there are none.

Cultivation practices have evolved since the 1960s when Efigenia Encinas planted quinoa

Fig. 12.14 Efigenia Encinas with her lamb. Photo credit: Diego Nuñez de Arco.

Fig. 12.15 Replanted quinoa in a field where earlier plantings had been eaten by rabbits. Photo credit: Stefan Jeremiah.

Fig. 12.16 Quinoa emerging before first rains. Photo credit: Diego Nuñez de Arco.

Fig. 12.17 Quinoa field prepared with tractors. Photo credit: Diego Nuñez de Arco.

manually with her grandfather. At present, manual planting is relegated to steep hillsides, and when farmers need to replant because of frost or localized pest attacks (Fig. 12.15).

The quinoa crop year starts with the last rains of January or February. During this time, fallow fields are overturned to allow rainfall to be captured underground. This water is essential for the seeds to sprout come September/October, when farmers start to sow. This way, the quinoa benefits not only from the rains that come from November to February but also from the previous rainy season. Therefore, each field of quinoa uses two rainy seasons' worth of rainfall (Fig. 12.16).

Farmers today rely on both natural indicators and modern technology, such as meteorological reports, to time the sowing. If planted too early, the quinoa will sprout but quickly run out of water and wither. The flowering of certain bushes and cacti, the presence of lizards, scorpions, and where birds lay their eggs are signs still used by farmers to predict frost or rain and serve to determine the types of seeds and the best locations in each field for planting them.

Today, most farmers hire tractors to assist in the overturning of lands and in sowing (Fig. 12.17). This is a major departure from traditional practices as Efigenia mentions in her interview. The use of tractors has benefits and also drawbacks, including soil erosion because of the strong winds in the highlands. Efigenia and agronomists in Bolivia point to the importance of maintaining live barriers against the wind. These indigenous "tholas" (short bushes) seem to be as important as the llamas that feed on them. For millennia, llamas have had a symbiotic relationship with quinoa. They provide what is widely recognized by all quinoa farmers in Bolivia as the very best of fertilizers. Most quinoa farmers will apply llama manure on their fields at least once every 3 years.

LIVING WELL, REVERSED MIGRATION, AND CULTURAL IDENTITY

The following excerpts from interviews illustrate how farmers view their success and their hopes for "living well," which, as Willy Choque explains, is synonymous to the farmers with sustainable development.

> We have been working with some communities on defining first of all what it means to 'live well.' We are using 'living well' as a development paradigm coming from within the communities. We have begun by realizing that our communities did not understand the meaning of sustainable development, for example. But it's easier (for the growers) to conceptualize or understand 'living well.' Living well is in our ancestral languages: Aymara, Quechua, Guaraní. It's easier for a producer to define what it is to live well and according to that define how he wants to improve his life.
>
> Willy Choque, Agronomist with the University of Oruro, during an interview for *Shared Interest* documentary, October 2013.

> Now you can see (the impact) in the country. I think it's an improvement in all the families, I think it is, that they start to buy motorcycles, they buy their jeeps, trucks, they have solid houses, comfortable families. I think it's success that mother earth has given us, to the region. Thanks to quinoa I think we in this region are not migrating. Before there was a lot of migration, and all those fellows who had migrated are coming back. So I think that quinoa is in the right place, and it's important. It's also important that this country is beginning to consume more and more quinoa than before. Now there is no more discrimination. Now nobody says 'I don't eat quinoa'.
>
> Alfredo Perez, President of Chamber of Quinoa Producers from Potosi (CADEQUIR) (Fig. 12.18)

> The producer is very excited with the high prices of quinoa at this moment, and they are buying a lot of things: houses, transportation; they have more comfort, their children are going to better schools, and they are learning to use their money on other things, on living well.
>
> Danny Mamani, Agronomist and Representative of the Quinoa Producer Association Sumaj Juira

> Living well, for me, means not only having machines. It means having a good house, good nutrition, good clothes, education, good health, that's for me 'living well'. So that's the aim. And to raise a family. Having the opportunity for the children to study.
>
> Efigenia Encinas, Representative of Sumaj Camaña Association of Women Farmers

What has been another valuable impact of quinoa is that community ties are also strengthening as producers return to their communities:

> When people leave the communities, the children little by little lose their cultural identity: their language, their dress, their nutrition, etc. But now that people are back in their communities, all this **cultural identity is alive.** If there are people in the communities, the culture is alive.
>
> Willy Choque, Agronomist with the University of Oruro, during an interview for *Shared Interest* documentary, October 2013

> Quinoa production contributes to social cohesion as people return to their communities and accept the responsibilities that these impose upon them. They participate in community events. The canvas of society is woven.
>
> Pablo Laguna, in "Is the commercialization of quinoa as negative as it is being said to be?" April 7, 2013, *La Razon* newspaper

It is evident that much good has come to Bolivian quinoa growers through the development of the quinoa market. It is to be hoped that this impact will continue to expand as growers increase acreage and improve their yields, marrying their traditional knowledge and production practices with modern adapted technologies.

OPPORTUNITIES FOR THE BOLIVIAN FARMER

> At the conference (International Quinoa Research Symposium At Washington State University, August 12, 2014), there was talk of creating a special brand for Andean quinoa, a little like the special recognition granted to other traditional foods, like Bordeaux wine. This brand would be top-quality quinoa, and consumers might also be willing to pay extra for it, just because it would be quinoa from the land and the communities that safeguarded it for thousands of years.
>
> "Can Quinoa Farming Go Global Without Leaving Andeans Behind?" by Dan Charles for *National Public Radio*

The year 2014 will mark the entry into the market of quinoa from nontraditional origins and mass agriculture. Already quinoa is being planted successfully in the United States (Pacific Northwest and Colorado), Canada (Saskatchewan), Western Australia, France, and India, to name a few. Peru is also increasing their farming at sea level along its coast, with outstanding yields that triple those obtained by smallholder farmers in Bolivia. Moreover, large producers have a distinct advantage: the food supply chain relies on future contracts from reliable sources. For example, large food manufacturers who use quinoa as an ingredient in their cereal (and this may be less than 5% of the total ingredients used) seek to have their purchase volumes and prices fixed for a year. In the current quinoa supply chain, this is very difficult because quinoa comes from thousands of smallholder farmers, who do not release all their product at once, so that supply trickles into the processing plants. Exporting quinoa-processing plants are the source of purchase contracts and are currently the suppliers to the market. The issue with these processing plants is that they can only assure prices on what inventory they hold at the moment. This inventory is limited by their ability to purchase quinoa stocks and process these stocks. The largest of quinoa-processing plants in the world can process 25 containers (500 tons) per month and their working capital is limited to 60 containers (1,200 tons).

Commodity traders in the quinoa supply chain play the important role of aggregating the supply from multiple quinoa exporters, assuring the requested 6-month to 1-year-long contracts with food manufacturers. To date, they have played a crucial role in the growth of the quinoa market.

As the market for quinoa grows, demand will flow toward sources that can offer the forward contracting that other mature commodities offer. Commodity traders who until today absorbed some of the inefficiencies of the current supply chain will favor the sources that are more reliable and pose lower risk.

The lives of the Bolivian farmers have been dramatically improved and no one wishes them to return to a state of poverty. To prevent this situation, many stakeholders agree that a niche should be created for the smallholder farmer (Fig. 12.18). This niche would seek to decommoditize their quinoa, giving it distinguishing marks such as Fair Trade, Appellation of Origin, and Certified Organic.

Fig. 12.18 Typical quinoa farmer family in the Potosí region of Bolivia. Photo credit: Vitaliy Prokopets.

We are obtaining Fair Trade Certifications with the organizations we work with, so we can in the future protect our producers under two main premises: organic quinoa and fair trade. What is coming in the future is strong competition from other emerging countries in quinoa production, but we think that the actions Andean Naturals and Jacha Inti are taking on behalf of our producers -- because we consider them our family -- will protect them. In those times of crisis, they will continue to receive a fair price for the quinoa they produce, and all their work will be acknowledged.

Yeris Peric, Jacha Inti Quinoa Processing Plant Agronomist, October 2008

Index

Quinoa: Improvement and Sustainable Production, First Edition. Edited by Kevin Murphy and Janet Matanguihan.